New Advances into Nanostructured Oxides, 2nd Edition

New Advances into Nanostructured Oxides, 2nd Edition

Guest Editors

Silvia Mostoni
Roberto Nisticò

Basel • Beijing • Wuhan • Barcelona • Belgrade • Novi Sad • Cluj • Manchester

Guest Editors

Silvia Mostoni
Department of Materials
Science, U5
University of Milano-Bicocca
Milano
Italy

Roberto Nisticò
Department of Materials
Science, U5
University of Milano-Bicocca
Milano
Italy

Editorial Office
MDPI AG
Grosspeteranlage 5
4052 Basel, Switzerland

This is a reprint of the Special Issue, published open access by the journal *Inorganics* (ISSN 2304-6740), freely accessible at: www.mdpi.com/journal/inorganics/special_issues/6063SX09DU.

For citation purposes, cite each article independently as indicated on the article page online and using the guide below:

Lastname, A.A.; Lastname, B.B. Article Title. *Journal Name* **Year**, *Volume Number*, Page Range.

ISBN 978-3-7258-3440-2 (Hbk)
ISBN 978-3-7258-3439-6 (PDF)
https://doi.org/10.3390/books978-3-7258-3439-6

© 2025 by the authors. Articles in this book are Open Access and distributed under the Creative Commons Attribution (CC BY) license. The book as a whole is distributed by MDPI under the terms and conditions of the Creative Commons Attribution-NonCommercial-NoDerivs (CC BY-NC-ND) license (https://creativecommons.org/licenses/by-nc-nd/4.0/).

Contents

About the Editors vii

Silvia Mostoni and Roberto Nisticò
New Advances into Nanostructured Oxides, 2nd Edition
Reprinted from: *Inorganics* 2025, 13, 60, https://doi.org/10.3390/inorganics13020060 1

Mariyem Abouri, Abdellah Benzaouak, Fatima Zaaboul, Aicha Sifou, Mohammed Dahhou and Mohammed Alaoui El Belghiti et al.
Efficient Catalytic Reduction of Organic Pollutants Using Nanostructured CuO/TiO_2 Catalysts: Synthesis, Characterization, and Reusability
Reprinted from: *Inorganics* 2024, 12, 297, https://doi.org/10.3390/inorganics12110297 7

Mahboubeh Dolatyari, Mehdi Tahmasebi, Sudabeh Dolatyari, Ali Rostami, Armin Zarghami and Ashish Yadav et al.
Core/Shell ZnO/TiO_2, SiO_2/TiO_2, Al_2O_3/TiO_2, and $Al_{1.9}Co_{0.1}O_3/TiO_2$ Nanoparticles for the Photodecomposition of Brilliant Blue E-4BA
Reprinted from: *Inorganics* 2024, 12, 281, https://doi.org/10.3390/inorganics12110281 23

Tiangui Zhao, Tihao Cao, Qifu Bao, Weixia Dong, Ping Li and Xingyong Gu et al.
Significantly Enhanced Self-Cleaning Capability in Anatase TiO_2 for the Bleaching of Organic Dyes and Glazes
Reprinted from: *Inorganics* 2023, 11, 341, https://doi.org/10.3390/inorganics11080341 37

Tan Mao, Junyan Zha, Ying Hu, Qian Chen, Jiaming Zhang and Xueke Luo
Research Progress of TiO_2 Modification and Photodegradation of Organic Pollutants
Reprinted from: *Inorganics* 2024, 12, 178, https://doi.org/10.3390/inorganics12070178 52

Ahmed H. Naggar, Abdelaal S. A. Ahmed, Tarek A. Seaf El-Nasr, N. F. Alotaibi, Kwok Feng Chong and Gomaa A. M. Ali
Morphological Dependence of Metal Oxide Photocatalysts for Dye Degradation
Reprinted from: *Inorganics* 2023, 11, 484, https://doi.org/10.3390/inorganics11120484 75

Adil Alshoaibi
The Influence of Annealing Temperature on the Microstructure and Electrical Properties of Sputtered ZnO Thin Films
Reprinted from: *Inorganics* 2024, 12, 236, https://doi.org/10.3390/inorganics12090236 110

Dmitrii Filimonov, Marina Rozova, Sergey Maksimov and Denis Pankratov
Synthesis and Redox Properties of Iron and Iron Oxide Nanoparticles Obtained by Exsolution from Perovskite Ferrites Promoted by Auxiliary Reactions
Reprinted from: *Inorganics* 2024, 12, 223, https://doi.org/10.3390/inorganics12080223 123

Anil Abduraman, Ana-Maria Brezoiu, Rodica Tatia, Andreea-Iulia Iorgu, Mihaela Deaconu and Raul-Augustin Mitran et al.
Mesoporous Titania Nanoparticles for a High-End Valorization of *Vitis vinifera* Grape Marc Extracts
Reprinted from: *Inorganics* 2024, 12, 263, https://doi.org/10.3390/inorganics12100263 147

Reo Kimura, Kota Shiba, Kanata Fujiwara, Yanni Zhou, Iori Yamada and Motohiro Tagaya
Precipitative Coating of Calcium Phosphate on Microporous Silica–Titania Hybrid Particles in Simulated Body Fluid
Reprinted from: *Inorganics* 2023, 11, 235, https://doi.org/10.3390/inorganics11060235 165

Dilshat U. Tulyaganov, Simeon Agathopoulos, Konstantinos Dimitriadis, Hugo R. Fernandes, Roberta Gabrieli and Francesco Baino
The Story, Properties and Applications of Bioactive Glass "1d": From Concept to Early Clinical Trials
Reprinted from: *Inorganics* **2024**, *12*, 224, https://doi.org/10.3390/inorganics12080224 . . . . . . **178**

About the Editors

Silvia Mostoni

Silvia Mostoni earned a MSc degree in Chemistry at the University of Milano in 2015 and received her PhD in Materials Science and Nanotechnology at the University of Milano–Bicocca in 2019. Her research interests focus on the synthesis, characterization and surface functionalization of inorganic nanomaterials mainly based on metal oxide nanoparticles for applications in the catalytic field, automotive sector, and materials with luminescent properties. Since 2023, she has been a Researcher with a fixed-term contract in General and Inorganic Chemistry at the University of Milano–Bicocca.

Roberto Nisticò

Roberto Nistico (RN) is Associate Professor in General and Inorganic Chemistry at the University of Milano-Bicocca (Department of Materials Science, Italy). Up until today, RN has published more than 90 papers in renowned international journals and 2 book chapters. Additionally, RN is a member of the Editorial Board of the journal Inorganics ("Inorganic Materials" Section). RN's research mainly focuses on several aspects at the interface between inorganic chemistry and materials science, always looking for novel and appealing solutions for a sustainable future. RN's principal field of interest is the development of magnet-responsive nanomaterials (i.e., iron oxides, ferrites) and other inorganic systems for the environmental remediation of contaminated (waste)water, (photo)catalysis, and energetic applications.

Editorial

New Advances into Nanostructured Oxides, 2nd Edition

Silvia Mostoni and Roberto Nisticò *

Department of Materials Science, INSTM, University of Milano-Bicocca, U5, Via R. Cozzi 55, 20125 Milano, Italy; silvia.mostoni@unimib.it
* Correspondence: roberto.nistico@unimib.it; Tel.: +39-02-6448-5111

1. Introduction

The interest in inorganic nanostructured oxides is growing extensively, thanks to their remarkable features and their wide range of applications, which include (photo)catalysis [1–4], controlled drug-delivery and chemical release [5–7], environmental remediation [8–10], energy and batteries [11], smart materials [12], and so on. One interesting aspect of inorganic nanostructured oxides is the high level of control that can be achieved over particle morphology, size, shape, and porosity [13], as well as their surface properties [14,15], which further broaden the possible application of these materials in a wide variety of fields. In fact, the use of suitable synthetic pathways and ad hoc surface functionalization procedures provides a powerful tool to obtain pure inorganic or hybrid inorganic–organic composite materials that represent strategic materials in the most recent research literature [16,17].

In this context, after the success of the first edition published in *Inorganics* in 2022, the second edition of the Special Issue, entitled "New Advances into Nanostructured Oxides, 2nd Edition" was launched to bring together the most recent developments on the class of inorganic materials. This Special Issue is inserted in the "Inorganic Materials" section that, since its birth, has rapidly grown thanks to its attention to advanced inorganic materials, as well as their high technological demand. The aim was to collect research papers and reviews focused on the synthesis and characterization of inorganic oxide nanomaterials through soft chemistry approaches, favoring a high control on the particle morphology, size, porosity and surface functionalities. In addition, in view of their application, the materials' performances are described with the ambitious goal of identifying the structure–property relations to connect the surface, morphological and structural features to the material activity. This represents a shared interest in the development of innovative materials, as the modulation of the key structural parameters may be used to tune and improve the materials' resulting performances.

Prior to proceeding with the overview of the contributions, the Guest Editors would like to thank all of the reviewers who spent their valuable time thoroughly reviewing and improving the articles published in this volume. We also sincerely thank all the authors for choosing *Inorganics*, and, in particular, the "Inorganic Materials" section, as the recipient of their excellent science.

2. An Overview of the Published Articles

Overall, this Special Issue collected 10 original papers (i.e., seven research articles and three reviews) and received more than 16,000 views, which paved the way for the further proposal of a third edition of the same Special Issue. The published papers can be categorized into three main subsections as reported in Table 1, including: (i) environmental

Received: 6 February 2025
Revised: 7 February 2025
Accepted: 13 February 2025
Published: 16 February 2025

Citation: Mostoni, S.; Nisticò, R. New Advances into Nanostructured Oxides, 2nd Edition. *Inorganics* **2025**, *13*, 60. https://doi.org/10.3390/inorganics13020060

Copyright: © 2025 by the authors. Licensee MDPI, Basel, Switzerland. This article is an open access article distributed under the terms and conditions of the Creative Commons Attribution (CC BY) license (https://creativecommons.org/licenses/by/4.0/).

remediation, (ii) development and optimization of new synthetic routes for metal oxides, and (iii) drug-delivery and biomedicine. This classification was made considering both the materials applications and the applied synthetic routes for the preparation of the materials.

Table 1. Correlation between subsections and contributions collected in the present Special Issue.

Subsections	Title	References
Environmental remediation	"Efficient catalytic reduction of organic pollutants using nanostructured CuO/TiO_2 catalysts: synthesis, characterization, and reusability"	[18]
	"Core/shell ZnO/TiO_2, SiO_2/TiO_2, Al_2O_3/TiO_2, and $Al_{1.9}Co_{0.1}O_3/TiO_2$ nanoparticles for the photodecomposition of Brilliant Blue E-4BA"	[19]
	"Significantly enhanced self-cleaning capability in anatase TiO_2 for the bleaching of organic dyes and glazes"	[20]
	"Research progress of TiO_2 modification and photodegradation of organic pollutants"	[21]
	"Morphological dependence of metal oxide photocatalysts for dye degradation"	[22]
Development and optimization of new synthetic routes for metal oxides	"The influence of annealing temperature on the microstructure and electrical properties of sputtered ZnO thin films"	[23]
	"Synthesis and redox properties of iron and iron oxide nanoparticles obtained by exsolution from perovskite ferrites promoted by auxiliary reactions"	[24]
	"Mesoporous titania nanoparticles for a high-end valorization of *Vitis vinifera* grape marc extracts"	[25]
Drug-delivery and biomedicine	"Precipitative coating of calcium phosphate on microporous silica–titania hybrid particles in simulated body fluid"	[26]
	"The story, properties and applications of bioactive glass "1d": from concept to early clinical trials"	[27]

2.1. Environmental Remediation

A large portion of the contributions reported in this Special Issue is focused on the exploitation of inorganic nanomaterials (mainly TiO_2) for environmental remediation. In fact, the urgency of solving environmental pollution caused by the anthropogenic activities both in gaseous and liquid phases has driven research interest from more than a decade [28–30]. In this field, the use of photocatalysis is one of the most studied techniques to reduce the concentration of organic pollutants in wastewater, by using inorganic nanomaterials as efficient photo-catalysts. At the same time, organic pollutants can be degraded by using other catalytic reactions, such as catalytic reduction reactions in the presence of suitable nanostructured catalysts. In this first section, three research papers and two reviews were published [18–22].

Abouri et al. [18] reported the synthesis of nanostructured CuO/TiO_2 catalysts via a combustion technique, followed by calcination at 700 °C to achieve a rutile-phase TiO_2 structure with varying copper loadings (in the 5–40 wt.% range). Tests were performed aiming to reduce 4-Nitrophenol and Methyl Orange as target molecules with sodium borohydride ($NaBH_4$) in the presence of the CuO/TiO_2 catalysts. Results indicated a 98% reduction of 4-Nitrophenol in 480 s and 98% reduction of Methyl Orange in 420 s. The catalysts exhibited high stability over 10 reuse cycles, maintaining over 96% efficiency for Methyl Orange and 94% efficiency for 4-Nitrophenol.

Dolatyari et al. [19] reported the synthesis of core/shell ZnO/TiO_2, SiO_2/TiO_2, and Al_2O_3/TiO_2 nanoparticles using ethylene glycol for governing the nanoparticle size, followed by calcination at 400 °C to decompose the organic residues at the nanoparticles' surface. Furthermore, Cobalt (Co^{3+}) was added during the Al_2O_3 nanoparticle synthesis to form Co-doped nanoparticles with formula $Al_{1.9}Co_{0.1}O_3/TiO_2$. All of the synthesized nanoparticles were tested for the photocatalytic degradation of Brilliant Blue E-4BA under UV and visible light irradiation. Photocatalytic tests revealed that both Al_2O_3/TiO_2 and $Al_{1.9}Co_{0.1}O_3/TiO_2$ showed superior degradation under UV and visible light compared to ZnO/TiO_2 and SiO_2/TiO_2 with complete photodecomposition of the target dye (20 ppm) in only 20 min using a 10 mg of photocatalyst. Furthermore, the "Co-doped" $Al_{1.9}Co_{0.1}O_3/TiO_2$ nanoparticles showed the best performance under visible light irradiation, due to the increased absorption in the visible range as Co-doping introduces additional energy levels into Al_2O_3, resulting in improved electron–hole pair generation.

Zhao et al. [20] reported the hydro/solvothermal synthesis of Mg-doped TiO_2 anatase samples in water/ethanol environment at 180 °C for 36 h without using any surfactants or templates. Subsequently, glaze samples were prepared by sintering the raw powders (95% Kaolin clay, 5% Mg-doped TiO_2) at ca. 1200 °C. Photocatalytic experiments revealed that Mg-doped TiO_2 samples have higher photocatalytic activities (99.5% in 80 min under visible light) against Rhodamine B compared with undoped ones, due to pure anatase phase formation. Moreover, ceramic glaze materials present self-cleaning properties, achieving a water contact angle of ca. 6° at room temperature.

Mao et al. [21] analyzed the scientific literature describing the latest advances in the synthesis procedures and strategies to produce and/or properly modify TiO_2-based nanomaterials, aiming at maximizing/improving the photocatalytic performances in environmental remediation processes for the abatement of organic pollutants.

Lastly, Naggar et al. [22] reviewed the latest scientific literature describing the recent progress regarding the use of metal oxides in photocatalysis, with a particular focus on the critical role played by their morphology in the overall degradation process. The state-of-the-art analysis revealed that non-spherical morphologies exhibit enhanced photocatalytic performance due to their unique crystal facets and surface areas, which can promote charge transfer and improve catalytic efficiency. Furthermore, porous design and substantial specific surface area are responsible for an increased photocatalytic activity, whereas flake-like structures exhibit comparatively lower performance.

2.2. Development and Optimization of New Synthetic Routes for Metal Oxides

Careful consideration has been dedicated in this Special Issue to the development of new synthetic routes for metal oxides, and to the optimization of the synthetic approaches for the preparation of metal oxide nanostructures. In fact, the use of specific experimental conditions can drive the structural and morphological properties of the materials, and the fine control of these parameters play a key role in the determination of the material performances. Moreover, the use of innovative and more recent methodologies is reported, as well as the preparation of hybrid materials [23–25].

Alshoaibi [23] reported a study involving the deposition of a hetero-structured (ZnO/Zn/ZnO) thin film on a glass substrate using the DC magnetron sputtering technique. Subsequently, samples were annealed at different temperatures in the 100–500 °C range. Characterization results indicated the formation of both metallic zinc and the hexagonal ZnO crystal structure for samples annealed below 200 °C, whereas pure hexagonal ZnO formed for samples annealed at 300 and 500 °C, with a slight crystallinity decrease for the sample annealed at the highest temperature. Since both roughness and particle size are inversely proportional to the annealing temperature (with the exception of the sample annealed at 500 °C), the optimal annealing temperature was determined to be 400 °C.

Filimonov et al. [24] reported an original approach to synthesize hollow and layered oxide magnetic nanoparticles (either Fe_3O_4, $\gamma\text{-}Fe_2O_3$, or $Fe_3O_4/La_{1-x}Ca_xFeO_{3-\gamma}$), by a solid-state exsolution process carried out in a reducing environment at elevated temperatures from Ca- and La-based unsubstituted (and substituted) perovskite-related ferrites, and using h-BN as a reducing agent.

Lastly, Abduraman et al. [25] reported a synthetic sol–gel protocol assisted by solvothermal treatment (100 °C, 24 h) using either a triblock copolymer (Pluronic P123) or a nonionic surfactant (Pluronic F127) as soft-templating agents, followed by purification through either Soxhlet extraction or calcination at 400 °C for the production of mesoporous titania nanoparticles. The results indicated that samples prepared using Pluronic F127 presented a higher surface area and less agglomeration than the sample synthesized with Pluronic P123. Furthermore, an extract from *Vitis vinifera* grape marc (*Feteasca neagra* cultivar) with high radical scavenging activity was encapsulated in mesoporous titania and compared with reference SBA-15 silica support. Both resulting materials showed biocompatibility and even better radical scavenging potential than the free extract. Furthermore, the titania encapsulated sample showed better cytocompatibility than the silica one, thus making it suitable for skin-care products.

2.3. Drug Delivery and Biomedicine

In this last part, inorganic nanomaterials for drug delivery systems and for biomedicine purposes are discussed [26,27].

Kimura et al. [26] reported the development of a calcium phosphate-coating method to homogeneously cover silica–titania porous nanoparticles (with a well-defined spherical shape, uniform size, and tunable nanoporous structure) in simulated body fluids. The results indicated that the pore size distribution is a fundamental parameter significantly affecting the coating formation, with surfaces with bimodal pore sizes becoming rough after the calcium phosphate precipitation, whereas those with a unimodal pore size remaining smooth, thus indicating that pore sizes serve as different nucleation sites leading to different surface morphologies.

Lastly, Tulyaganov et al. [27] reviewed and critically discussed the genesis, development, properties, and applications of the bioactive glass "1d" (i.e., from the primary crystallization field of pseudo-wollastonite in the $CaO\text{–}MgO\text{–}SiO_2$ ternary system, after addition of P_2O_5, Na_2O and CaF_2) and its relevant glass-ceramic derivative products (i.e., diopside, fluorapatite, and wollastonite crystalline phases, formed by performing a thermal treatment), which are extremely appealing inorganic biomaterials alternatives to the reference 45S5 Bioglass® exploitable in a variety of bone-regenerative clinical applications, such as the repair of periodontal defects, ridge preservation and sinus augmentation.

3. Conclusions

With this Special Issue "New Advances into Nanostructured Oxides, 2nd Edition" published in the "Inorganics Materials" section, and also published as a book, the Editors hope that the high quality of the contributions collected here will receive the visibility and attention they deserve. These would help readers to increase their knowledge in the field of inorganic materials, and be a new source of inspiration for novel, focused investigations.

Acknowledgments: The Editors would like to thank all authors, reviewers, and the entire editorial staff of *Inorganics* who provided their new science, constructive recommendations, and assisted in the realization of the present Special Issue.

Conflicts of Interest: The authors declare no conflicts of interest.

References

1. Innocenzi, P.; Malfatti, L. Mesoporous ordered titania films: An advanced platform for photocatalysis. *J. Photochem. Photobiol. C* **2024**, *58*, 100646. [CrossRef]
2. Wang, L.; Zhang, J.; Zhang, Y.; Yu, H.; Qu, Y.; Yu, J. Inorganic metal-oxide photocatalyst for H_2O_2 production. *Small* **2022**, *18*, 2104561. [CrossRef] [PubMed]
3. Kumar, A.; Priyanka, C.; Kumar, A.; Camargo, P.H.C.; Venkata, K. Recent advances in plasmonic photocatalysis based on TiO_2 and noble metal nanoparticles for energy conversion, environmental remediation, and organic synthesis. *Small* **2022**, *18*, 2101638. [CrossRef] [PubMed]
4. Zhang, Q.; Gao, S.; Yu, J. Metal sites in zeolites: Synthesis, characterization, and catalysis. *Chem. Rev.* **2023**, *123*, 6039–6106. [CrossRef] [PubMed]
5. Qiao, R.; Fu, C.; Forgham, H.; Javed, I.; Huang, X.; Zhu, J.; Whittaker, A.K.; Davis, T.P. Magnetic iron oxide nanoparticles for brain imaging and drug delivery. *Adv. Drug Deliv. Rev.* **2023**, *197*, 114822. [CrossRef]
6. Wagner, J.; Goßl, D.; Ustyanovska, N.; Xiong, M.; Hauser, D.; Zhuzhgova, O.; Hočevar, S.; Taskoparan, B.; Poller, L.; Datz, S.; et al. Mesoporous silica nanoparticles as pH-responsive carrier for the immune-activating drug resiquimod enhance the local immune response in mice. *ACS Nano* **2021**, *15*, 4450–4466. [CrossRef] [PubMed]
7. Caldera, F.; Nisticò, R.; Magnacca, G.; Matencio, A.; Khazaei Monfared, Y.; Trotta, F. Magnetic composites of dextrin-based carbonate nanosponges and iron oxide nanoparticles with potential application in targeted drug delivery. *Nanomaterials* **2022**, *12*, 754. [CrossRef] [PubMed]
8. Chairungsri, W.; Subkomkaew, A.; Kijjanapanich, P.; Chimupala, Y. Direct dye wastewater photocatalysis using immobilized titanium dioxide on fixed substrate. *Chemosphere* **2022**, *286*, 131762. [CrossRef] [PubMed]
9. Danish, M.S.S.; Estrella, L.L.; Alemaida, I.M.A.; Lisin, A.; Moiseev, N.; Ahmadi, M.; Nazari, M.; Wali, M.; Zaheb, H.; Senjyu, T. Photocatalytic applications of metal oxides for sustainable environmental remediation. *Metals* **2021**, *11*, 80. [CrossRef]
10. Abdullah, F.H.; Abu Bakar, N.H.H.; Abu Bakar, M. Current advancements on the fabrication, modification, and industrial application of zinc oxide as photocatalyst in the removal of organic and inorganic contaminants in aquatic systems. *J. Hazard. Mater.* **2022**, *424*, 127416. [CrossRef]
11. Mezzomo, L.; Bonato, S.; Mostoni, S.; Di Credico, B.; Scotti, R.; D'Arienzo, M.; Mustarelli, P.; Ruffo, R. Composite solid-state electrolyte based on hybrid poly (ethylene glycol)-silica fillers enabling long-life lithium metal batteries. *Electrochim. Acta* **2022**, *411*, 140060. [CrossRef]
12. Sun, L.; Liu, H.; Ye, Y.; Lei, Y.; Islam, R.; Tan, S.; Tong, R.; Miao, Y.-B.; Cai, L. Smart nanoparticles for cancer therapy. *Sig. Transduct. Target. Ther.* **2023**, *8*, 418. [CrossRef] [PubMed]
13. Nisticò, R.; Scalarone, D.; Magnacca, G. Sol-gel chemistry, templating and spin-coating deposition: A combined approach to control in a simple way the porosity of inorganic thin films/coatings. *Microporous Mesoporous Mater.* **2017**, *248*, 18–29. [CrossRef]
14. Zhou, Y.; Liu, L.; Li, G.; Hu, G. Insights into the influence of ZrO_2 crystal structures on methyl laurate hydrogenation over Co/ZrO_2 catalysts. *ACS Catal.* **2021**, *11*, 7099–7113. [CrossRef]
15. Liccardo, L.; Bordin, M.; Sheverdyaeva, P.M.; Belli, M.; Moras, P.; Vomiero, A.; Moretti, E. Surface defect engineering in colored TiO_2 hollow spheres toward efficient photocatalysis. *Adv. Funct. Mater.* **2023**, *33*, 2212486. [CrossRef]
16. Masoud, M.; Khodamorady, M.; Tahmasbi, B.; Bahrami, K.; Ghorbani-Choghamarani, A. Boehmite nanoparticles as versatile support for organic–inorganic hybrid materials: Synthesis, functionalization, and applications in eco-friendly catalysis. *J. Ind. Eng. Chem.* **2021**, *97*, 1–78. [CrossRef]
17. Bona, G.; Viganò, L.; Cantoni, M.; Mantovan, R.; Di Credico, B.; Mostoni, S.; Scotti, R.; Nisticò, R. An experimental demonstration on the recyclability of hybrid magnetite-humic acid nanoparticles. *Sustain. Mater. Technol.* **2025**, *43*, e01275. [CrossRef]
18. Abouri, M.; Benzaouak, A.; Zaaboul, F.; Sifou, A.; Dahhou, M.; El Belghiti, M.A.; Azzaoui, K.; Hammouti, B.; Rhazi, L.; Sabbahi, R.; et al. Efficient Catalytic Reduction of Organic Pollutants Using Nanostructured CuO/TiO_2 Catalysts: Synthesis, Characterization, and Reusability. *Inorganics* **2024**, *12*, 297. [CrossRef]
19. Dolatyari, M.; Tahmasebi, M.; Dolatyari, S.; Rostami, A.; Zarghami, A.; Yadav, A.; Klein, A. Core/Shell ZnO/TiO_2, SiO_2/TiO_2, Al_2O_3/TiO_2, and $Al_{1.9}Co_{0.1}O_3/TiO_2$ Nanoparticles for the Photodecomposition of Brilliant Blue E-4BA. *Inorganics* **2024**, *12*, 281. [CrossRef]
20. Zhao, T.; Cao, T.; Bao, Q.; Dong, W.; Li, P.; Gu, X.; Liang, Y.; Zhou, J. Significantly Enhanced Self-Cleaning Capability in Anatase TiO_2 for the Bleaching of Organic Dyes and Glazes. *Inorganics* **2023**, *11*, 341. [CrossRef]
21. Mao, T.; Zha, J.; Hu, Y.; Chen, Q.; Zhang, J.; Luo, X. Research Progress of TiO_2 Modification and Photodegradation of Organic Pollutants. *Inorganics* **2024**, *12*, 178. [CrossRef]
22. Naggar, A.H.; Ahmed, A.S.A.; El-Nasr, T.A.S.; Alotaibi, N.F.; Chong, K.F.; Ali, G.A.M. Morphological Dependence of Metal Oxide Photocatalysts for Dye Degradation. *Inorganics* **2023**, *11*, 484. [CrossRef]
23. Alshoaibi, A. The Influence of Annealing Temperature on the Microstructure and Electrical Properties of Sputtered ZnO Thin Films. *Inorganics* **2024**, *12*, 236. [CrossRef]

24. Filimonov, D.; Rozova, M.; Maksimov, S.; Pankratov, D. Synthesis and Redox Properties of Iron and Iron Oxide Nanoparticles Obtained by Exsolution from Perovskite Ferrites Promoted by Auxiliary Reactions. *Inorganics* **2024**, *12*, 223. [CrossRef]
25. Abduraman, A.; Brezoiu, A.-M.; Tatia, R.; Iorgu, A.-I.; Deaconu, M.; Mitran, R.-A.; Matei, C.; Berger, D. Mesoporous Titania Nanoparticles for a High-End Valorization of *Vitis vinifera* Grape Marc Extracts. *Inorganics* **2024**, *12*, 263. [CrossRef]
26. Kimura, R.; Shiba, K.; Fujiwara, K.; Zhou, Y.; Yamada, I.; Tagaya, M. Precipitative Coating of Calcium Phosphate on Microporous Silica–Titania Hybrid Particles in Simulated Body Fluid. *Inorganics* **2023**, *11*, 235. [CrossRef]
27. Tulyaganov, D.U.; Agathopoulos, S.; Dimitriadis, K.; Fernandes, H.R.; Gabrieli, R.; Baino, F. The Story, Properties and Applications of Bioactive Glass "1d": From Concept to Early Clinical Trials. *Inorganics* **2024**, *12*, 224. [CrossRef]
28. Zamora-Ledezma, C.; Negrete-Bolagay, D.; Figuroa, F.; Zamora-Ledezma, E.; Ni, M.; Alexis, F.; Guerrero, V.H. Heavy metal water pollution: A fresh look about hazards, novel and conventional remediation methods. *Environ. Technol. Innov.* **2021**, *21*, 101504. [CrossRef]
29. Rahman, A.; Sarkar, A.; Yadav, O.P.; Achari, G.; Slobodnik, J. Potential human health risks due to environmental exposure to nano- and microplastics and knowledge gaps: A scoping review. *Sci. Total Environ.* **2021**, *757*, 143872. [CrossRef] [PubMed]
30. Siddiqua, A.; Hahladakis, J.N.; Al-Attiya, W.A.K.A. An overview of the environmental pollution and health effects associated with waste landfilling and open dumping. *Environ. Sci. Pollut. Res.* **2022**, *29*, 58514–58536. [CrossRef] [PubMed]

Disclaimer/Publisher's Note: The statements, opinions and data contained in all publications are solely those of the individual author(s) and contributor(s) and not of MDPI and/or the editor(s). MDPI and/or the editor(s) disclaim responsibility for any injury to people or property resulting from any ideas, methods, instructions or products referred to in the content.

Article

Efficient Catalytic Reduction of Organic Pollutants Using Nanostructured CuO/TiO$_2$ Catalysts: Synthesis, Characterization, and Reusability

Mariyem Abouri [1,2], Abdellah Benzaouak [3], Fatima Zaaboul [2], Aicha Sifou [2], Mohammed Dahhou [2], Mohammed Alaoui El Belghiti [1], Khalil Azzaoui [4,5], Belkheir Hammouti [5], Larbi Rhazi [6], Rachid Sabbahi [7,*], Mohammed M. Alanazi [8] and Adnane El Hamidi [2,*]

[1] Laboratory of Spectroscopy, Molecular Modeling, Materials, Nanomaterial, Water and Environment, Faculty of Sciences, Mohammed V University in Rabat, Avenue Ibn Battouta, Rabat BP1014, Morocco; mariyemab@gmail.com (M.A.); m.elbelghiti@um5r.ac.ma (M.A.E.B.)

[2] Laboratory of Materials, Nanotechnologies and Environment, Faculty of Sciences, Mohammed V University in Rabat, Avenue Ibn Battouta, Rabat BP1014, Morocco; zaaboul.fatima00@gmail.com (F.Z.); aichasifou@gmail.com (A.S.); m.dahhou@um5r.ac.ma (M.D.)

[3] Laboratory of Spectroscopy, Molecular Modeling, Materials, Nanomaterials, Water and Environment, Environmental Materials Team, ENSAM, Mohammed V University, B.P. 765, Rabat 10090, Morocco; a.benzaouak@um5r.ac.ma

[4] Engineering Laboratory of Organometallic, Molecular Materials and Environment, Faculty of Sciences, Sidi Mohammed Ben Abdellah University, UEMF, Fes 30000, Morocco; k.azzaoui@yahoo.com

[5] Euromed Research Center, Euromed Polytechnic School, Euromed University of Fes, Eco-Campus, Fes Meknes Road, Fes 30030, Morocco; hammoutib@gmail.com

[6] Institut Polytechnique UniLaSalle, Université d'Artois, ULR 7519, 19 rue Pierre Waguet, BP 30313, 60026 Beauvais, France; larbi.rhazi@unilasalle.fr

[7] Research Team in Science and Technology, Higher School of Technology, Ibn Zohr University, Quartier 25 Mars, P.O. Box 3007, Laayoune 70000, Morocco

[8] Department of Pharmaceutical Chemistry, College of Pharmacy, King Saud University, Riyadh 11451, Saudi Arabia; mmalanazi@ksu.edu.sa

* Correspondence: r.sabbahi@uiz.ac.ma (R.S.); adnane.elhamidi@fsr.um5.ac.ma (A.E.H.)

Citation: Abouri, M.; Benzaouak, A.; Zaaboul, F.; Sifou, A.; Dahhou, M.; El Belghiti, M.A.; Azzaoui, K.; Hammouti, B.; Rhazi, L.; Sabbahi, R.; et al. Efficient Catalytic Reduction of Organic Pollutants Using Nanostructured CuO/TiO$_2$ Catalysts: Synthesis, Characterization, and Reusability. *Inorganics* 2024, 12, 297. https://doi.org/10.3390/inorganics12110297

Academic Editors: Roberto Nisticò and Silvia Mostoni

Received: 17 October 2024
Revised: 15 November 2024
Accepted: 15 November 2024
Published: 19 November 2024

Copyright: © 2024 by the authors. Licensee MDPI, Basel, Switzerland. This article is an open access article distributed under the terms and conditions of the Creative Commons Attribution (CC BY) license (https://creativecommons.org/licenses/by/4.0/).

Abstract: The catalytic reduction of organic pollutants in water is a critical environmental challenge due to the persistent and hazardous nature of compounds like azo dyes and nitrophenols. In this study, we synthesized nanostructured CuO/TiO$_2$ catalysts via a combustion technique, followed by calcination at 700 °C to achieve a rutile-phase TiO$_2$ structure with varying copper loadings (5–40 wt.%). The catalysts were characterized using X-ray diffraction (XRD), attenuated total reflectance-Fourier transform infrared (ATR–FTIR) spectroscopy, thermogravimetric analysis-differential thermal analysis (TGA–DTA), UV-visible diffuse reflectance spectroscopy (DRS), and scanning electron microscopy with energy-dispersive X-ray spectroscopy (SEM–EDS). The XRD results confirmed the presence of the crystalline rutile phase in the CuO/TiO$_2$ catalysts, with additional peaks indicating successful copper oxide loading onto TiO$_2$. The FTIR spectra confirmed the presence of all the functional groups in the prepared samples. SEM images revealed irregularly shaped copper oxide and agglomerated TiO$_2$ particles. The DRS results revealed improved optical properties and a decreased bandgap with increased Cu content, and 4-Nitrophenol (4-NP) and methyl orange (MO), which were chosen for their carcinogenic, mutagenic, and nonbiodegradable properties, were used as model organic pollutants. Catalytic activities were tested by reducing 4-NP and MO with sodium borohydride (NaBH$_4$) in the presence of a CuO/TiO$_2$ catalyst. Following the in situ reduction of CuO/TiO$_2$, Cu (NPs)/TiO$_2$ was formed, achieving 98% reduction of 4-NP in 480 s and 98% reduction of MO in 420 s. The effects of the NaBH$_4$ concentration and catalyst mass were investigated. The catalysts exhibited high stability over 10 reuse cycles, maintaining over 96% efficiency for MO and 94% efficiency for 4-NP. These findings demonstrate the potential of nanostructured CuO/TiO$_2$ catalysts for environmental remediation through efficient catalytic reduction of organic pollutants.

Keywords: titanium dioxide; catalysts; catalytic reduction; 4-nitrophenol; methyl orange; copper oxide

1. Introduction

Across the world, water and energy shortages represent urgent and increasingly severe global challenges. On a global level, over one billion people face the critical issue of not having access to clean drinking water, which is a necessity for human existence. Furthermore, more than two billion individuals lack proper sanitation facilities, resulting in almost two million annual fatalities due to diseases transmitted through impure water sources or insufficient sewage systems [1]. Water pollution occurs when the concentration of harmful chemicals or biological substances in a water body surpasses established standards, leading to adverse impacts on both human health and the natural environment [2].

Azo dyes and nitroaromatic compounds exhibit high stability and are known for their carcinogenic, mutagenic, and nonbiodegradable properties, posing a significant threat to human life, water quality, and the environment. These substances, such as Congo red, rhodamine B (RhB), and methyl orange (MO), are extensively used in numerous chemical sectors, including paper, textiles, paint, and plastics. These industrial processes produce significant quantities of dyes, leading to the emergence of dye-laden particles that are introduced directly into the environment [3,4]. MO, for example, is a highly toxic and nonbiodegradable azo. The adverse effects of this toxic dye may disrupt the balance of water within ecosystems and pose health risks, including symptoms such as vomiting, diarrhea, breathing difficulties, and nausea [5]. In particular, 4-Nitrophenol (4-NP) is a highly toxic compound that is challenging to degrade and treat effectively [6].

Compared with alternative methods, nitrophenol and dye reduction in the presence of a suitable catalyst and sodium borohydride ($NaBH_4$) is a recommended protocol, since it is relatively more affordable, secure, and environmentally friendly [7–9]. Therefore, the exploration of techniques such as chemical reduction to efficiently degrade diverse chemical pollutants is of considerable interest. However, this method commonly faces a challenge in achieving a fast degradation rate at lower concentrations of reducing agents, such as sodium borohydride. Consequently, to minimize the quantity of reducing agent required and increase the reaction rate, the development of an efficient catalyst for this chemical reduction becomes crucial [10,11]. Owing to their exceptional physicochemical proprieties, there has been significant interest in the field of catalysis regarding transition and noble metal nanoparticles in recent years [12–16]. Among these materials, copper nanoparticles (CuNPs) have gained recognition as excellent materials that exhibit novel physiochemical characteristics. This has prompted their investigation in various applications, including catalysis, sensors, photocatalysis, diverse biological activities, energy storage, and organic synthesis applications [17–21]. The catalytic activity is closely related to the degree of dispersion of the Cu(NPs), and a superior performance has been noted with smaller-sized nanoparticles [22,23].

One significant challenge in employing nanoparticles as catalysts involves their agglomeration and accumulation, resulting in a decrease in their catalytic efficiency. A viable solution to address this concern is the utilization of solid supports to stabilize the nanoparticles [24]. For this purpose, a series of solid supports, such as polymers, metal oxides, and carbon materials [25–28], have been adopted to avoid sintering, enhance stability, and facilitate the optimal mobilization/dispersion of nanoparticles, aiming to maximize catalytic activities across a wide variety of applications. Among these supports, the immobilization of CuNPs on TiO_2, which is a representative n-type semiconductor material, has been extensively used as a photocatalyst, catalyst support, and cocatalyst because of its high oxidation ability, environmental friendliness, physicochemical stability, and cost effectiveness [29,30]. Research on the use of Cu/TiO_2 for catalytic reduction has gained significant attention because of its promising applications in environmental remediation. One study focused on synthesizing $Cu-TiO_2$ nanoparticles from the extract of Phoenix dactylifera. These catalysts effectively degraded dyes, such as RhB, in 11 min and MO in 25 min, achieving reductions of 89.8 and 95.3%, respectively, with 21.3 mg/mL of the catalyst [31]. Other investigations explored various shapes and sizes of Cu nanostructures and revealed that smaller particles and specific morphologies enhanced the catalytic activity for nitroaromatic reduction [32]. Additionally, Cu/TiO_2 catalyst synthesized from *Chimonanthus praecox* extract exhibited superior performance in degrading pollutants, such as 4-nitrophenol and other organic dyes, with 10 mg of catalyst [33]. CuO/TiO_2 composite created using *Tilia platyphyllos* extract also demonstrated high catalytic activity, reducing MO in 10 min and methyl blue in 9 min with 3 mg of the catalyst [34]. Recent studies of CuO/TiO_2 nanocomposite have further highlighted their exceptional catalytic performance, particularly under direct sunlight, where these photocatalysts effectively drive the selective hydrogenation of 4-NP to 4-AP in the presence of $NaBH_4$ [35]. Moreover, binary CuO/TiO_2 composites have demonstrated excellent reactivity in tandem hydrogenation processes for nitro compounds [36].

In the present study, different wt.% CuO/TiO_2 (rutile) were synthesized through the combustion technique followed by calcination at 700 °C to ensure a complete transition from anatase to rutile. The prepared catalysts were characterized via X-ray diffraction (XRD), UV–vis diffuse reflectance spectroscopy, attenuated total reflectance–Fourier transform infrared (ATR–FTIR) spectroscopy, thermogravimetric analysis-differential thermal analysis (TGA–DTA), and scanning electron microscopy with energy-dispersive X-ray spectroscopy (SEM–EDS). The catalytic performance of CuO/TiO_2 was evaluated afterwards via the reduction of MO and 4-nitophenol as models of organic contaminants in the presence of $NaBH_4$. The influence of parameters, including $NaBH_4$ concentration, catalyst mass and reusability of the catalyst, were examined.

2. Results and Discussion

2.1. X-Ray Diffraction

The XRD patterns of the CuO/TiO_2 catalyst series are presented in Figure 1. The results obtained reveal that the materials are primarily composed of the rutile phase of TiO_2 in accordance with ICDD 21-1276. The presence of CuO in its monoclinic form was also verified in accordance with the ICDD 48-1548 reference pattern, confirming the effective loading of copper oxide in TiO_2. Figure 1 also reveals a correlation between the deposited quantity of Cu(II) ions and the intensities of the characteristic CuO peaks. These findings collectively indicate the successful preparation of the CuO/TiO_2 catalysts with varying percentages of CuO deposition. The average crystallite size of CuO was determined to be 26, 27, 32, 35, 36 nm for 40, 30, 20, 10 and 5 wt.% CuO/TiO_2 catalysts, respectively. The crystallite size was determined using the Scherrer equation [37]:

$$L = \frac{0.9\lambda}{\beta \cos \theta}$$

where λ is the X-ray wavelength (0.1540 nm), β is the full width at half maximum (FWHM), θ is Bragg's angle, and L represents the average crystallite size.

On the other hand, the absence of the anatase phase of TiO_2 is due to its transformation to rutile upon heating at 700 °C, as observed in Figure 2, which shows the XRD curves for the 40 wt.% CuO/TiO_2 catalysts subjected to various calcination temperatures. The primary difference observed was the phase transition of TiO_2 from anatase (ICDD 21-1272) to rutile (ICDD 21-1276), with complete transformation occurring at 700 °C. The same results were observed in many previous studies [38].

Following the reduction experiment using $NaBH_4$, the rutile structure of TiO_2 remained unchanged, indicating its stability under the reaction conditions. However, the monoclinic structure of CuO significantly decreased, leading to the formation of Cu(NPs), as depicted in Figure 3. Notably, two additional peaks emerged at 43.28° and 50.37°, corresponding to the (111) and (200) planes, respectively. These peaks are characteristic of the face-centered cubic (FCC) crystal structure of metallic copper nanoparticles (ICDD No. 04-0836). The average Cu(NP) size is about 15 nm.

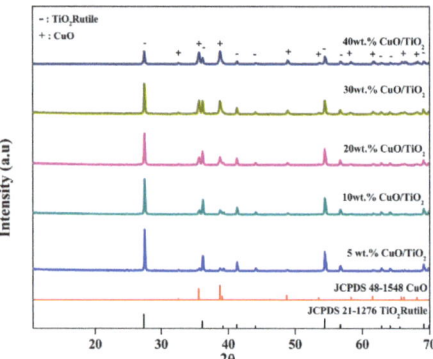

Figure 1. XRD patterns of different wt.% CuO/TiO_2 samples.

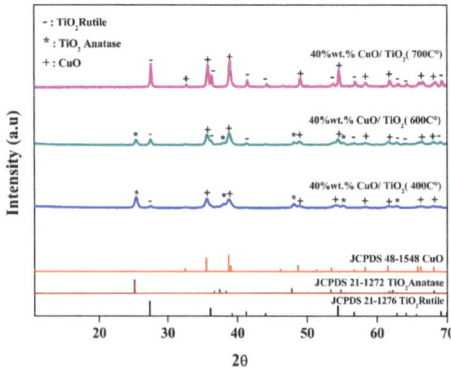

Figure 2. XRD patterns of 40 wt.% CuO/TiO_2 samples calcined at different temperatures.

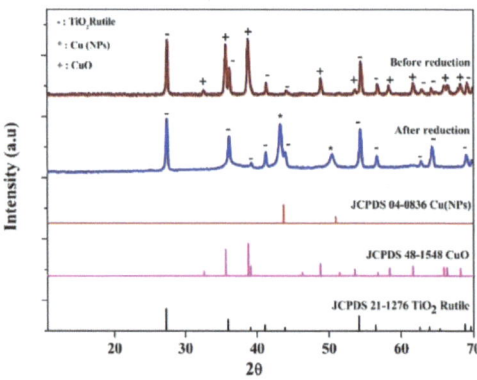

Figure 3. XRD patterns of 40 wt.% CuO/TiO_2 before and after the reduction experiment.

2.2. Thermal Analysis Measurement

To examine the thermal behavior of CuO/TiO_2, thermal experiments involving thermogravimetric and differential measurements were performed. The results are shown in Figure 4. The initial step, characterized by a weight loss of 3% within the temperature range of 30 to 150 °C, can be attributed to the vaporization of water [39]. The notable weight reduction occurring between 200 and 350 °C, as evaluated at 30%, appears to be associated with the decomposition and combustion of the organic constituents. The latter was confirmed by an intense exothermic peak observed in the DTA curve and two intense peaks in the DTG curve [40]. On the other hand, no weight loss was observed at temperatures above 450 °C, and the anatase to rutile phase transition that should occur above 600 °C [38] was not observed in the DTA curve because of its reduced sensitivity compared with that of differential scanning calorimetry (DSC).

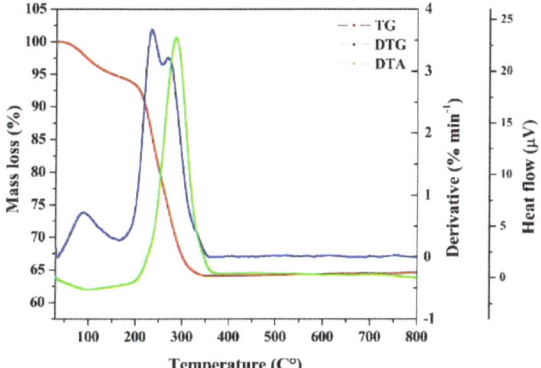

Figure 4. Thermal curves of the as-prepared 40 wt.% CuO/TiO_2 without calcination under air flow at 10 °C min^{-1}.

2.3. ATR–FTIR Spectroscopy

ATR–FTIR spectroscopy was performed on the catalysts calcined at 700 °C. The measurements were conducted across the spectrum ranging from 400 to 4000 cm^{-1}, and the results are shown in Figure 5. The analysis revealed similar results for the different catalysts with different percentages of CuO. The absence of water molecules and other contaminants in the catalyst was confirmed by the observation of strong bands in the infrared spectrum just below 1000 cm^{-1}. These bands typically correspond to lattice vibrations within metal oxides. The peak observed at approximately 650 cm^{-1} is associated with the presence of the Ti-O-O stretching vibration bond. Additionally, the spectra revealed two other distinct bonds below 500 cm^{-1} that were attributed to metal–oxygen (M–O) bond vibrations. All the observed bands suggest the existence of both Ti–O vibrational stretching and Cu–O stretching vibration bonds [41], which confirms the successful incorporation of CuO into TiO_2.

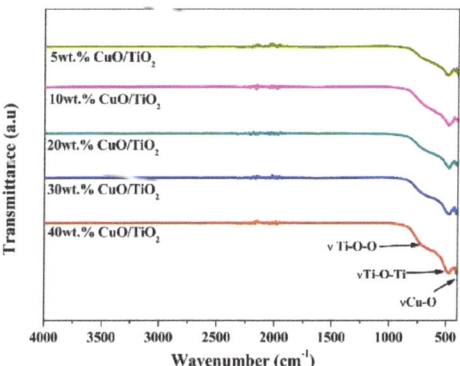

Figure 5. Infrared spectra of the wt.% CuO/TiO_2 samples.

2.4. UV–Vis Diffuse Reflectance Spectroscopy

Figure 6a presents the diffuse reflectance spectroscopy (DRS) results for TiO_2 and the various wt.% CuO/TiO_2 catalysts, revealing their optical properties in the 200–800 nm range. TiO_2 exhibits a pronounced absorption peak below 400 nm, which is attributed to its intrinsic interbond absorption [42]. In contrast, the spectrum of the wt.% CuO/TiO_2 catalysts displays additional absorption bands in the 400–800 nm range. The absorption band between 400 and 600 nm is attributed to charge transfer from the valence band (VB) of TiO_2 to CuO. Additionally, the lower absorption band within the 500–800 nm range is associated with the intrinsic exciton band of CuO and the d–d transition of Cu^{2+} species [43]. The band gap energies (Eg) of TiO_2, CuO, and the different wt.% CuO/TiO_2 catalysts were determined by identifying the intersection of the linear portion of the $(\alpha h\nu)^2$ vs. energy (eV) curves, as detailed in Figure 6b. The Eg values for TiO_2 and wt.% CuO/TiO_2 are summarized in Table 1. These results indicate that increasing the CuO content leads to a progressive narrowing of the band gap. This trend can be attributed to the incorporation of Cu, which introduces additional energy states within the band structure, thereby lowering the bandgap. The reduced bandgap in the CuO/TiO_2 catalysts enhances their visible light absorption, which is beneficial for photocatalytic applications [44].

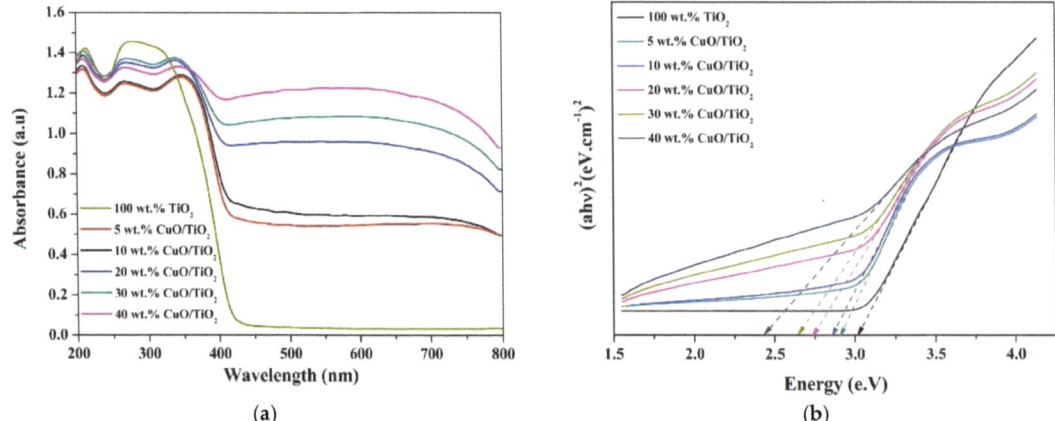

Figure 6. (a) UV–vis diffuse reflectance spectra of TiO_2 and wt.% CuO/TiO_2 and (b) Tauc plot.

Table 1. Bandgaps of different catalysts.

Catalyst	TiO_2	5 wt.% CuO	10 wt.% CuO	20 wt.% CuO	30 wt.% CuO	40 wt.% CuO
Band gap (eV)	3.02	2.91	2.86	2.73	2.65	2.43

2.5. SEM–EDS Analysis

Morphological studies via SEM–EDS were carried out on the CuO/TiO_2 catalyst. Figure 7 shows representative SEM images of synthesized CuO/TiO_2, which reveal a homogeneous spherical morphology with varying diameters and numerous micrometer-sized particles. Additionally, the catalyst has a porous structure and some aggregation of spheres. The analysis of the CuO/TiO_2 catalyst via energy dispersive X-ray (EDX) spectroscopy aimed to examine the presence and distribution of copper, titanium, and oxygen within the material. The EDX spectrum of CuO/TiO_2 is shown in Figure 8, and the results indicate the successful incorporation of copper into the TiO_2 sample. Additionally, the elemental mapping analysis revealed a uniform distribution of copper (depicted in yellow), titanium (represented in pink), and oxygen (shown in blue) in the samples, as illustrated in Figure 9.

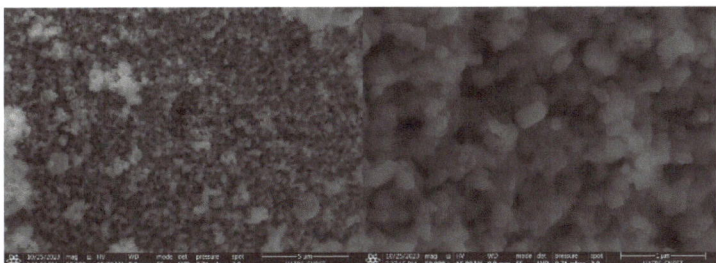

Figure 7. SEM micrographs of 40 wt.% CuO/TiO$_2$.

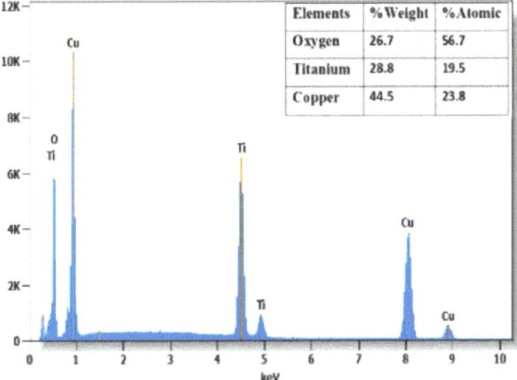

Figure 8. EDS Spectrum of 40 wt.% CuO/TiO$_2$.

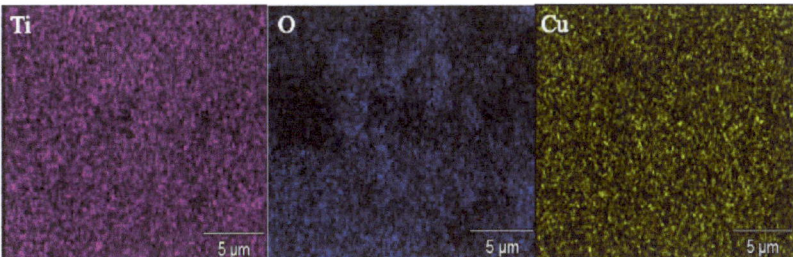

Figure 9. EDX mapping of the 40 wt.% CuO/TiO$_2$ catalyst.

2.6. Catalytic Reduction of 4-NP and MO by a CuO/TiO$_2$ Catalyst

The catalytic performance of the CuO/TiO$_2$ samples with different CuO loadings (5 to 40 wt.%) was evaluated for the reduction of organic pollutants via NaBH$_4$, as shown in Figure 10a,b. As shown in Figure 10a, which represents the reduction of 4-nitrophenol (4-NP), the 40 wt.% CuO/TiO$_2$ catalyst exhibited the fastest catalytic activity, reaching nearly 100% removal within a shorter reaction time than the other catalysts and pure CuO. Similarly, in Figure 10b, which shows the reduction of MO, 40 wt.% CuO/TiO$_2$ again demonstrates the fastest catalytic performance. The pure CuO sample exhibited lower catalytic activity than the 40 wt.% CuO/TiO$_2$ sample did, likely due to the agglomeration of copper oxide nanoparticles, which reduced the number of accessible active sites. On the other hand, at lower CuO loadings (e.g., 5 wt.%), the catalytic activity is the slowest, which can be attributed to the lower number of active CuO/TiO$_2$ sites available for the reduction process.

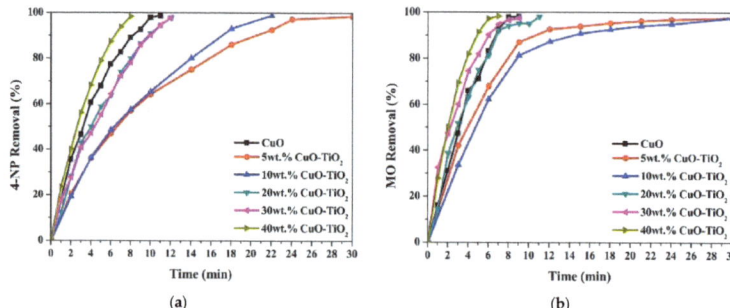

Figure 10. Reduction performance of the wt.% CuO/TiO$_2$ catalysts for 4-NP (**a**) and MO (**b**).

Figure 11a,b display the absorption spectra of 4-NP and MO in the presence of NaBH$_4$ but without the use of a catalyst. Under these conditions, the peak intensities show little change even after 180 min, indicating a negligible reduction rate, as no significant degradation occurred during this time. However, when CuO/TiO$_2$ is introduced alongside NaBH$_4$ (Figure 11c,d) and forming Cu(NPs)/TiO$_2$, a rapid and noticeable reduction in the characteristic absorption peaks of both 4-NP and MO is observed. For 4-NP, the peak at ~400 nm diminished progressively, vanishing completely within 480 s (Figure 11c). Similarly, the absorption peak of MO at approximately 464 nm disappears within 420 s (Figure 11d). This demonstrates a much faster and more efficient reduction process in the presence of the CuO/TiO$_2$ catalyst than in the presence of NaBH$_4$ alone. Simultaneously, a pair of new peaks emerged at 297 nm and 231 nm for the reduction of 4-NP and at 242 nm during the reduction of MO. These new peaks can be attributed to the characteristic absorption bands of the colorless 4-aminophenol (4-AP) from 4-NP and sulfanilic acid along with dimentyl-4-phenylenediamine from OM [45,46].

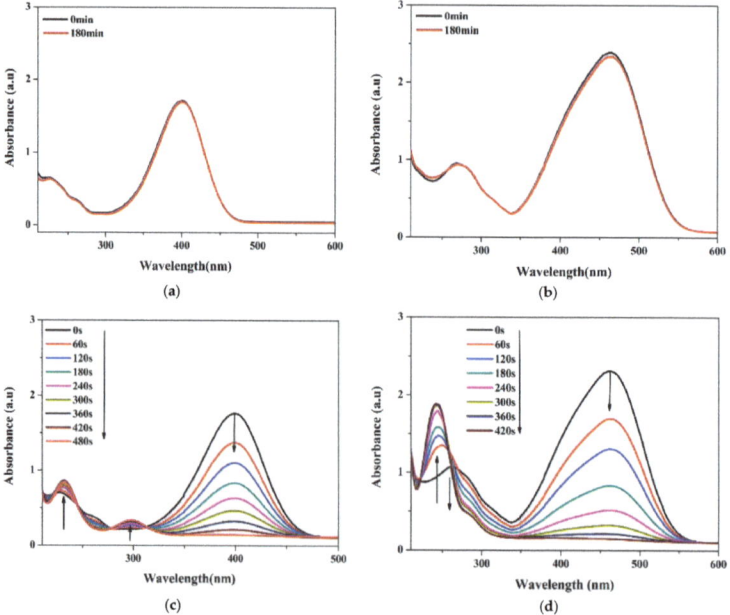

Figure 11. UV–vis spectra of the reduction of (**a**) 4-NP and (**b**) MO by NaBH$_4$ without the catalyst and absorbance spectra of (**c**) 4-NP and (**d**) MO in the presence of NaBH$_4$ and CuO/TiO$_2$.

2.6.1. Kinetic Study

The pseudo-first-order kinetic model of Langmuir–Hinshelwood was employed to analyze the kinetics data due to the substantial excess of NaBH$_4$ compared with the organic dyes, denoted as ([NaBH$_4$]/[Pollutant]) = 40. The rate constants (k$_{app}$) governing the conversion reactions for both 4-NP and MO were determined via the following equation:

$$\ln\frac{C_t}{C_0} = \ln\frac{A_t}{A_0} = -k_{app}t \quad (1)$$

As depicted in Figure 12a,b, the Ln (A$_t$/A$_0$) vs. time of reaction was determined by analyzing the absorption peak intensities at 400 nm (4-NP) and 464 nm (MO). Remarkably, the reduction process was initiated promptly, devoid of any induction time requirement. The graphical representation of Ln (A$_t$/A$_0$) against time exhibited a robust linear correlation, aligning well with the expectations of pseudo-first-order kinetics. Consequently, the calculated rate constants were determined to be 0.00517 s^{-1} for 4-NP and 0.00727 s^{-1} for MO. To enable a comparative assessment of CuO/TiO$_2$ performance, an activity factor denoted as k$'$ was computed via the following equation:

$$k' = \frac{k_{app}}{m} \quad (2)$$

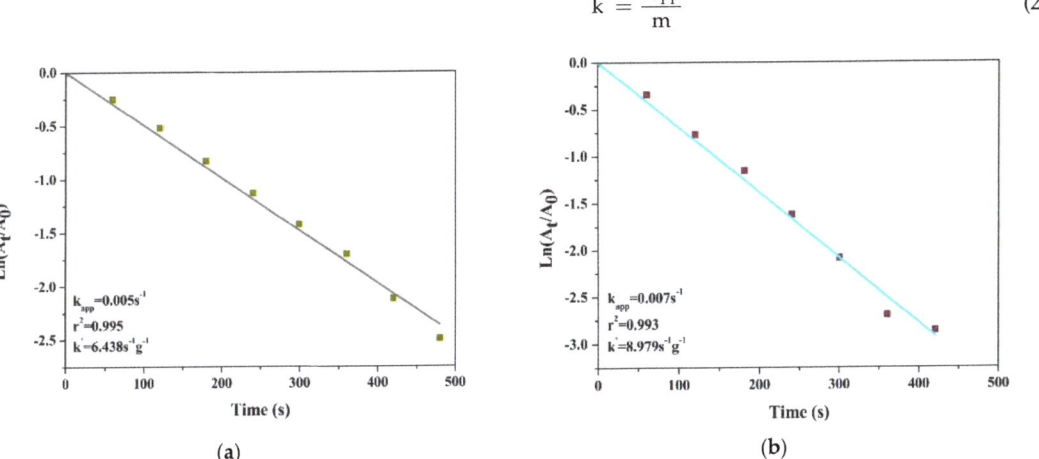

Figure 12. First-order plot for the catalytic reduction of (**a**) 4-NP and (**b**) MO. Experimental conditions: 40 wt.% CuO/TiO$_2$ dose = 0.04 g/L, [NaBH$_4$] = 14 mM, and [4-NP or MO] = 0.35 mM.

In this equation, m(g) represents the mass of the copper in the catalyst involved in the reaction.

In this context, the derived k$'$ values were 6.46 s^{-1} g^{-1} for 4-NP and 8.98 s^{-1} g^{-1} for MO. For a comprehensive comparison, Table 2 illustrates the varying efficacies of Cu(NPs) catalysts documented in the literature.

Tables 3 and 4 show an increase in the observed rate constants (k$_{app}$) with increasing CuO percentage in the catalyst. The 40 wt.% CuO/TiO$_2$ catalyst exhibits the highest rate constant (0.005 s^{-1}), indicating that a higher copper loading favors faster reduction kinetics for 4-NP and MO. However, when considering the mass of active copper in the catalyst, the normalized values (k$'$) indicate that the 5 wt.% CuO/TiO$_2$ catalyst exhibits the highest activity (17.5 s^{-1}g^{-1}). This suggests that, although increasing the CuO loading accelerates the overall reaction, the efficiency per unit mass of copper is maximized at lower loadings.

After the in situ reductive reaction of CuO/TiO$_2$ to Cu (NPs)/TiO$_2$ with NaBH$_4$, an underlying mechanism enables the catalytic reduction of 4-NP and MO using Cu(NPs)/TiO$_2$ in the presence of NaBH$_4$ (Figure 13). In this process, the BH^{4-} ions adhere to the surface of the catalyst, thereby initiating the creation of BO^{2-} through the self-hydrolysis of NaBH$_4$. Moreover, BH^{4-} interacts with the catalyst, facilitating the transfer of active hydrogen species, leading to the formation of an energetically charged hydrogen layer on the catalyst surface [47]. The organic pollutants subsequently attach themselves to the catalyst surface, undergoing reduction into 4-AP for 4-NP and sulfanilic acid along with dimentyl-4-phenylenediamine for MO during the step that governs the reaction rate. The produced compounds are then desorbed from the surface of the Cu(NPs)/TiO$_2$ catalyst. As a result, Cu(NPs)/TiO$_2$ exhibits highly efficient catalytic reduction due to the supply of electrons to the catalyst by BH$_4^-$ ions, a mechanism that enables 4-NP and MO to bond with the catalyst surface, consequently yielding enhanced catalytic activity [48].

Table 2. Comparison of the catalytic properties of catalysts with those of Cu(NPs) reported in the literature and this work.

Pollutant	Catalyst	k' (s^{-1}g^{-1})	Time (min)	Conditions			Reference
				NaBH$_4$ (M)	Pollutant (M)	Catalyst Mass (mg)	
4-Nitrophenol (4-NP)	Cu10/MZ	5.00	10	0.0045	0.0000675	1	[22]
	Cu10/ZSM-5	1.17	50	0.0045	0.0000675	1	[22]
	CuO/TiO$_2$	6.46	8	0.014	0.00035	2	This work
	Cu NPs@Fe$_3$O$_4$-LS	1.87	3	0.125	0.00125	7	[4]
	MnO@Cu/C	17.33	1.5	0.0193	0.0008	4	[16]
	C@Cu	59.00	1	0.033	0.00014	1	[49]
	Cu-Ag/GP	0.40	10	0.0714	0.000857	10	[50]
	Cu-Ni/GP	0.60	7	0.0714	0.000857	10	[50]
	CuVOS@SiO$_2$-3	1.57	2	0.00528	0.00014377	5	[51]
	CuVOS-3	8.20	2	0.00582	0.000138	5	[52]
Methyl orange (MO)	C@Cu	62.00	1	0.033	0.000061	1	[49]
	CuO/TiO$_2$	8.98	7	0.014	0.00035	2	This work
	Cu-Ag/GP	0.77	4	0.0714	0.00004286	10	[50]
	Cu-Ni/GP	0.38	5	0.0714	0.00004286	10	[50]
	CuVOS@SiO$_2$-3	1.37	4	0.00528	0.0003055	5	[51]
	CuVOS-3	6.47	2	0.00582	0.00029	5	[52]

Table 3. Rate constants obtained for 4-NP reduction by wt.% CuO/TiO$_2$.

Catalyst	k$_{app}$ (s^{-1})	k' (s^{-1}g^{-1})
5 wt.% CuO/TiO$_2$	0.00175	17.5
10 wt.% CuO/TiO$_2$	0.00175	8.7
20 wt.% CuO/TiO$_2$	0.00328	8.2
30 wt.% CuO/TiO$_2$	0.00308	5.1
40 wt.% CuO/TiO$_2$	0.00512	6.4

Table 4. Rate constants obtained for MO reduction by wt.% CuO/TiO$_2$.

Catalyst	k$_{app}$ (s^{-1})	k' (s^{-1}g^{-1})
5 wt.% CuO/TiO$_2$	0.00150	15.0
10 wt.% CuO/TiO$_2$	0.00126	6.3
20 wt.% CuO/TiO$_2$	0.00479	12.0
30 wt.% CuO/TiO$_2$	0.00546	9.1
40 wt.% CuO/TiO$_2$	0.00727	8.9

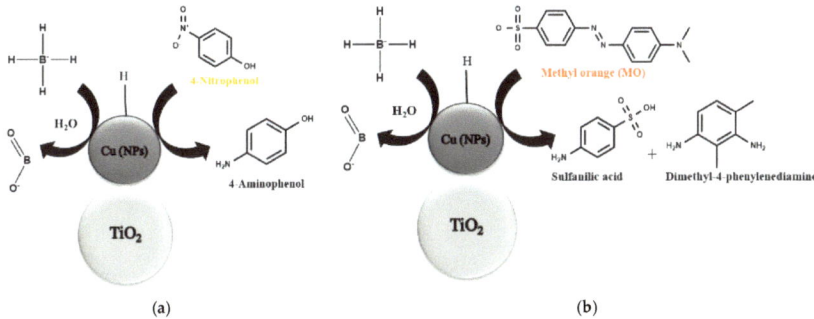

Figure 13. Schematic reactions of (**a**) 4-NP and (**b**) MO, by NaBH$_4$ in the presence of Cu(NPs)/TiO$_2$.

2.6.2. Effects of Parameters and Reusability

Multiple factors were investigated to understand their influence on the catalyzed reduction of 4-NP and MO using 40 wt.% CuO/TiO$_2$ catalysts. The parameters examined included the catalyst dosage and concentration of NaBH$_4$. Figure 14a,b show the impact of varying catalyst amounts on the rate of pollutant reduction. By making more active sites accessible for the reaction, the large amount of catalyst led to improved efficiency. The performance constantly increased as the catalyst amount increased from 0.02 g/L to 0.06 g/L, which confirms the principle of Sabatier [53]. The variation in NaBH$_4$ concentration had a pronounced influence on the catalytic reduction of 4-NP and MO (Figure 14c,d). Because of the limited quantity of hydrogen emitted, the catalytic reduction rate decreased as the NaBH$_4$ concentration decreased [54].

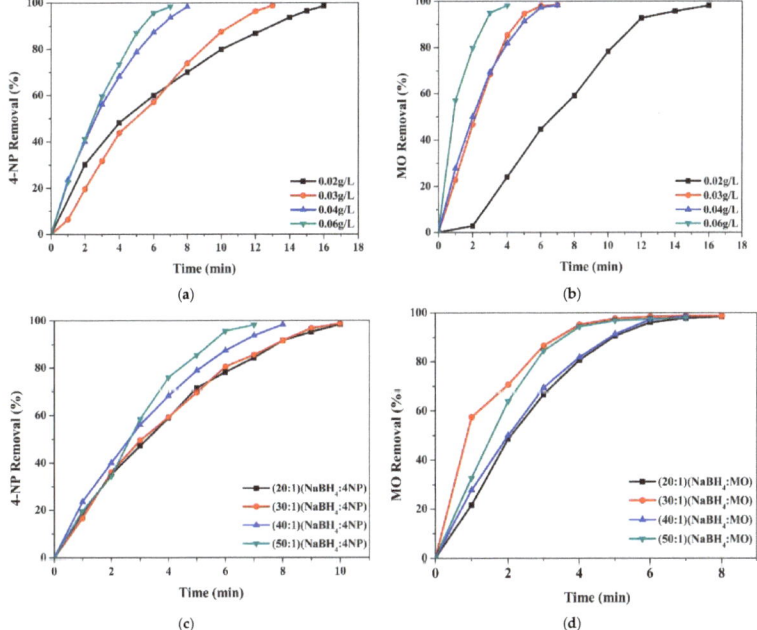

Figure 14. Pollutant removal versus time plots for catalytic reduction using 40 wt.% CuO/TiO$_2$, showing the effect of catalyst dose on the reduction of (**a**) 4-NP and (**b**) MO, and the effect of NaBH$_4$ concentration on the reduction of (**c**) 4-NP and (**d**) MO.

An essential characteristic of catalysts is their stability and reusability. To assess this, a fresh solution containing pollutants was introduced after each catalytic reduction cycle. The initial concentration remained constant at 0.35 mM throughout. As depicted in Figure 15a,b, the catalyst consistently maintained over 96% MO capacity and 94% 4-NP capacity for reduction of the dyes across 10 successive reaction cycles. This outcome highlights the exceptional reusability of the CuO/TiO$_2$ catalyst. This excellent reusability can be attributed to the even distribution and robust stability of Cu on the TiO$_2$ surface, which in turn provides a greater number of active sites for catalytic processes.

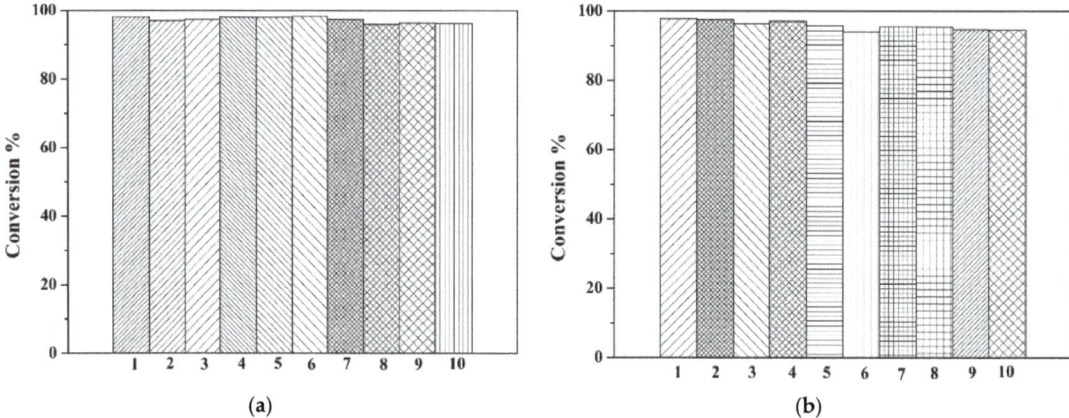

Figure 15. Recycling of CuO/TiO$_2$ in the hydrogenation of (a) 4-NP and (b) MO.

3. Materials and Methods

3.1. Synthesis of the CuO/TiO$_2$ Catalysts

The combustion technique was used to prepare different percentages of wt.% CuO/TiO$_2$ (wt = 5, 10, 20, 30, and 40%). Analytical grade Cu(NO$_3$)$_2$·6H$_2$O (Sigma Aldrich, ≥99.9%), along with TiO$_2$-P25 (Degussa Aeroxide P25) and citric acid as a fuel (Sigma Aldrich, ≥99.5%), served as the starting materials. Initially, these materials were added to deionized water with a fixed weight percentage of metal on TiO$_2$ and an excess amount of citric acid. The resulting mixtures were then stirred to achieve a well-dispersed suspension. The suspension subsequently underwent evaporation from 80 to 130 °C to yield a thick gel. This gel was subsequently converted into a dry form. Finally, the dry gel was calcined in an air atmosphere at 700 °C for 4 h with a ramp rate of 10 °C/min. Figure 16 represents the synthesis method for the catalysts.

Figure 16. Synthesis procedure for the catalysts.

3.2. Catalytic Reduction of Organic Pollutants

The catalytic activity of CuO/TiO$_2$ was evaluated by studying the reduction of MO (Sigma Aldrich, ≥98.0%) and 4-nitrophenylphenol (Sigma Aldrich, ≥99.0%) as the target pollutants for the experiments. The experimental setup involved a beaker (50 mL) placed at room temperature. The catalytic reaction proceeded as follows: 2 mg of CuO/TiO$_2$ was uniformly dispersed in 35 mL of deionized water, followed by the addition of 10 mL of freshly prepared aqueous NaBH$_4$ (Sigma Aldrich, ≥99.0%) (70 mM). The solution was sonicated for 10 min, resulting in a color change from a black CuO/TiO$_2$ suspension to a gray color, indicating the formation of Cu(NPs)/TiO$_2$. Next, 5 mL of MO (3.5 mM) or 4-NP (3.5 mM) was added to the mixture. The overall concentrations of the organic dyes and NaBH$_4$ were 0.35 mM and 14 mM, respectively. The progress of the reaction was monitored by recording the time-dependent UV–vis absorption spectra of the reaction mixture in the wavelength range between 210 and 600 nm via a spectrophotometer (Techcomp UV 2300). The catalytic reduction experiment is shown in Figure 17.

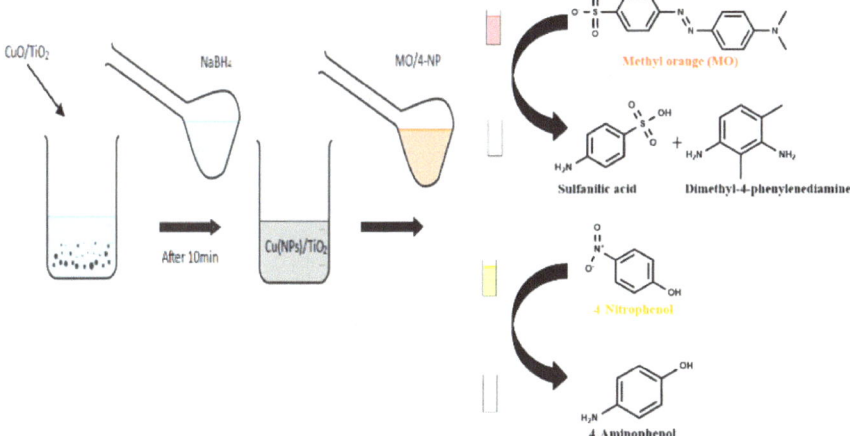

Figure 17. Catalytic reduction experiments of organic pollutants.

3.3. Characterization Techniques

The identification of the mineral crystalline phase of the catalysts was carried out via X-ray diffraction, via a Schimadzu 6100 powder diffractometer with a monochromatic beam ($\lambda_{Cuk\alpha}$ = 1.541838°). These measurements were taken at room temperature over a 2θ range of 10° to 70°, with a scanning rate of 2°/min. A UV–vis spectrophotometer, namely, a PerkinElmer Lamda 900 UV/Vis/NIR spectrometer, was used to obtain diffuse reflectance spectra of the catalysts in the wavelength range of 200–800 nm. To identify the main functional groups of the catalyst, ATR–FTIR spectroscopy was used with a Nicolet iS50 instrument with a resolution of 4 cm^{-1} in the spectral range of 400–4000 cm^{-1}. The thermal stability and degradation behavior of the catalysts were verified via simultaneous TGA and DTA under an air flow rate of 30 mL/min via a Labsys^{TM} (1F) Setaram instrument. An 8 mg sample was placed in an alumina crucible and heated from 30 °C to 800 °C at a heating rate of 10 °C/min. The surface morphology and chemical analysis of the samples were performed by SEM-EDX on a Quattros S-FEG-Thermofisher scientific instrument.

4. Conclusions

In this study, CuO/TiO$_2$ catalysts were prepared through a combustion technique. The catalysts were characterized via XRD, ATR–FTIR spectroscopy, TGA–DTA, UV–vis DRS, and SEM-EDX techniques. Compared with TiO$_2$ and CuO, the presence of Cu in the CuO/TiO$_2$ catalysts significantly enhanced the reduction of 4-NP and MO. The fastest dye reduction was obtained with 40 wt.% CuO, followed by 40 > 30 > 20 > 10 > 5 wt.% CuO/TiO$_2$ in the presence of NaBH$_4$ and 2 mg of catalyst. The kinetic data were successfully modeled via the pseudo-first-order Langmuir–Hinshelwood mechanism, yielding rate constants of 0.000517 s^{-1} for 4-NP and 0.00727 s^{-1} for MO. The efficiency of 4-NP and MO reduction with the CuO/TiO$_2$ catalyst improved as the catalyst dosage increased from 0.02 g/L to 0.06 g/L, offering more active sites. However, a decrease in NaBH$_4$ concentration led to a decrease in the catalytic reduction rate due to reduced hydrogen production. The presence of TiO$_2$ rutile functions as a protective agent to slow catalyst damage and agglomeration of particles. This material can be reused for several cycles without significant loss of catalytic activity.

Author Contributions: Conceptualization, M.A.; methodology, M.A.; software, M.A. and B.H.; validation, A.B. and A.E.H.; formal analysis, M.A. and A.B.; investigation, M.A.; resources, A.E.H. and A.B.; data curation, M.A.; writing—original draft preparation, M.A.; writing—review and editing, A.B., F.Z., A.E.H., A.S., M.D., K.A., R.S. and L.R.; visualization F.Z., M.D. and L.R.; supervision, A.E.H., A.B., M.A.E.B. and M.M.A. All authors have read and agreed to the published version of the manuscript.

Funding: This research was supported by the Researchers Supporting Project number (RSPD2024R628), King Saud University, Riyadh, Saudi Arabia.

Institutional Review Board Statement: Not applicable.

Data Availability Statement: The data presented in this study are available upon request from the corresponding author.

Acknowledgments: The authors extend their appreciation to the Researchers Supporting Project number (RSPD2024R628), King Saud University, Riyadh, Saudi Arabia for supporting this research.

Conflicts of Interest: The authors declare no conflicts of interest.

References

1. Szymczyk, A.; van der Bruggen, B.; Ulbricht, M. Surface Modification of Water Purification Membranes. In *Surface Modification of Polymers*; John Wiley & Sons, Ltd.: Hoboken, NJ, USA, 2019; pp. 363–398. ISBN 978-3-527-81924-9.
2. Bilal, M.; Khan, S.; Ali, J.; Ismail, M.; Khan, M.I.; Asiri, A.M.; Khan, S.B. Biosynthesized Silver Supported Catalysts for Disinfection of Escherichia Coli and Organic Pollutant from Drinking Water. *J. Mol. Liq.* **2019**, *281*, 295–306. [CrossRef]
3. Ismail, M.; Khan, M.I.; Khan, S.B.; Akhtar, K.; Khan, M.A.; Asiri, A.M. Catalytic Reduction of Picric Acid, Nitrophenols and Organic Azo Dyes via Green Synthesized Plant Supported Ag Nanoparticles. *J. Mol. Liq.* **2018**, *268*, 87–101. [CrossRef]
4. Nezafat, Z.; Karimkhani, M.M.; Nasrollahzadeh, M.; Javanshir, S.; Jamshidi, A.; Orooji, Y.; Jang, H.W.; Shokouhimehr, M. Facile Synthesis of Cu NPs@Fe$_3$O$_4$-Lignosulfonate: Study of Catalytic and Antibacterial/Antioxidant Activities. *Food Chem. Toxicol.* **2022**, *168*, 113310. [CrossRef] [PubMed]
5. Munagapati, V.S.; Yarramuthi, V.; Kim, D.-S. Methyl Orange Removal from Aqueous Solution Using Goethite, Chitosan Beads and Goethite Impregnated with Chitosan Beads. *J. Mol. Liq.* **2017**, *240*, 329–339. [CrossRef]
6. Albukhari, S.M.; Ismail, M.; Akhtar, K.; Danish, E.Y. Catalytic Reduction of Nitrophenols and Dyes Using Silver Nanoparticles @ Cellulose Polymer Paper for the Resolution of Waste Water Treatment Challenges. *Colloids Surf. A Physicochem. Eng. Asp.* **2019**, *577*, 548–561. [CrossRef]
7. El-Aal, M.A.; Said, A.E.-A.A.; Goda, M.N.; Abo Zeid, E.F.; Ibrahim, S.M. Fe$_3$O$_4$@CMC-Cu Magnetic Nanocomposite as an Efficient Catalyst for Reduction of Toxic Pollutants in Water. *J. Mol. Liq.* **2023**, *385*, 122317. [CrossRef]
8. Ghorbani-Vaghei, R.; Veisi, H.; Aliani, M.H.; Mohammadi, P.; Karmakar, B. Alginate Modified Magnetic Nanoparticles to Immobilization of Gold Nanoparticles as an Efficient Magnetic Nanocatalyst for Reduction of 4-Nitrophenol in Water. *J. Mol. Liq.* **2021**, *327*, 114868. [CrossRef]
9. Kaid, M.; Ali, A.E.; Shamsan, A.Q.S.; Younes, S.M.; Abdel-Raheem, S.A.A.; Abdul-Malik, M.A.; Salem, W.M. Efficiency of Maturation Oxidation Ponds as a Post-Treatment Technique of Wastewater. *Curr. Chem. Lett.* **2022**, *11*, 415–422. [CrossRef]
10. Lajevardi, A.; Tavakkoli Yaraki, M.; Masjedi, A.; Nouri, A.; Hossaini Sadr, M. Green Synthesis of MOF@Ag Nanocomposites for Catalytic Reduction of Methylene Blue. *J. Mol. Liq.* **2019**, *276*, 371–378. [CrossRef]
11. Zhang, K.; Suh, J.M.; Choi, J.-W.; Jang, H.W.; Shokouhimehr, M.; Varma, R.S. Recent Advances in the Nanocatalyst-Assisted NaBH$_4$ Reduction of Nitroaromatics in Water. *ACS Omega* **2019**, *4*, 483–495. [CrossRef]
12. Landge, V.K.; Sonawane, S.H.; Manickam, S.; Bhaskar Babu, G.U.; Boczkaj, G. Ultrasound-Assisted Wet-Impregnation of Ag–Co Nanoparticles on Cellulose Nanofibers: Enhanced Catalytic Hydrogenation of 4-Nitrophenol. *J. Environ. Chem. Eng.* **2021**, *9*, 105719. [CrossRef]

13. Mekki, A.; Mokhtar, A.; Hachemaoui, M.; Beldjilali, M.; Meliani, M.F.; Zahmani, H.H.; Hacini, S.; Boukoussa, B. Fe and Ni Nanoparticles-Loaded Zeolites as Effective Catalysts for Catalytic Reduction of Organic Pollutants. *Microporous Mesoporous Mater.* **2021**, *310*, 110597. [CrossRef]
14. Omole, M.A.; K'Owino, I.O.; Sadik, O.A. Palladium Nanoparticles for Catalytic Reduction of Cr(VI) Using Formic Acid. *Appl. Catal. B Environ.* **2007**, *76*, 158–167. [CrossRef]
15. Saravanakumar, K.; Priya, V.S.; Balakumar, V.; Prabavathi, S.L.; Muthuraj, V. Noble Metal Nanoparticles (Mx = Ag, Au, Pd) Decorated Graphitic Carbon Nitride Nanosheets for Ultrafast Catalytic Reduction of Anthropogenic Pollutant, 4-Nitrophenol. *Environ. Res.* **2022**, *212*, 113185. [CrossRef]
16. Shokry, R.; Abd El Salam, H.M.; Aman, D.; Mikhail, S.; Zaki, T.; El Rouby, W.M.A.; Farghali, A.A.; Al Zoubi, W.; Ko, Y.G. MOF-Derived Core–Shell MnO@Cu/C as High-Efficiency Catalyst for Reduction of Nitroarenes. *Chem. Eng. J.* **2023**, *459*, 141554. [CrossRef]
17. Alencar, L.M.; Silva, A.W.B.N.; Trindade, M.A.G.; Salvatierra, R.V.; Martins, C.A.; Souza, V.H.R. One-Step Synthesis of Crumpled Graphene Fully Decorated by Copper-Based Nanoparticles: Application in H_2O_2 Sensing. *Sens. Actuators B Chem.* **2022**, *360*, 131649. [CrossRef]
18. Chaerun, S.K.; Prabowo, B.A.; Winarko, R. Bionanotechnology: The Formation of Copper Nanoparticles Assisted by Biological Agents and Their Applications as Antimicrobial and Antiviral Agents. *Environ. Nanotechnol. Monit. Manag.* **2022**, *18*, 100703. [CrossRef]
19. Cuong, H.N.; Pansambal, S.; Ghotekar, S.; Oza, R.; Thanh Hai, N.T.; Viet, N.M.; Nguyen, V.-H. New Frontiers in the Plant Extract Mediated Biosynthesis of Copper Oxide (CuO) Nanoparticles and Their Potential Applications: A Review. *Environ. Res.* **2022**, *203*, 111858. [CrossRef]
20. Godiya, C.B.; Kumar, S.; Park, B.J. Superior Catalytic Reduction of Methylene Blue and 4-Nitrophenol by Copper Nanoparticles-Templated Chitosan Nanocatalyst. *Carbohydr. Polym. Technol. Appl.* **2023**, *5*, 100267. [CrossRef]
21. Patil, D.; Manjanna, J.; Chikkamath, S.; Uppar, V.; Chougala, M. Facile Synthesis of Stable Cu and CuO Particles for 4-Nitrophenol Reduction, Methylene Blue Photodegradation and Antibacterial Activity. *J. Hazard. Mater. Adv.* **2021**, *4*, 100032. [CrossRef]
22. Aslam, S.; Subhan, F.; Waqas, M.; Zifeng, Y.; Yaseen, M.; Naeem, M. Cu Nanoparticles Confined within ZSM-5 Derived Mesoporous Silica (MZ) with Enhanced Stability for Catalytic Hydrogenation of 4-Nitrophenol and Degradation of Azo Dye. *Microporous Mesoporous Mater.* **2023**, *354*, 112547. [CrossRef]
23. Patra, A.K.; Dutta, A.; Bhaumik, A. Cu Nanorods and Nanospheres and Their Excellent Catalytic Activity in Chemoselective Reduction of Nitrobenzenes. *Catal. Commun.* **2010**, *11*, 651–655. [CrossRef]
24. Taherian, Z.; Khataee, A.; Han, N.; Orooji, Y. Hydrogen Production through Methane Reforming Processes Using Promoted-Ni/Mesoporous Silica: A Review. *J. Ind. Eng. Chem.* **2022**, *107*, 20–30. [CrossRef]
25. Wang, C.; Liu, Y.; Cui, Z.; Yu, X.; Zhang, X.; Orooji, Y.; Zhang, Q.; Chen, L.; Ma, L. In Situ Synthesis of Cu Nanoparticles on Carbon for Highly Selective Hydrogenation of Furfural to Furfuryl Alcohol by Using Pomelo Peel as the Carbon Source. *ACS Sustain. Chem. Eng.* **2020**, *8*, 12944–12955. [CrossRef]
26. Ningsih, L.A.; Yoshida, M.; Sakai, A.; Andrew Lin, K.-Y.; Wu, K.C.W.; Catherine, H.N.; Ahamad, T.; Hu, C. Ag-Modified $TiO_2/SiO_2/Fe_3O_4$ Sphere with Core-Shell Structure for Photo-Assisted Reduction of 4-Nitrophenol. *Environ. Res.* **2022**, *214*, 113690. [CrossRef]
27. Tran, X.T.; Hussain, M.; Kim, H.T. Facile and Fast Synthesis of a Reduced Graphene Oxide/Carbon Nanotube/Iron/Silver Hybrid and Its Enhanced Performance in Catalytic Reduction of 4–Nitrophenol. *Solid State Sci.* **2020**, *100*, 106107. [CrossRef]
28. Zhao, X.; Lv, L.; Pan, B.; Zhang, W.; Zhang, S.; Zhang, Q. Polymer-Supported Nanocomposites for Environmental Application: A Review. *Chem. Eng. J.* **2011**, *170*, 381–394. [CrossRef]
29. Bakhsh, E.M.; Khan, S.B.; Maslamani, N.; Danish, E.Y.; Akhtar, K.; Asiri, A.M. Carboxymethyl Cellulose/Copper Oxide–Titanium Oxide Based Nanocatalyst Beads for the Reduction of Organic and Inorganic Pollutants. *Polymers* **2023**, *15*, 1502. [CrossRef]
30. Liao, X.; Zheng, L.; He, Q.; Li, G.; Zheng, L.; Li, H.; Tian, T. Fabrication of Ag/TiO_2 Membrane on Ti Substrate with Integral Structure for Catalytic Reduction of 4-Nitrophenol. *Process Saf. Environ. Prot.* **2022**, *168*, 792–799. [CrossRef]
31. Hassan, F.; Abbas, A.; Ali, F.; Nazir, A.; Al Huwayz, M.; Alwadai, N.; Iqbal, M.; Ali, Z. Bio-Mediated Synthesis of $Cu-TiO_2$ Nanoparticles Using *Phoenix Dactylifera* Lignocellulose as Capping and Reducing Agent for the Catalytic Degradation of Toxic Dyes. *Desalination Water Treat.* **2023**, *298*, 53–60. [CrossRef]
32. Kaur, R.; Pal, B. Cu Nanostructures of Various Shapes and Sizes as Superior Catalysts for Nitro-Aromatic Reduction and Co-Catalyst for Cu/TiO_2 Photocatalysis. *Appl. Catal. A Gen.* **2015**, *491*, 28–36. [CrossRef]
33. Akbari, R. Green Synthesis and Catalytic Activity of Copper Nanoparticles Supported on TiO_2 as a Highly Active and Recyclable Catalyst for the Reduction of Nitro-Compounds and Degradation of Organic Dyes. *J. Mater. Sci. Mater. Electron.* **2021**, *32*, 15801–15813. [CrossRef]
34. Khodadadi, B.; Yeganeh Faal, A.; Shahvarughi, A. Tilia Platyphyllos Extract Assisted Green Synthesis of $CuO/TiO2$ Nanocomposite: Application as a Reusable Catalyst for the Reduction of Organic Dyes in Water. *Journal of Applied Chemical Research* **2019**, *13*, 51–65.
35. Tian, X.; Dong, Y.; Zahid, M. One-Pot Synthesis of CuO/TiO Nanocomposites for Improved Photocatalytic Hydrogenation of 4-Nitrophenol to 4-Aminophenol under Direct Sunlight. *J. Chin. Chem. Soc.* **2023**, *70*, 848–856. [CrossRef]

36. Li, P.; Zhang, X.; Wang, J.; Xu, B.; Zhang, X.; Fan, G.; Zhou, L.; Liu, X.; Zhang, K.; Jiang, W. Binary CuO/TiO$_2$ Nanocomposites as High-Performance Catalysts for Tandem Hydrogenation of Nitroaromatics. *Colloids Surf. A Physicochem. Eng. Asp.* **2021**, *629*, 127383. [CrossRef]
37. Monshi, A.; Foroughi, M.R.; Monshi, M. Modified Scherrer Equation to Estimate More Accurately Nano-Crystallite Size Using XRD. *World J. Nano Sci. Eng.* **2012**, *2*, 154–160. [CrossRef]
38. Hanaor, D.A.H.; Sorrell, C.C. Review of the Anatase to Rutile Phase Transformation. *J. Mater. Sci.* **2011**, *46*, 855–874. [CrossRef]
39. Ha, C.A.; Nguyen, D.T.; Nguyen, T. Green Fabrication of Heterostructured CoTiO$_3$/TiO$_2$ Nanocatalysts for Efficient Photocatalytic Degradation of Cinnamic Acid. *ACS Omega* **2022**, *7*, 40163–40175. [CrossRef]
40. Lin, Y.-J.; Chang, Y.-H.; Yang, W.-D.; Tsai, B.-S. Synthesis and Characterization of Ilmenite NiTiO$_3$ and CoTiO$_3$ Prepared by a Modified Pechini Method. *J. Non-Cryst. Solids* **2006**, *352*, 789–794. [CrossRef]
41. Bopape, D.A.; Mathobela, S.; Matinise, N.; Motaung, D.E.; Hintsho-Mbita, N.C. Green Synthesis of CuO-TiO$_2$ Nanoparticles for the Degradation of Organic Pollutants: Physical, Optical and Electrochemical Properties. *Catalysts* **2023**, *13*, 163. [CrossRef]
42. Shi, Q.; Li, Y.; Zhan, E.; Ta, N.; Shen, W. Anatase TiO$_2$ Hollow Nanosheets: Dual Roles of F−, Formation Mechanism, and Thermal Stability. *CrystEngComm* **2014**, *16*, 3431–3437. [CrossRef]
43. Li, B.; Hao, Y.; Zhang, B.; Shao, X.; Hu, L. A Multifunctional Noble-Metal-Free Catalyst of CuO/TiO$_2$ Hybrid Nanofibers. *Appl. Catal. A Gen.* **2017**, *531*, 1–12. [CrossRef]
44. Angel, R.D.; Durán-Álvarez, J.C.; Zanella, R.; Angel, R.D.; Durán-Álvarez, J.C.; Zanella, R. TiO2-Low Band Gap Semiconductor Heterostructures for Water Treatment Using Sunlight-Driven Photocatalysis. In *Titanium Dioxide—Material for a Sustainable Environment*; IntechOpen: London, UK, 2018; ISBN 978-1-78923-327-8.
45. Sirohi, S.; Mittal, A.; Nain, R.; Jain, N.; Singh, R.; Dobhal, S.; Pani, B.; Parida, D. Effect of Nanoparticles Shape on Conductivity of Ag Nanoparticle PVA Composite Films. *Polym. Int.* **2019**, *68*, 1961–1967. [CrossRef]
46. Vidhu, V.K.; Philip, D. Catalytic Degradation of Organic Dyes Using Biosynthesized Silver Nanoparticles. *Micron* **2014**, *56*, 54–62. [CrossRef] [PubMed]
47. Li, X.; Ma, Y.; Yang, Z.; Huang, D.; Xu, S.; Wang, T.; Su, Y.; Hu, N.; Zhang, Y. In Situ Preparation of Magnetic Ni-Au/Graphene Nanocomposites with Electron-Enhanced Catalytic Performance. *J. Alloys Compd.* **2017**, *706*, 377–386. [CrossRef]
48. Lin, F.; Doong, R. Bifunctional Au−Fe3O4 Heterostructures for Magnetically Recyclable Catalysis of Nitrophenol Reduction. *J. Phys. Chem. C* **2011**, *115*, 6591–6598. [CrossRef]
49. Ahsan, M.A.; Jabbari, V.; El-Gendy, A.A.; Curry, M.L.; Noveron, J.C. Ultrafast Catalytic Reduction of Environmental Pollutants in Water via MOF-Derived Magnetic Ni and Cu Nanoparticles Encapsulated in Porous Carbon. *Appl. Surf. Sci.* **2019**, *497*, 143608. [CrossRef]
50. Ismail, M.; Khan, M.I.; Khan, S.B.; Khan, M.A.; Akhtar, K.; Asiri, A.M. Green Synthesis of Plant Supported CuAg and CuNi Bimetallic Nanoparticles in the Reduction of Nitrophenols and Organic Dyes for Water Treatment. *J. Mol. Liq.* **2018**, *260*, 78–91. [CrossRef]
51. Sun, H.; Abdeta, A.B.; Zelekew, O.A.; Guo, Y.; Zhang, J.; Kuo, D.-H.; Lin, J.; Chen, X. Spherical Porous SiO$_2$ Supported CuVOS Catalyst with an Efficient Catalytic Reduction of Pollutants under Dark Condition. *J. Mol. Liq.* **2020**, *313*, 113567. [CrossRef]
52. Sun, H.; Zelekew, O.A.; Chen, X.; Guo, Y.; Kuo, D.-H.; Lu, Q.; Lin, J. A Noble Bimetal Oxysulfide CuVOS Catalyst for Highly Efficient Catalytic Reduction of 4-Nitrophenol and Organic Dyes. *RSC Adv.* **2019**, *9*, 31828–31839. [CrossRef]
53. Ahmad, S.; Khan, S.B.; Asiri, A.M. Catalytic Efficiency of Copper Nanoparticles Modified Silica-Alginate Hydrogel Nanocomposite towards Reduction of Water Pollutants and H$_2$ Generation. *Int. J. Hydrogen Energy* **2023**, *48*, 6399–6417. [CrossRef]
54. Yang, K.; Yan, Y.; Wang, H.; Sun, Z.; Chen, W.; Kang, H.; Han, Y.; Zahng, W.; Sun, X.; Li, Z. Monodisperse Cu/Cu$_2$O@C Core–Shell Nanocomposite Supported on RGO Layers as an Efficient Catalyst Derived from a Cu-Based MOF/GO Structure. *Nanoscale* **2018**, *10*, 17647–17655. [CrossRef] [PubMed]

Disclaimer/Publisher's Note: The statements, opinions and data contained in all publications are solely those of the individual author(s) and contributor(s) and not of MDPI and/or the editor(s). MDPI and/or the editor(s) disclaim responsibility for any injury to people or property resulting from any ideas, methods, instructions or products referred to in the content.

Article

Core/Shell ZnO/TiO$_2$, SiO$_2$/TiO$_2$, Al$_2$O$_3$/TiO$_2$, and Al$_{1.9}$Co$_{0.1}$O$_3$/TiO$_2$ Nanoparticles for the Photodecomposition of Brilliant Blue E-4BA

Mahboubeh Dolatyari [1], Mehdi Tahmasebi [2], Sudabeh Dolatyari [2], Ali Rostami [1,3,*], Armin Zarghami [3], Ashish Yadav [4] and Axel Klein [5,*]

[1] SP-EPT Labs, ASEPE Company, Industrial Park of Advanced Technologies, Tabriz 5364196795, Iran; mdolatya@uni-koeln.de

[2] QC Lab, Alborz Farmad Company, Eshtehard Industrial Town, Karaj 3188115475, Iran; niksirat2000@yahoo.com (M.T.); sudabeh.dolatyari@gmail.com (S.D.)

[3] Photonics and Nanocrystal Research Lab (PNRL Lab), Faculty of Electrical and Computer Engineering, University of Tabriz, Tabriz 5166614761, Iran; zarghamiarmin@gmail.com

[4] Center for Advanced Laser Manufacturing (CALM), Shandong University of Technology, Zibo 255000, China; ashish@sdut.edu.cn

[5] University of Cologne, Faculty of Mathematics and Natural Sciences, Department of Chemistry and Biochemistry, Institute for Inorganic and Materials Chemistry, Greinstraße 6, D-50939 Cologne, Germany

* Correspondence: rostami@tabrizu.ac.ir (A.R.); axel.klein@uni-koeln.de (A.K.); Tel.: +49-221-470-4006 (A.K.)

Citation: Dolatyari, M.; Tahmasebi, M.; Dolatyari, S.; Rostami, A.; Zarghami, A.; Yadav, A.; Klein, A. Core/Shell ZnO/TiO$_2$, SiO$_2$/TiO$_2$, Al$_2$O$_3$/TiO$_2$, and Al$_{1.9}$Co$_{0.1}$O$_3$/TiO$_2$ Nanoparticles for the Photodecomposition of Brilliant Blue E-4BA. *Inorganics* **2024**, *12*, 281. https://doi.org/10.3390/inorganics12110281

Academic Editor: Antonino Gulino

Received: 29 August 2024
Revised: 15 October 2024
Accepted: 25 October 2024
Published: 30 October 2024

Copyright: © 2024 by the authors. Licensee MDPI, Basel, Switzerland. This article is an open access article distributed under the terms and conditions of the Creative Commons Attribution (CC BY) license (https://creativecommons.org/licenses/by/4.0/).

Abstract: The synthesis and characterization of ZnO/TiO$_2$, SiO$_2$/TiO$_2$, Al$_2$O$_3$/TiO$_2$, and Al$_{1.9}$Co$_{0.1}$O$_3$/TiO$_2$ core/shell nanoparticles (NPs) is reported. The NPs were used for photocatalytic degradation of brilliant blue E-4BA under UV and visible light irradiation, monitored by high-performance liquid chromatography and UV-vis absorption spectroscopy. The size of the NPs ranged from 10 to 30 nm for the core and an additional 3 nm for the TiO$_2$ shell. Al$_2$O$_3$/TiO$_2$ and Al$_{1.9}$Co$_{0.1}$O$_3$/TiO$_2$ showed superior degradation under UV and visible light compared to ZnO/TiO$_2$ and SiO$_2$/TiO$_2$ with complete photodecomposition of 20 ppm dye in 20 min using a 10 mg/100 mL photocatalyst. The "Co-doped" Al$_{1.9}$Co$_{0.1}$O$_3$/TiO$_2$ NPs show the best performance under visible light irradiation, which is due to increased absorption in the visible range. DFT-calculated band structure calculations confirm the generation of additional electronic levels in the band gap of γ-Al$_2$O$_3$ through Co^{3+} ions. This indicates that Co-doping enhances the generation of electron–hole pairs after visible light irradiation.

Keywords: photocatalyst; nanomaterials; heterojunction; dye degradation; DFT calculations

1. Introduction

Organic dyes represent an important class of environmental pollutants, and their decomposition and removal are of technical and social importance [1–4]. Photocatalytic advanced oxidation processes (AOPs) based on TiO$_2$ are technically available [1–4], but are established almost exclusively in developed countries [5]. Photosemiconducting anatase (β-TiO$_2$, tetragonal) is the base of the photocatalytic process [3–7] and is technically preferred due to its robust oxidizing capability, good photostability, and the non-toxic nature of TiO$_2$ materials [1–4]. However, conventional TiO$_2$ photocatalysis faces limitations such as rapid electron–hole recombination, low solar light absorption, and low activity on conventional crystal facets [3–8]. To overcome these obstacles and enhance photocatalytic efficiency, the creation of heterojunctions with other photosemiconductors has been explored [1,6,8–11]. Amongst such TiO$_2$-based heterojunction materials, core/shell nanoparticles (NPs) combine the large surface with an intimate contact of the two photosemiconductors [1]. Basically, two different approaches are possible. The first is to cover various functional oxides (core) with TiO$_2$ as the shell (Approach A, Scheme 1) [7,8,11–29]. Alternatively, a TiO$_2$ core can be supported by an oxide shell (Approach B) [7,8,17,22,30–35].

TiO$_2$ shell
core:
ZnO [13-17], SiO$_2$ [18-22], Al$_2$O$_3$ [23-25], Fe$_3$O$_4$ [12], rGO [26], BiFeO$_3$ [27], CuO [28], WO$_3$ [29]

Approach A

this work

TiO$_2$ core
shell:
ZnO [17], SiO$_2$ [22,30-33], Al$_2$O$_3$ [34], Fe$_3$O$_4$ [35], WO$_3$ [29]

Approach B - inverse approach

Scheme 1. Two different approaches for the use of TiO$_2$ in core/shell NPs.

Recent work on core/shell NPs for the photodecomposition of dyes, following approach A (Scheme 1), included magnetically separable Fe$_3$O$_4$ (core)/TiO$_2$ (shell) [12], NPs ZnO/TiO$_2$ core/shell nanorods [16], SiO$_2$/TiO$_2$ NPs [18], or rGO/TiO$_2$ NPs (rGO = reduced graphene oxide) [26] for the photodegradation of methylene blue (MB). ZnO@TiO$_2$ core/shell NPs were reported for the photodecomposition of rhodamine B (RhB) along with antimicrobial activity [13]. Al$_2$O$_3$/TiO$_2$ NPs were used for the photodecomposition of RhB and methyl orange [23], SiO$_2$/TiO$_2$ NPs were applied in the photodecomposition of crystal violet [19,20], and BiFeO$_3$/TiO$_2$ NPs, including the ferroelectric BiFeO$_3$, were used for methyl violet degradation [27]. These examples have in common that the TiO$_2$ shell was chemically built-up around the core, and characterization of the materials showed that the TiO$_2$ shell fully covers the core and thus very likely forms an intimate heterojunction between the core and shell (Scheme 1, Approach A). This is not necessarily the case in examples such as the reported CuO/TiO$_2$ heterojunction NPs that were synthesized by ball-milling Cu$_2$O and TiO$_2$ [28] or the Al$_2$O$_3$/TiO$_2$ (or TiO$_2$/Al$_2$O$_3$) NPs that were obtained through oxidation of bimetallic Ti/Al NPs [34].

The above examples differ also from the inverse approach (Approach B, Scheme 1) of using TiO$_2$ as the core, as in the TiO$_2$/SiO$_2$ NP photocatalysts used for singlet oxygen generation or the decomposition of MB [30,32] or RhB [33], the TiO$_2$/Al$_2$O$_3$ NPs for photodecomposition of the typical textile dye reactive brilliant red, the dye X-3B [24], the TiO$_2$/WO$_3$ photocatalyst reported for photocatalytic ozonation of organic contaminants [29], or the TiO$_2$/Fe$_3$O$_4$ NPs used for photodecomposition of X-3B [35]. The approach of modifying the TiO$_2$ surface with SiO$_2$, WO$_3$, or Fe$_3$O$_4$ (Scheme 1, Approach A) was driven by the idea to optimize the Fenton process, which is the production of the reactive superoxide O$_2^{\bullet-}$ and hydroxy $^\bullet$OH radicals [1,3,4,7,24,29-35] in addition to the electronic effect of a heterojunction. The approach of using TiO$_2$ as the reactive shell (Scheme 1, Approach B) relies on the TiO$_2$ photochemistry in combination with the electronic modification of the surface TiO$_2$ through the formation of the heterojunction and faster charge transport, leading to decreased recombination rates [1,3,12,13,18-20,23,26-28].

Starting from the idea that cheaper and environmentally benign oxides like ZnO, SiO$_2$, and Al$_2$O$_3$ as core materials would allow for producing metal oxide/TiO$_2$ core/shell NPs also for large-scale applications, we embarked on synthesizing core/shell ZnO/TiO$_2$, SiO$_2$/TiO$_2$, and Al$_2$O$_3$/TiO$_2$ NPs and studying their behavior as photocatalysts for photodegradation of the dye brilliant blue E-4BA, thus following the approach of modifying the electronic inventory of TiO$_2$ for high photoreactivity (Scheme 1). Similar materials with ZnO, SiO$_2$, or Al$_2$O$_3$ cores and TiO$_2$ shells have been reported as photocatalysts for such applications [13-25]. Cobalt (Co^{3+}) was added during the Al$_2$O$_3$ NP synthesis to generate Al$_{1.9}$Co$_{0.1}$O$_3$/TiO$_2$, what we will further call "Co-doped" [36]. From the fractional replacement of Al^{3+} through Co^{3+}, we expected extension of the absorption into the visible range and thus substantial improvements in the photocatalytic properties. A similar idea of Co doping for visible photocatalytic applications has been reported for materials such as LuFeO$_3$ [37], ZnO [38-40], and Eu(OH)$_3$ [41].

2. Results and Discussion

2.1. Synthesis and Characterization of the NPs

In the synthesis of the NPs, ethylene glycol serves as the key agent for governing the nanoparticle size. Upon heating the glycol-covered as-prepared NPs to 400 °C, ethylene glycol on the NP surface decomposed and led to an increased surface-to-volume ratio. The decomposition of ethylene glycol begins at 110 °C, resulting in a color change in the sample from white to gray. After one hour at 400 °C, the obtained powder turns yellowish, indicating that the degradation of ethylene glycol is complete.

Transmission electron micrographs (TEMs) show NPs of spherical shape for SiO_2 and ZnO, while the TiO_2 particles display a cubic morphology, and Al_2O_3 appears flake-like (Figure S1, Supplementary Materials). Their approximate diameters are 12, 10, 10, and 30 nm for Al_2O_3, TiO_2, SiO_2, and ZnO, respectively. After covering the NPs with TiO_2, the TEM shows sizes that are about 3 nm larger than those of the pure core materials (Figure S2). Previously, it was suggested that NPs for efficient photocatalysis should have sizes below 30 nm with a preferred range from 10 to 20 nm [42,43].

Brunauer–Emmett–Teller (BET) analysis shows a surface area of 583 m^2/g for Al_2O_3/TiO_2, 160 m^2/g for ZnO/TiO_2, and 87 m^2/g for SiO_2/TiO_2. The surface area for $Al_{1.9}Co_{0.1}O_3$/TiO_2 is 618 m^2/g (Figure S3). The high surface areas for Al_2O_3/TiO_2 and $Al_{1.9}Co_{0.1}O_3$/TiO_2 are very promising for the adsorption of the dye on the surface for photodegradation.

Powder XRD (PXRD) patterns of the synthesized Al_2O_3 NPs (Figure S4A) confirm the γ-Al_2O_3 (cubic, spinel, Fd-3m) structure with observed peaks at 31.92°, 37.60°, 45.86°, and 67.03° corresponding to the (220), (311), (222), (400), and (440) planes, respectively. The crystallinity is low, as can be seen from the broad reflections. The $Al_{1.9}Co_{0.1}O_3$ NPs have retained the γ-Al_2O_3 structure (Figure S4B). However, the reflections exhibit a shift to lower angles in line with increased lattice parameters caused by the larger Co^{3+} (r = 75 pm) [44] ion partially replacing the smaller Al^{3+} ions (r = 67 pm) [44], assuming that large parts of the Al^{3+} ions residing in octahedral positions [45] were replaced by high-spin Co^{3+}. The ZnO particles show higher crystallinity, with peaks located at 31.84°, 34.52°, 36.33°, 47.63°, 56.71°, 62.96°, 68.13°, and 69.18° corresponding to the (100), (002), (101), (102), (110), (103), and (112) planes (Figure S4C) of hexagonal ZnO crystallizing in the space group P63mc (wurtzite type). The SiO_2 NPs show only a broad peak at 23.36° (Figure S4D), which is probably due to the largely amorphous character of the material.

The PXRD of the TiO_2-coated particles show the peaks corresponding to the core materials together with additional signals agreeing with the anatase (β-TiO_2, tetragonal, $I4_1$/amd) structure with peaks at 25.39°, 37.89°, 48.1°, and 55.18° corresponding to the (101), (004), (200), and (211) planes, respectively (Figure 1, compare Figure S4E).

The sharp signals of the core materials in the core/shell NPs are in line with the small but highly crystalline TiO_2 shell of only a few nm concluded from the TEM. This is supported by previously reported Al_2O_3/TiO_2 core/shell NPs for which the PXRD exclusively showed the signals for TiO_2 at an estimated thickness of the TiO_2 shell of 17 nm [23]. Furthermore, for SiO_2/TiO_2 core/shell NPs of 20 to 30 nm core and about 3 nm shell size, broad signals for the TiO_2 shell were found along with one broad reflection for SiO_2 [19]. The crystallinity of the TiO_2 shell is thus strongly dependent on the preparation method, and our method seems very suitable to produce high crystallinity.

Energy-dispersive X-ray spectroscopy (EDX) of the core/shell Al_2O_3/TiO_2, ZnO/TiO_2, and SiO_2/TiO_2 NPs shows the core elements Al, Si, Zn, and O together with smaller amounts of Ti from the shell (Figure 2) in line with a thin TiO_2 shell. For $Al_{1.9}Co_{0.1}O_3$, cobalt was found with an Al/Co ratio in keeping with 5 to 10% Co in the structure.

The UV-vis absorption spectrum of the pristine Al_2O_3 NPs (Figure 3A) shows the typical absorption band at around 200 nm [46]. For ZnO, a characteristic maximum at 378 nm was recorded in keeping with previous reports [47,48], and for SiO_2 and TiO_2 absorption, the UV-vis range shows only tails down from 170 to 1000 nm (Figure S5). The $Al_{1.9}Co_{0.1}O_3$ NPs show an additional broad absorption band in the visible range at 520 nm (2.38 eV) (Figure 3B,C). A very similar absorption band at 2.58 eV was reported for ZnO

NPs containing 3, 5, and 7 mol% Co [38] and for $LaAl_{0.98}Co_{0.02}O_3$ containing 2% Co(III) and corresponds to the $^5T_2 \rightarrow {}^5E$ transition [49]. Thus, the spectrum of $Al_{1.9}Co_{0.1}O_3$ is in line with high-spin Co^{3+} ions replacing Al^{3+} at the octahedral sites, supporting our assumption from PXRD.

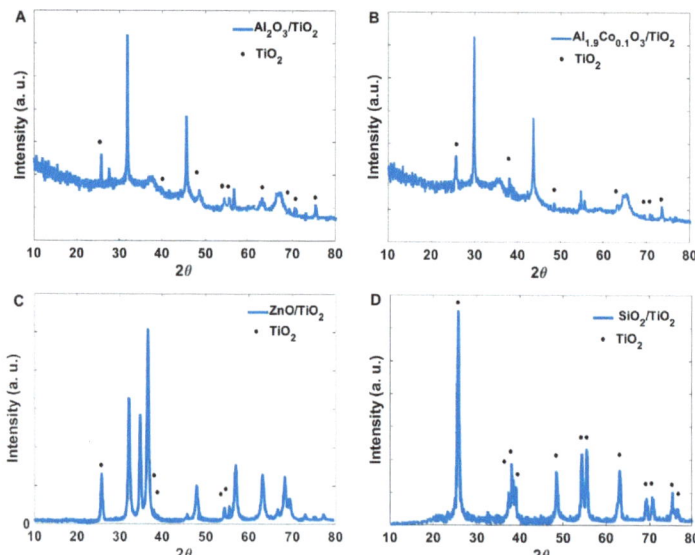

Figure 1. XRD patterns of the core/shell nanoparticles (**A**) Al_2O_3/TiO_2, (**B**) $Al_{1.9}Co_{0.1}O_3/TiO_2$, (**C**) ZnO/TiO_2, and (**D**) SiO_2/TiO_2. The signals corresponding to TiO_2 are marked with black dots.

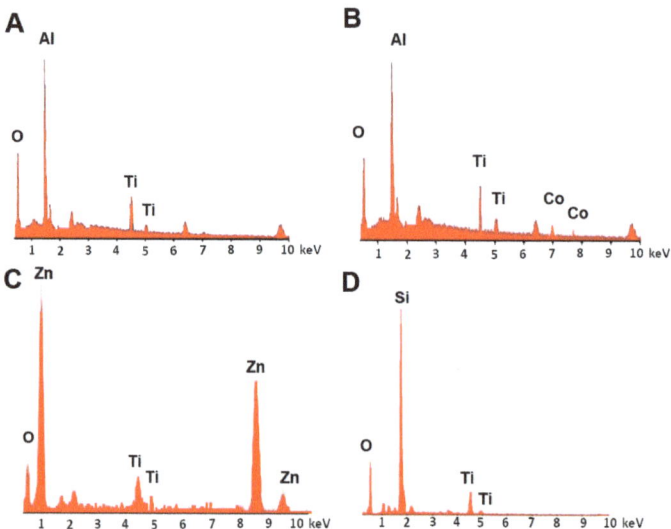

Figure 2. EDX images of (**A**) Al_2O_3/TiO_2, (**B**) $Al_{1.9}Co_{0.1}O_3/TiO_2$, (**C**) ZnO/TiO_2, and (**D**) SiO_2/TiO_2 NPs.

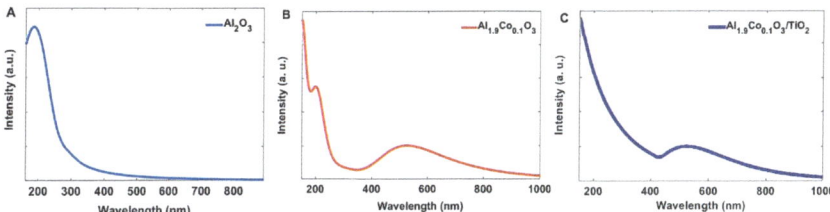

Figure 3. UV-vis absorption spectra for the NPs (**A**) Al$_2$O$_3$, (**B**) Al$_{1.9}$Co$_{0.1}$O$_3$, and (**C**) Al$_{1.9}$Co$_{0.1}$O$_3$/TiO$_2$ NPs.

The tiny red shift in the peak observed at around 200 nm for Co^{3+}-doped nanoparticles compared to undoped Al$_2$O$_3$ can be attributed to oxygen defects in the structure of alumina, known as F centers [50]. Two types of F centers can be created in the structure of alumina. F and F$_2$ centers can be reversibly interconverted by irradiating them with different wavelengths of light. For example, irradiating with 300 nm light can convert F$_2$ and F$_2^+$ centers, and irradiating with about 200 nm light can convert F centers [51,52]. Previous studies [53,54] report that the impact of electronic defects on optical properties is more significant than that of the electronic levels introduced by the doped ions.

The UV-Vis spectra of similarly reported Al$_2$O$_3$/TiO$_2$ [23], ZnO$_2$/TiO$_2$ [15], and SiO$_2$/TiO$_2$ [16] NPs show a broad UV absorption tailing down till 400 nm (3.1 eV) originating from the TiO$_2$ shells. This is fully in line with our results (Figure 3C).

2.2. Photocatalysis

The photocatalytic degradation of brilliant blue E-4BA (Scheme 2) under both UV and visible light irradiation was studied for pure TiO$_2$, Al$_2$O$_3$/TiO$_2$, ZnO/TiO$_2$, SiO$_2$/TiO$_2$, and Al$_{1.9}$Co$_{0.1}$O$_3$/TiO$_2$ NPs.

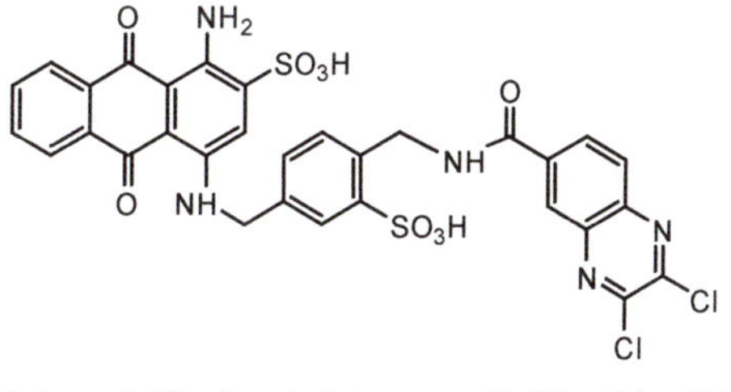

Scheme 2. The chemical structure of brilliant blue E-4BA.

The photodegradation was first monitored using UV-vis absorption spectroscopy with the visible and UV bands of the dye with progressing reactions (Figures 4 and 5). In the first experiments, the optimum amount of catalyst was found at around 10 mg/100 mL, and the time for complete photodecomposition ranged from 15 to 200 min. Without any catalyst, no photodecomposition was observed, neither under UV nor under visible light irradiation.

When comparing identical amounts of catalyst (10 mg) under the same UV irradiation conditions, the reactivity increased within the series–no catalyst (no reaction) < pure TiO$_2$ < SiO$_2$/TiO$_2$ < ZnO/TiO$_2$ < Al$_2$O$_3$/TiO$_2$ < Al$_{1.9}$Co$_{0.1}$O$_3$/TiO$_2$–(Figure 4). The time for complete conversion under UV irradiation was 70 min for pure TiO$_2$, 60 min for SiO$_2$/TiO$_2$, 30 min for ZnO/TiO$_2$, 15 to 20 min for Al$_2$O$_3$/TiO$_2$, and 15 to 20 min for Al$_{1.9}$Co$_{0.1}$O$_3$/TiO$_2$.

No significant degradation of the dye under visible light irradiation was found for Al$_2$O$_3$/TiO$_2$, ZnO/TiO$_2$, and SiO$_2$/TiO$_2$ NPs within 90 min. The catalysts seem to absorb a small amount of dye after adding it, but the solution retains the blue color of the dye. In contrast to this, more than 95% of the dye can be degraded in the solution containing Al$_{1.9}$Co$_{0.1}$O$_3$/TiO$_2$ NPs under visible light irradiation within 90 min (Figure 5).

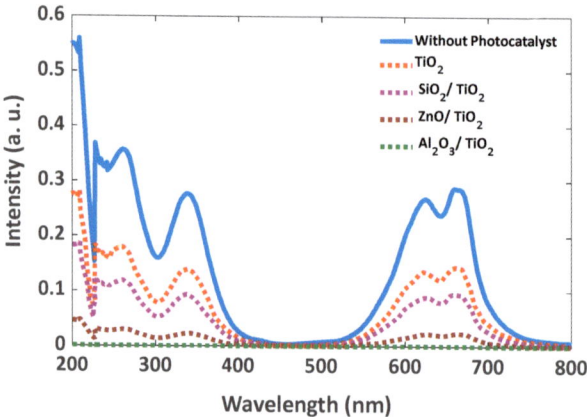

Figure 4. Absorption spectra of brilliant blue E-4BA in H_2O after addition of 10 mg catalyst (or none) to a 20 ppm dye solution and UV light irradiation for 15 min.

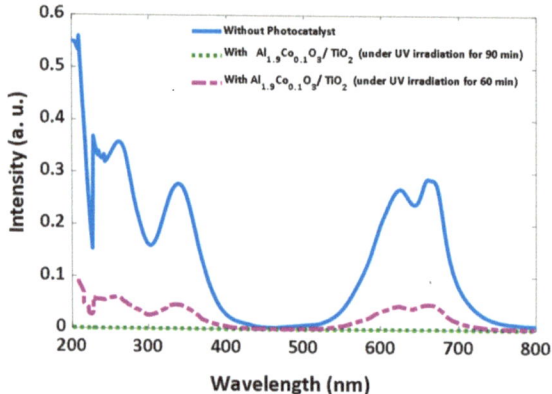

Figure 5. Absorption spectra of brilliant blue E-4BA in H_2O after addition of 10 mg $Al_{1.9}Co_{0.1}O_3/TiO_2$ (or none) to a 20 ppm dye solution and visible light irradiation for 90 min.

HPLC analysis of the reaction mixtures (Figure 6 and Figure S6) gave the same sequence for the reactivity – no catalyst < pure TiO_2 < SiO_2/TiO_2 < ZnO/TiO_2 < Al_2O_3/TiO_2 < $Al_{1.9}Co_{0.1}O_3/TiO_2$.

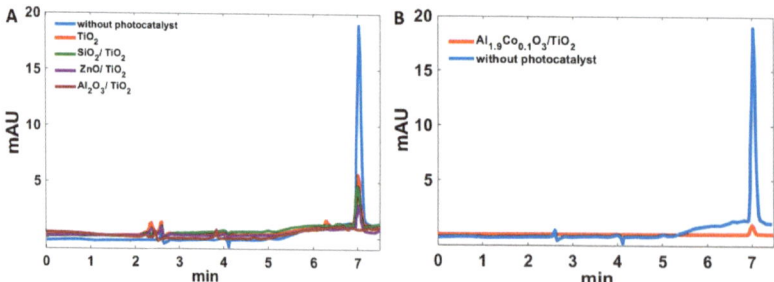

Figure 6. (**A**): HPLC plots of reaction solutions containing 20 ppm brilliant blue E-4BA in H_2O after addition of 10 mg catalyst (or none) and UV irradiation for 15 min. (**B**): HPLC plots of products of a 20 ppm dye solution after visible irradiation for 90 min in the absence and presence of $Al_{1.9}Co_{0.1}O_3/TiO_2$.

Thus, all NP catalysts were active, and the core/shell NPs were more active than the pure TiO_2 particles. For the core/shell NPs, reactivity increased along the series $SiO_2/TiO_2 <$ $ZnO/TiO_2 < Al_2O_3/TiO_2$. The increase from SiO_2 to ZnO is roughly in line with the decreasing band gap, with reported values around 4 eV for SiO_2 [54] and around 3 eV for ZnO [55] NPs. The high activity of the Al_2O_3/TiO_2 NPs does not agree with this idea, as γ-Al_2O_3 has the highest band gap in the series, with reported values ranging from 5 to 8 eV [56,57]. The 195 nm absorption found in the UV-vis absorption spectrum of Al_2O_3 (Figure 3) recalculates to 6.36 eV, in agreement with these reports. The superior properties of Al_2O_3/TiO_2 over the other two materials can be explained in terms of the heterojunction effect.

On the other hand, the activities correlate quite well with the BET analysis showing far larger surface areas for Al_2O_3/TiO_2 (583 m^2/g) and $Al_{1.9}Co_{0.1}O_3/TiO_2$ (618 m^2/g) than for ZnO/TiO_2 (160 m^2/g) and SiO_2/TiO_2 (87 m^2/g). This underlines that adsorption of the dye onto the surface is crucial for the photodegradation, which is supported by other reports [2,7,13,18–20,23,26]. The reactivity of the SiO_2/TiO_2 NPs in our study compares well to a recent report on the photodecomposition of methylene blue (MB) [18]. The therein reported rather amorphous core/shell SiO_2/TiO_2 NPs with sizes around 330 nm (shell about 35 nm) and BET surfaces ranging from 300 to 400 m^2/g allowed for complete photodecomposition using UV light within 30 min. Crystal violet was photodecomposed using SiO_2/TiO_2 NPs within 60 min, while the BET surface for these particles was smaller (50 to 100 m^2/g) [20]. Another study using Al_2O_3/TiO_2 NPs for the photodecomposition of methyl orange and rhodamine B reported more than 120 min till completeness under similar conditions [23].

The higher activity found for the $Al_{1.9}Co_{0.1}O_3/TiO_2$ compared with the non-doped Al_2O_3/TiO_2 NPs can additionally be rationalized with the introduction of Co(III)-based traps into the Al_2O_3 electronic structure. To shed more light on this, a theoretical study based on density functional theory (DFT) was started, comparing Al_2O_3 and $Al_{1.9}Co_{0.1}O_3$. While the band structure of γ-Al_2O_3 has previously been calculated on a high level of sophistication [57,58], a comparison of Al_2O_3 and Al-doped $Al_{1.9}Co_{0.1}O_3$ has not been reported.

2.3. DFT-Calculated Electronic Structures of Al_2O_3 and $Al_{1.9}Co_{0.1}O_3$

The DFT-calculated band structure for Al_2O_3 (Figure 7A) shows a direct band gap of 6.106 eV, in good agreement with reported values [57,58], and the experimental value of 6.36 eV we derived from UV-vis absorption measurements.

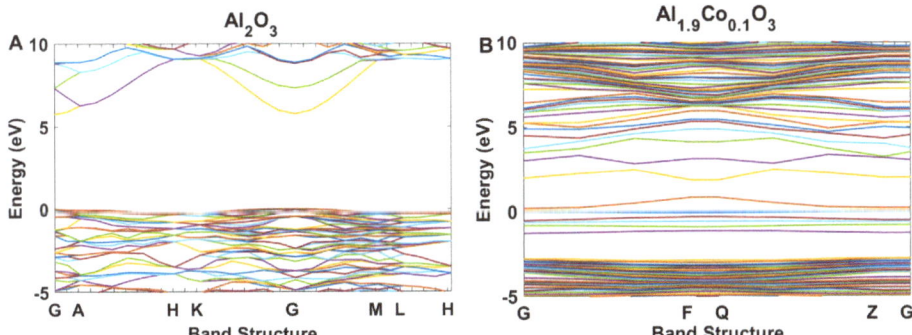

Figure 7. Calculated band structure for (**A**) Al_2O_3 and (**B**) $Al_{1.9}Co_{0.1}O_3$ NPs with the highest valence band = Fermi level set to $E = 0$ eV.

The band structure of $Al_{1.9}Co_{0.1}O_3$ contains a number of additional donor (below 0 eV) and acceptor (above 0 eV) levels (Figure 7B). When converting the DFT-calculated band structure into a calculated UV-vis absorption spectrum (Figure 8), the experimentally observed long-wavelength absorption at 520 nm was well reproduced by the calculations. The calculated minimum band gap energy of 1.35 eV agrees reasonably well with the end of the broad absorption from 400 to 900 nm, which recalculates to 3.10 to 1.38 eV. The calculations predict also a band at 280 nm and strong absorptions in the UV range around 200 nm. While the 200 nm band was found in the experimental structure and is attributed to the Al_2O_3 host material, the 280 nm band was not experimentally found and might lie below the very intense band centered at 200 nm.

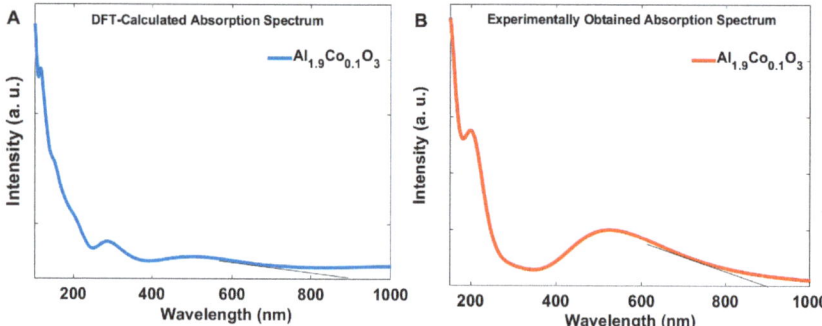

Figure 8. (**A**) DFT-calculated and (**B**) experimentally obtained UV-vis absorption spectra for $Al_{1.9}Co_{0.1}O_3$.

Thus, while the generation of electrons and holes in nano-Al_2O_3/TiO_2 requires UV irradiation (Figure 9), the incorporation of Co^{3+} ions in the Al_2O_3 lattice generates additional donor and acceptor levels and shifts the energy into the visible range.

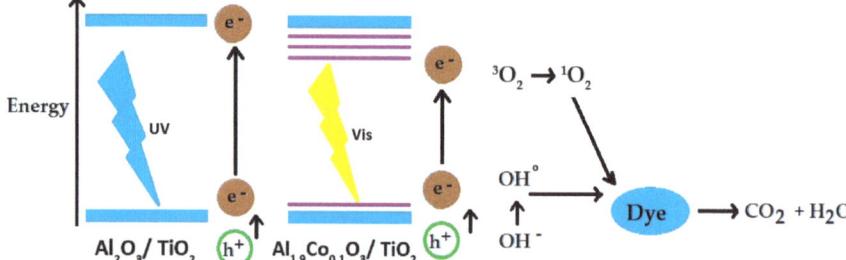

Figure 9. Proposed mechanism for dye photodegradation by Al_2O_3 and $Al_{1.9}Co_{0.1}O_3$ NPs under UV and visible light. $^3O_2$ and $^1O_2$ represent triplet and singlet oxygen, OH• the OH radical.

2.4. FT-IR Stability Tests on Al_2O_3/TiO_2 and $Al_{1.9}Co_{0.1}O_3$/TiO_2

Previous reports indicated that gamma alumina exhibits limited stability in aqueous environments, eventually becoming hydrated over time. To assess the stability of the synthesized Al_2O_3/TiO_2 nanoparticles, we used FT-IR spectroscopy. Within one week of storage in water, the synthesized Al_2O_3/TiO_2 nanoparticles are seemingly stable (Figure S7). After two weeks, peaks at 3551, 3460, and 3400 cm^{-1} indicate hydrolysis (formation of OH-functions). The results for the Co^{3+}-doped particles are essentially the same (Figure S8).

3. Materials and Methods
3.1. Materials

The starting materials, tetraethyl orthosilicate $Si(OEt)_4$, titanium tetraisopropoxide $Ti(OiPr)_4$, $AlCl_3$, ethylene glycol, ethylene diamine, EtOH, cobalt nitrate hexahydrate $Co(NO_3)_2 \cdot 6H_2O$, zinc nitrate dihydrate $Zn(NO_3)_2 \cdot 2H_2O$, Na_2CO_3, and ethylamine ($EtNH_2$) were all provided from Merck (Darmstadt, Germany) and used without further purification.

3.2. Preparation of γ-Al$_2$O$_3$ NPs

A total of 1 g of AlCl$_3$ and 3 g of ethylene glycol were dissolved in 100 mL of deionized water. Subsequently, 10 mL of ethylene diamine diluted with 50 mL of water were added dropwise to the reactor, and the reaction mixture was stirred for 4 h. The resulting NPs were separated and washed with deionized water and EtOH using centrifugation. The final material was calcinated at 400 °C for 1 h.

3.3. Preparation of Al$_2$O$_3$/TiO$_2$ NPs

The Al$_2$O$_3$ particles were dispersed in 100 mL of deionized water. In a separate beaker, 1 mL of Ti(OiPr)$_4$ and 5 mL of ethylene glycol were mixed and added dropwise to the first beaker. The mixture was stirred for 5 h, and the resulting precipitated NPs were separated and washed with deionized water and EtOH using centrifugation. The final material was calcinated at 400 °C for 1 h.

3.4. Preparation of Al$_{1.9}$Co$_{0.1}$O$_3$/TiO$_2$ NPs

A total of 0.95 g of AlCl$_3$, 0.09 g of Co(NO$_3$)$_2$·6H$_2$O, and 3 g of ethylene glycol were dissolved in 100 mL of deionized water. Subsequently, 10 mL of ethylene diamine, diluted with 50 mL of water, was added dropwise to the reaction mixture. The resulting Al$_{1.9}$Co$_{0.1}$O$_3$ NPs were isolated and washed several times with deionized water and EtOH using centrifugation. The NPs were then dispersed in 100 mL of deionized water. In a separate beaker, 1 mL of Ti(OiPr)$_4$ and 5 mL of ethylene glycol were mixed and added dropwise to the first beaker. The mixture was stirred for 5 h, and the resulting precipitated NPs were separated and washed with deionized water and EtOH using centrifugation. The final material was calcinated at 400 °C for 1 h.

3.5. Preparation of ZnO NPs

A total of 1 g of Zn(NO$_3$)$_2$·2H$_2$O and 3 g of ethylene glycol were dissolved in a beaker. Subsequently, 1 g of dissolved Na$_2$CO$_3$ was added dropwise to the first beaker. The precipitated NPs were separated and washed with deionized water and EtOH using centrifugation. The final material was calcinated at 400 °C for 1 h.

3.6. Preparation of ZnO/TiO$_2$ NPs

The ZnO particles were dispersed in 100 mL of deionized water. In another beaker, 1 mL of Ti(OiPr)$_4$ and 5 mL of ethylene glycol were mixed and added dropwise to the first beaker. The resulting mixture was stirred for 5 h, and the precipitated NPs were separated and washed with deionized water and EtOH using centrifugation. The final material was calcinated at 400 °C for 1 h.

3.7. Preparation of SiO$_2$ NPs

A total of 10 mL of Si(OEt)$_4$ and 50 mL of EtOH were mixed in a beaker and stirred for 15 min. Then, 50 mL of deionized water and EtNH$_2$ were added immediately under stirring. After 12 h, the SiO$_2$ NPs were separated and washed with deionized water and EtOH using centrifugation. The final material was calcinated at 400 °C for 1 h.

3.8. Preparation of SiO$_2$/TiO$_2$ NPs

The SiO$_2$ particles were dispersed in 100 mL of deionized water. In another beaker, 1 mL of Ti(OiPr)$_4$ and 5 mL of ethylene glycol were mixed and added dropwise to the first beaker. The resulting mixture was stirred for 5 h, and the precipitated NPs were separated and washed with deionized water and EtOH, using centrifugation. The final material was calcinated at 400 °C for 1 h.

3.9. Photocatalysis

A solution was prepared by dissolving 0.2 g of brilliant blue E-4BA (C$_{31}$H$_{19}$O$_9$N$_5$S$_2$Cl$_2$Na$_2$) in 100 mL deionized water (20 ppm). Subsequently, 10 mg of the NPs was added to 100 mL of the prepared solution. The solution was irradiated at 475 nm using an Osram, 3.8 mW/cm^2 UV light-emitting diode (LED)(ams-OSRAM International, Regensburg, Germany) or at 525 nm using an Osram, 4 mW/cm^2 visible light LED. The light sources were kept at a distance of 10 cm from the samples. During irradiation, samples were taken in time intervals of 5 to 200 min, and UV-vis absorption spectra were recorded. After the complete irradiation time, the solutions were filtered and analyzed using HPLC.

3.10. Chromatography

The mobile phase consisted of Solution A: 0.1 M ammonium acetate adjusted to pH = 6.7, and Solution B: MeCN. Gradient elution was set as summarized in Table 1.

Table 1. Details for the gradient elution [a].

Time	A	B	Flow
0	97	3	1
18	40	60	1
20	40	60	1

[a] Flow: 1 mL/min, wavelength: 261 nm, injection volume: 20 µL. Calibration curves were generated from five different concentrations between 25 and 75 ppm of brilliant blue E-4BA with three replicates, and the correlation coefficient was assessed.

3.11. Instrumentation

Transmission electron micrographs (TEMs) were recorded using a TEM Philips EM 208S (Philips, Beaverton, OR, USA). Powder XRD (PXRD) was carried out using a PW 1730 (Philips, Eindhoven, The Netherland) diffractometer in the range of 2θ = 4 to 100° and using Cu-Kα (λ = 1.5406 Å) radiation. Energy-dispersive X-ray spectroscopy (EDX) was measured on a FE-SEM TESCAN MIRA2 (Tescan, Kohoutovice, Czech Republic). UV-vis absorption spectra of the NPs were measured on powder samples in transmission mode using a T70 UV-vis photospectrophotometer (PG Instruments, Wibtoft, UK). HPLC was carried out on an Agilent instrument AGI1200-120013780 (Agilent, Santa Clara, CA, USA). Brunauer–Emmett–Teller (BET) analysis was measured using a BET BELSORP Mini II instrument (Microtrac Retsch, Haan/Duesseldorf, Germany). FT-IR measurements were recorded using an FT-IR spectrometer MB3000 (ABB, Zürich, Switzerland).

3.12. DFT Calculations

Electronic band structures and optical properties of Al_2O_3 and $Al_{1.9}Co_{0.1}O_3$ nanoparticles were computed using density functional theory (DFT) calculations. The CASTEP code was employed for the calculations, and optimization was achieved through the BFGS (Broyden-Fletcher-Goldfarb-Shanno) geometry optimization method (CASTEP version 16.11) [59]. The calculations utilized the generalized gradient approximation (GGA) and the non-local gradient-corrected exchange-correlation functional, parameterized using the Perdew-Burke-Ernzerhof (PBE) approach [60]. The plane-wave cutoff energy was set to 550 eV. The optical characteristics of Al_2O_3 and $Al_{1.9}Co_{0.1}O_3$ were investigated using the CASTEP frequency computation results at the gamma point (GP). Brillouin zone summation employed k-point sampling with a Monkhorst-Pack grid at a 2 × 2 × 2 parameter set. The "2 × 2 × 2 parameter" refers to the grid used for k-point sampling within the Brillouin zone when performing calculations on a material's electronic band structure. In this case, the Monkhorst-Pack grid with a 2 × 2 × 2 parameter indicates that the Brillouin zone is divided into a grid of k-points in each direction (x, y, z) with two points in the x-direction, two points in the y-direction, and two points in the z-direction. This grid defines the discretization of the Brillouin zone for numerical calculation. Additionally, geometry optimization under applied hydrostatic pressure was utilized to determine the material modulus (B) and its pressure derivative, B′ = dB/dP [59,61].

4. Conclusions

Metal oxide core/shell nanoparticles (NPs), Al_2O_3/TiO_2, ZnO/TiO_2, SiO_2/TiO_2, and $Al_{1.9}Co_{0.1}O_3/TiO_2$, were synthesized, characterized, and investigated for their photocatalytic activity in the degradation of the dye brilliant blue E-4BA. The size of the cores ranged from 10 to 30 nm, while the TiO_2 shells were about 3 nm thick. This thin shell was confirmed by powder X-ray diffraction (PXRD) showing reflections for both the core and the shell material and makes sure that both the core and shell can absorb photons. Furthermore, both parts are highly crystalline, except for the SiO_2 core. The optimal catalyst dosage for the photodegradation was identified at 10 mg per 100 mL of a 20 ppm dye solution using UV-vis absorption spectroscopy and HPLC for reaction monitoring. No reactivity was found without a catalyst, while for the studied NPs, the reactivity increased along the series–pure TiO_2 < SiO_2/TiO_2 < ZnO/TiO_2 < Al_2O_3/TiO_2 < $Al_{1.9}Co_{0.1}O_3/TiO_2$– under UV irradiation, which translates into times for complete conversion of 70 min for pure TiO_2 and 15 min for $Al_{1.9}Co_{0.1}O_3/TiO_2$. This aligns well with the increasing BET surfaces SiO_2/TiO_2 (87 m^2/g) < ZnO/TiO_2 (160 m^2/g) < Al_2O_3/TiO_2 (583 m^2/g) and $Al_{1.9}Co_{0.1}O_3/TiO_2$ (618 m^2/g), underpinning that the adsorption capacity of a material is crucial for photodegradation. Under visible light irradiation, only $Al_{1.9}Co_{0.1}O_3/TiO_2$ was found to be active in photodecomposition of the dye, decomposing the dye within 90 min. DFT-based band structure calculations showed that Co-doping introduces additional energy levels into Al_2O_3, resulting in improved electron–hole pair generation. The DFT-calculated absorption spectra of Al_2O_3/TiO_2 and $Al_{1.9}Co_{0.1}O_3/TiO_2$ agree very well with the experimental spectra, thus supporting the band structure calculations.

On the other hand, from this preliminary study, it is not possible to identify the "best material" from our series or the "most important parameter" for determining this. Size, layer thickness, specific surface area, band gap, and photon absorptions are strongly interconnected. Especially, we were not able to quantitatively assess the heterojunction effect of combining ZnO, SiO_2, and Al_2O_3 with TiO_2. The experimentally observed increasing reactivity along the series SiO_2/TiO_2 < ZnO/TiO_2 < Al_2O_3/TiO_2 correlates with the BET surfaces, but additionally, Al_2O_3 might show a pronounced heterojunction effect decreasing the band gap of pure Al_2O_3 around 6 eV, thus reaching those of ZnO (around 3 eV) and SiO_2 (around 4 eV). Thus, manufacturing of core/shell NPs with thin TiO_2 shells using these basically cheap and highly abundant metal oxides allows for generating potent photocatalysts for UV photocatalysis. The introduction of transition metals such as Co turns out to be suitable to shift the irradiation wavelength into the visible range.

Supplementary Materials: The following supporting information can be downloaded at: https://www.mdpi.com/article/10.3390/inorganics12110281/s1, Figure S1: TEM images of synthesized (A) Al_2O_3, (B) TiO_2, (C) SiO_2, and (D) ZnO nanoparticles; Figure S2: TEM images of the synthesized (A) Al_2O_3/TiO_2, (B) $Al_{1.9}Co_{0.1}O_3/TiO_2$, (C) ZnO/TiO_2, and (D) SiO_2/TiO_2 nanoparticles; Figure S3: BET plots of the ZnO/TiO_2 (top left), SiO_2/TiO_2 (top right), Al_2O_3/TiO_2 (bottom, left), and $Al_{1.9}Co_{0.1}O_3/TiO_2$ (bottom right) nanoparticles; Figure S4: XRD patterns of the synthesized NPs (A) Al_2O_3, (B) ZnO, (C) TiO_2, (D) SiO_2, and (E) $Al_{1.9}Co_{0.1}O_3$; Figure S5: UV-vis absorption spectra for the synthesized NPs (A) SiO_2, (B) TiO_2, and (C) ZnO; Figure S6: HPLC from reaction mixtures containing 20 ppm brilliant blue E-4BA in H_2O after addition of 10 mg bare SiO_2 or TiO_2 NPs as catalyst (or none) and UV irradiation for 15 min; Figure S7: FT-IR spectra of Al_2O_3/TiO_2 nanoparticles stored in water and after synthesis; Figure S8: FT-IR spectra of $Al_{1.9}Co_{0.1}O_3/TiO_2$ nanoparticles stored in water and after synthesis.

Author Contributions: M.D. synthesized and characterized the nanoparticles and evaluated the photocatalytic reactions. The synthesized nanoparticles were characterized by A.R. A.Z. set up the UV and visible reaction box for the photocatalytic reaction. The UV-vis absorption and chromatographic analysis of reaction mixtures were carried out by S.D. and M.T. A.Y. edited the manuscript and evaluated the photocatalytic reactions. A.R. and A.K. supervised the project. A.K. wrote and edited the manuscript. All authors contributed to revising the manuscript and have agreed with the last version. All authors have read and agreed to the published version of the manuscript.

Funding: This research received no external funding.

Data Availability Statement: The data supporting the findings of this study are available within this paper. Additional data are available from the corresponding author upon reasonable request.

Acknowledgments: The authors acknowledge Alborz Farmed Company for HPLC and UV-vis absorption analysis.

Conflicts of Interest: Authors Mahboubeh Dolatyari and Ali Rostami were employed by the ASEPE company. Authors Mehdi Tahmasebi and Sudabeh Dolatyari are employees of the company Alborz Farmad. The remaining authors declare that the research was conducted in the absence of any commercial or financial relationships that could be construed as a potential conflict of interest.

References

1. Thambiliyagodage, C. Activity enhanced TiO_2 nanomaterials for photodegradation of dyes—A review. *Environ. Nanotechnol. Monit. Manag.* **2021**, *16*, 100592. [CrossRef]
2. Kumari, H.; Suman, S.; Ranga, R.; Chahal, S.; Devi, S.; Sharma, S.; Kumar, S.; Kumar, P.; Kumar, S.; Kumar, A.; et al. A Review on Photocatalysis Used For Wastewater Treatment: Dye Degradation. *Water Air Soil Pollut.* **2023**, *234*, 349. [CrossRef] [PubMed]
3. Navidpour, A.H.; Abbasi, S.; Li, D.; Mojiri, A.; Zhou, J.L. Investigation of Advanced Oxidation Process in the Presence of TiO_2 Semiconductor as Photocatalyst: Property, Principle, Kinetic Analysis, and Photocatalytic Activity. *Catalysts* **2023**, *13*, 232. [CrossRef]
4. Reza, K.M.; Kurny, A.S.W.; Gulshan, F. Parameters affecting the photocatalytic degradation of dyes using TiO_2: A review. *Appl. Water Sci.* **2017**, *7*, 1569–1578. [CrossRef]
5. Liu, H.; Wang, C.; Wang, G. Photocatalytic Advanced Oxidation Processes for Water Treatment: Recent Advances and Perspective. *Chem. Asian J.* **2020**, *15*, 3239–3253. [CrossRef]
6. Hassaan, M.A.; El-Nemr, M.A.; Elkatory, M.R.; Ragab, S.; Niculescu, V.-C.; El Nemr, A. Principles of Photocatalysts and Their Different Applications: A Review. *Top. Curr. Chem.* **2023**, *381*, 31. [CrossRef]
7. Wen, J.; Li, X.; Liu, W.; Fang, Y.; Xie, J.; Xu, Y. Photocatalysis fundamentals and surface modification of TiO_2 nanomaterials. *Chin. J. Catal.* **2015**, *36*, 2049–2070. [CrossRef]
8. Lettieri, S.; Pavone, M.; Fioravanti, A.; Santamaria Amato, L.; Maddalena, P. Charge Carrier Processes and Optical Properties in TiO_2 and TiO_2-Based Heterojunction Photocatalysts: A Review. *Materials* **2021**, *14*, 1645. [CrossRef]
9. Rehman, Z.U.; Bilal, M.; Hou, J.; Butt, F.K.; Ahmad, J.; Ali, S.; Hussain, A. Photocatalytic CO_2 Reduction Using TiO_2-Based Photocatalysts and TiO_2 Z-Scheme Heterojunction Composites: A Review. *Molecules* **2022**, *27*, 2069. [CrossRef]
10. Li, F.; Zhu, G.; Jiang, J.; Yang, L.; Deng, F.; Arramel; Li, X. A review of updated S-scheme heterojunction photocatalysts. *J. Mater. Sci. Technol.* **2024**, *177*, 142–180. [CrossRef]
11. Hezam, A.; Peppel, T.; Strunk, J. Pathways towards a systematic development of Z-scheme photocatalysts for CO_2 reduction. *Curr. Opin. Green Sustain. Chem.* **2023**, *41*, 100789. [CrossRef]
12. Zhang, Q.; Yu, L.; Xu, C.; Zhang, W.; Chen, M.; Xu, Q.; Diao, G. A novel method for facile preparation of recoverable $Fe_3O_4@TiO_2$ core-shell nanospheres and their advanced photocatalytic application. *Chem. Phys. Lett.* **2020**, *761*, 138073. [CrossRef]
13. Selva Arasu, K.A.; Raja, A.G.; Rajaram, R. Photocatalysis and antimicrobial activity of $ZnO@TiO_2$ core-shell nanoparticles. *Inorg. Nano-Met. Chem.* **2024**, 1–12. [CrossRef]
14. Ali, M.M.; Haque, M.J.; Kabir, M.H.; Kaiyum, M.A.; Rahman, M.S. Nano synthesis of $ZnO–TiO_2$ composites by sol-gel method and evaluation of their antibacterial, optical and photocatalytic activities. *Results Mater.* **2021**, *11*, 100199. [CrossRef]
15. Karthikeyan, K.; Chandraprabh, M.N.; Hari Krishna, R.; Samrat, K.; Sakunthal, A.; Sasikumar, M. Optical and antibacterial activity of biogenic core-shell $ZnO@TiO_2$ nanoparticles. *J. Ind. Chem. Soc.* **2022**, *99*, 100361. [CrossRef]
16. Kwiatkowski, M.; Chassagnon, R.; Heintz, O.; Geoffroy, N.; Skompska, M.; Bezverkhyy, I. Improvement of photocatalytic and photoelectrochemical activity of ZnO/TiO_2 core/shell system through additional calcination: Insight into the mechanism. *Appl. Catal. B Environ.* **2017**, *204*, 200–208. [CrossRef]
17. Zhou, M.; Wu, B.; Zhang, X.; Cao, S.; Ma, P.; Wang, K.; Fan, Z.; Su, M. Preparation and UV Photoelectric Properties of Aligned $ZnO–TiO_2$ and $TiO_2–ZnO$ Core–Shell Structured Heterojunction Nanotubes. *ACS Appl. Mater. Interfaces* **2020**, *12*, 38490–38498. [CrossRef]
18. Gomes, B.R.; Lopes, J.L.; Coelho, L.; Ligonzo, M.; Rigoletto, M.; Magnacca, G.; Deganello, F. Development and Upscaling of $SiO_2@TiO_2$ Core-Shell Nanoparticles for Methylene Blue Removal. *Nanomaterials* **2023**, *13*, 2276. [CrossRef]
19. Ullah, S.; Ferreira-Neto, E.P.; Pasa, A.A.; Alcântara, C.C.J.; Acuna, J.J.S.; Bilmes, S.A.; Martínez Ricci, M.L.; Landers, R.; Fermino, T.Z.; Rodrigues-Filho, U.P. Enhanced photocatalytic properties of core@shell $SiO_2@TiO_2$ nanoparticles. *Appl. Catal. B Environ.* **2015**, *179*, 333–343. [CrossRef]
20. Ferreira-Neto, E.P.; Ullah, S.; Simões, M.B.; Perissinotto, A.P.; de Vicente, F.S.; Noeske, P.-L.M.; Ribeiro, S.J.L.; Rodrigues-Filho, U.P. Solvent-controlled deposition of titania on silica spheres for the preparation of $SiO_2@TiO_2$ core@shell nanoparticles with enhanced photocatalytic activity. *Colloids Surf. A Physicochem. Eng. Asp.* **2019**, *570*, 293–305. [CrossRef]

21. Kitsou, I.; Panagopoulos, P.; Maggos, T.; Arkas, M.; Tsetsekou, A. Development of SiO_2@TiO_2 core-shell nanospheres for catalytic applications. *Appl. Surf. Sci.* **2018**, *441*, 223–231. [CrossRef]
22. Budiarti, H.A.; Puspitasari, R.N.; Hatta, A.M.; Sekartedjo; Risanti, D.D. Synthesis and Characterization of TiO_2@SiO_2 and SiO_2@TiO_2 Core-Shell Structure Using Lapindo Mud Extract via Sol-Gel Method. *Proc. Eng.* **2017**, *170*, 65–71. [CrossRef]
23. Karunakaran, C.; Magesan, P.; Gomathisankar, P.; Vinayagamoorthy, P. Absorption, emission, charge transfer resistance and photocatalytic activity of Al_2O_3/TiO_2 core/shell nanoparticles. *Superlattices Microstruct.* **2015**, *83*, 659–667. [CrossRef]
24. Deng, H.; Zhang, M.; Cao, Y.; Lin, Y. Decolorization of Reactive Black 5 by Mesoporous Al_2O_3@TiO_2 Nanocomposites. *Environ. Prog. Sustain.* **2019**, *38*, S230–S242. [CrossRef]
25. Logar, M.; Kocjan, A.; Kocjan, A.; Aleš, D. Photocatalytic Activity of Nanostructured γ-Al_2O_3/TiO_2 Composite Powder Formed via a Polyelectrolyte-Multilayer-Assisted Sol–Gel Reaction. *Mater. Res. Bull.* **2012**, *47*, 12–17. [CrossRef]
26. Kocijan, M.; Curkovic, L.; Vengust, D.; Radoševic, T.; Shvalya, V.; Gonçalves, G.; Podlogar, M. Synergistic Remediation of Organic Dye by Titanium Dioxide/Reduced Graphene Oxide Nanocomposite. *Molecules* **2023**, *28*, 7326. [CrossRef] [PubMed]
27. Liu, Y.-L.; Wu, J.M. Synergistically catalytic activities of $BiFeO_3$/TiO_2 core-shell nanomaterials for degradation of organic dye molecule through piezo-phototronic effect. *Nano Energy* **2019**, *56*, 74–81. [CrossRef]
28. Hamad, H.; Elsenety, M.M.; Sadik, W.; El-Demerdash, A.-G.; Nashed, A.; Mostafa, A.; Elyamny, S. The superior Photo-catalytic performance and DFT insights of S-scheme CuO@TiO_2 heterojunction materials for simultaneous degradation of organics. *Sci. Rep.* **2022**, *12*, 2217. [CrossRef]
29. Rey, A.; Garcia-Munoz, P.; Hernandez-Alonso, M.D.; Mena, E.; Garcia-Rodriguez, S.; Beltran, F.J. WO_3–TiO_2 based catalysts for the simulated solar radiation assisted photocatalytic ozonation of emerging contaminants in a municipal wastewater treatment plant. *Appl. Catal. B* **2014**, *154*, 274–284. [CrossRef]
30. Shilova, O.A.; Kovalenko, A.S.; Nikolaev, A.M.; Mjakin, S.V.; Sinel'nikov, A.A.; Chelibanov, V.P.; Gorshkova, Y.E.; Tsvigun, N.V.; Ruzimuradov, O.N.; Kopitsa, G.P. Surface and photocatalytic properties of sol–gel derived TiO_2@SiO_2 core-shell nanoparticles. *J. Sol-Gel Sci. Technol.* **2023**, *108*, 263–273. [CrossRef]
31. Ren, Y.; Li, W.; Cao, Z.; Jiao, Y.; Xu, J.; Liu, P.; Li, S.; Li, X. Robust TiO_2 nanorods-SiO_2 core-shell coating with high-performance self-cleaning properties under visible light. *App. Surf. Sci.* **2020**, *509*, 145377. [CrossRef]
32. Babyszko, A.; Wanag, A.; Sadłowski, M.; Kusiak-Nejman, E.; Morawski, A.W. Synthesis and Characterization of SiO_2/TiO_2 as Photocatalyst on Methylene Blue Degradation. *Catalysts* **2022**, *12*, 1372. [CrossRef]
33. Nadrah, P.; Gaberšček, M.; Sever, Š.A. Selective degradation of model pollutants in the presence of core@shell TiO_2@SiO_2 photocatalyst. *App. Surf. Sci.* **2017**, *405*, 389–394. [CrossRef]
34. Lozhkomoev, A.S.; Kazantsev, S.O.; Bakina, O.V.; Pervikov, A.V.; Chzhou, V.R.; Rodkevich, N.G.; Lerner, M.I. Investigation of the Peculiarities of Oxidation of Ti/Al Nanoparticles on Heating to Obtain TiO_2/Al_2O_3 Composite Nanoparticles. *J. Clust. Sci.* **2023**, *34*, 2167–2176. [CrossRef]
35. Sun, Q.; Hong, Y.; Liu, Q.H.; Dong, L.F. Synergistic operation of photocatalytic degradation and Fenton process by magnetic Fe_3O_4 loaded TiO_2. *Appl. Surf. Sci.* **2018**, *430*, 399–406. [CrossRef]
36. Haber, J. Doping in catalysis: The action of adding a small amount of foreign atoms to form a solid solution in the lattice of a non-metallic catalyst. *Pure Appl. Chem.* **1991**, *63*, 1227. [CrossRef]
37. Wang, Z.; Shi, C.; Li, P.; Wang, W.; Xiao, W.; Sun, T.; Zhang, J. Optical and Photocatalytic Properties of Cobalt-Doped $LuFeO_3$ Powders Prepared by Oxalic Acid Assistance. *Molecules* **2023**, *28*, 5730. [CrossRef]
38. Roza, L.; Febrianti, Y.; Iwan, S.; Fauzia, V. The role of cobalt doping on the photocatalytic activity enhancement of ZnO nanorods under UV light irradiation. *Surf. Interfaces* **2020**, *18*, 1004. [CrossRef]
39. Ceng, C.Y.; Sum, J.Y.; Lai, L.S.; Toh, P.Y.; Chang, Z.H. Visible light-driven dye degradation by magnetic cobalt-doped zinc oxide/iron oxide photocatalyst. *Next Mater.* **2024**, *2*, 100074. [CrossRef]
40. Lu, Y.; Lin, Y.; Wang, D.; Wang, L.; Xie, T.; Jiang, T. A high performance cobalt-doped ZnO visible light photocatalyst and its photogenerated charge transfer properties. *Nano Res.* **2011**, *4*, 1144–1152. [CrossRef]
41. Matussin, S.N.; Khan, F.; Harunsani, M.H.; Kim, Y.M.; Khan, M.M. Impact of Co-Doping on the Visible Light-Driven Photocatalytic and Photoelectrochemical Activities of $Eu(OH)_3$. *ACS Omega* **2024**, *9*, 16420–16428. [CrossRef] [PubMed]
42. Ranno, L.; Dal Forno, S.; Lischner, J. Computational design of bimetallic core-shell nanoparticles for hot-carrier photocatalysis. *NPJ Comput. Mater.* **2018**, *4*, 31. [CrossRef]
43. Khanchandani, S.; Kumar, S.; Ganguli, A.K. Comparative Study of TiO_2/CuS Core/Shell and Composite Nanostructures for Efficient Visible Light Photocatalysis. *ACS Sustain. Chem. Eng.* **2016**, *4*, 1487–1499. [CrossRef]
44. Shannon, R.D. Revised Effective Ionic Radii and Systematic Studies of Interatomic Distances in Halides and Chalcogenides. *Acta Cryst. A* **1976**, *32*, 751–767. [CrossRef]
45. Prins, R. On the structure of γ-Al_2O_3. *J. Catal.* **2020**, *392*, 336–346. [CrossRef]
46. Dehghani, Z.; Nezamdoost, S.; Noghreiyan, A.V.; Nadafan, M. The influence of γ-irradiation on molecular structure and mass attenuation coefficients of γ-Al_2O_3 nanoparticles. *AIP Adv.* **2023**, *13*, 035120. [CrossRef]
47. Pudukudy, M.; Zahira Yaakob, Z. Facile Synthesis of Quasi Spherical ZnO Nanoparticles with Excellent Photocatalytic Activity. *J. Clust. Sci.* **2015**, *26*, 1187–1201. [CrossRef]
48. Hirano, T.; Kaseda, S.; Cao, K.L.A.; Iskandar, F.; Tanabe, E.; Ogi, T. Multiple ZnO Core Nanoparticles Embedded in TiO_2 Nanoparticles as Agents for Acid Resistance and UV Protection. *ACS Appl. Nano Mater.* **2022**, *5*, 15449–15456. [CrossRef]

49. Sanz-Ortiz, M.N.; Rodríguez, F.; Rodríguez, J.; Demazeau, G. Optical and magnetic characterisation of Co^{3+} and Ni^{3+} in $LaAlO_3$: Interplay between the spin state and Jahn–Teller effect. *J. Phys. Condens. Matter* **2011**, *23*, 415501. [CrossRef]
50. Surdo, A.I.; Kortov, V.S.; Pustovarov, V.A.; Yakovlev, V.Y. UV luminescence of F-centers in aluminum oxide. *Phys. Status Solidi C* **2005**, *2*, 527–530. [CrossRef]
51. Matysiak, W.; Tański, T. Analysis of the morphology, structure and optical properties of 1D SiO_2 nanostructures obtained with sol-gel and electrospinning methods. *Appl. Surf. Sci.* **2019**, *489*, 34–43. [CrossRef]
52. Shablonin, E.; Popov, A.I.; Prieditis, G.; Vasil'chenko, E.; Lushchik, A. Thermal annealing and transformation of dimer F centers in neutron-irradiated Al_2O_3 single crystals. *J. Nucl. Mater.* **2021**, *543*, 152600. [CrossRef]
53. Usseinov, A.B.; Gryaznov, D.; Popov, A.I.; Kotomin, E.A.; Seitov, D.; Abuova, F.; Nekrasov, K.A.; Akilbekov, A.T. Ab initio calculations of pure and Co^{+2}-doped MgF_2 crystals. *Nucl. Instrum. Methods Phys. Res. B* **2020**, *470*, 10–14. [CrossRef]
54. Brik, M.G.; Srivastava, A.M.; Popov, A.I. A few common misconceptions in the interpretation of experimental spectroscopic data. *Opt. Mater.* **2022**, *127*, 112276. [CrossRef]
55. Jafarova, V.N.; Orudzhev, G.S. Structural and electronic properties of ZnO: A first-principles density-functional theory study within LDA(GGA) and LDA(GGA)+U methods. *Solid State Commun.* **2021**, *325*, 114166. [CrossRef]
56. Filatova, E.O.; Konashuk, A.S. Interpretation of the Changing the Band Gap of Al_2O_3 Depending on its Crystalline Form: Connection with Different Local Symmetries. *J. Phys. Chem. C* **2015**, *119*, 20755–20761. [CrossRef]
57. Yazdanmehr, M.; Asadabadi, S.J.; Nourmohammadi, A.; Ghasemzadeh, M.; Rezvanian, M. Electronic structure and bandgap of γ-Al_2O_3 compound using mBJ exchange potential. *Nanoscale Res. Lett.* **2012**, *7*, 488. [CrossRef]
58. Papi, H.; Jalali-Asadabadi, S.; Nourmohammadi, A.; Ahmad, I.; Nematollahia, J.; Yazdanmehr, M. Optical properties of ideal gamma-Al_2O_3 and with oxygen point defects: An ab initio study. *RSC Adv.* **2015**, *5*, 55088. [CrossRef]
59. Clark, S.J.; Segall, M.D.; Pickard, C.J.; Hasnip, P.J.; Probert, M.J.; Refson, K.; Payne, M.C. *Materials Studio, CASTEP*, Version 5.0; Accelrys: San Diego, CA, USA, 2009; More information on the official CASTEP website at http://www.castep.org/.
60. Hamann, D.R.; Schlüter, M.; Chiang, C. Norm-Conserving Pseudopotentials. *Phys. Rev. Lett.* **1979**, *43*, 1494–1497. [CrossRef]
61. Dolatyari, M.; Alidoust, F.; Zarghami, A.; Rostami, A.; Mirtaheri, P.; Mirtagioglu, H. High-Resolution Color Transparent Display Using Superimposed Quantum Dots. *Nanomaterials* **2022**, *12*, 1423. [CrossRef]

Disclaimer/Publisher's Note: The statements, opinions and data contained in all publications are solely those of the individual author(s) and contributor(s) and not of MDPI and/or the editor(s). MDPI and/or the editor(s) disclaim responsibility for any injury to people or property resulting from any ideas, methods, instructions or products referred to in the content.

Article

Significantly Enhanced Self-Cleaning Capability in Anatase TiO$_2$ for the Bleaching of Organic Dyes and Glazes

Tiangui Zhao, Tihao Cao, Qifu Bao *, Weixia Dong *, Ping Li, Xingyong Gu, Yunzi Liang and Jianer Zhou

School of Materials Science and Engineering, Jingdezhen Ceramic University, Jingdezhen 333403, China
* Correspondence: bqf2002@hotmail.com (Q.B.); dongweixia@jci.edu.cn (W.D.)

Abstract: In this study, the Mg^{2+}-doped anatase TiO$_2$ phase was synthesized via the solvothermal method by changing the ratio of deionized water and absolute ethanol $V_{water}/V_{ethanol}$). This enhances the bleaching efficiency under visible light. The crystal structure, morphology, and photocatalytic properties of Mg-doped TiO$_2$ were characterized by X-ray diffraction, scanning electron microscopy, high-resolution transmission electron microscopy, N$_2$ adsorption-desorption, UV-Vis spectroscopy analysis, etc. Results showed that the photocatalytic activity of the Mg^{2+}-doped TiO$_2$ sample was effectively improved, and the morphology, specific surface area, and porosity of TiO$_2$ could be controlled by $V_{water}/V_{ethanol}$. Compared with the Mg-undoped TiO$_2$ sample, Mg-doped TiO$_2$ samples have higher photocatalytic properties due to pure anatase phase formation. The Mg-doped TiO$_2$ sample was synthesized at $V_{water}/V_{ethanol}$ of 12.5:2.5, which has the highest bleaching rate of 99.5% for the rhodamine B dye during 80 min under visible light. Adding Mg^{2+}-doped TiO$_2$ into the phase-separated glaze is an essential factor for enhancing the self-cleaning capability. The glaze samples fired at 1180 °C achieved a water contact angle of 5.623° at room temperature and had high stain resistance (the blot floats as a whole after meeting the water).

Keywords: solvothermal method; Mg-doped TiO$_2$; $V_{water}/V_{ethanol}$; self-cleaning properties; visible light

Citation: Zhao, T.; Cao, T.; Bao, Q.; Dong, W.; Li, P.; Gu, X.; Liang, Y.; Zhou, J. Significantly Enhanced Self-Cleaning Capability in Anatase TiO$_2$ for the Bleaching of Organic Dyes and Glazes. *Inorganics* **2023**, *11*, 341. https://doi.org/10.3390/inorganics11080341

Academic Editors: Roberto Nisticò and Silvia Mostoni

Received: 25 June 2023
Revised: 15 August 2023
Accepted: 15 August 2023
Published: 18 August 2023

Copyright: © 2023 by the authors. Licensee MDPI, Basel, Switzerland. This article is an open access article distributed under the terms and conditions of the Creative Commons Attribution (CC BY) license (https://creativecommons.org/licenses/by/4.0/).

1. Introduction

With the deterioration of environmental pollution, low-consumption and high-efficiency pollution technologies have received more attention [1,2]. As the durative utilizes clean energy, solar energy has vast potential for exploitation and application. Titanium dioxide is an important photocatalyst that has been widely studied because of its high activity, non-toxic characteristics, environmental friendliness, and good chemical stability [3–6]. As the energy barrier of the metastable phase was less than that of the stability phase, it was more likely to excite electrons and holes for the metastable phase [7,8]. Hence, anatase TiO$_2$ is considered to be the best photocatalyst of all of the structures of TiO$_2$ [9,10]. It can fully effectively utilize UV light from sunlight [11–13]. Several factors affect anatase TiO$_2$ photocatalytic activity, such as crystal size, specific surface area, and crystallinity [14–16]. The performance of the TiO$_2$ was optimized by doping [17–20], loading [21,22], and thin-film preparation [23,24]. Available studies indicated that some ions could enter the lattice as substitutional or interstitial; the titanium ions are substituted by metal ions in the crystal lattices. Some studies illustrate that rare-metal-ion-doped titania nanoparticles were prepared by the hydrothermal method, and their photocatalytic performance was greatly improved under UV irradiation [25,26]. At present, there exist a few studies concerning magnesium-ion-doped TiO$_2$ obtained by the sol-gel reaction synthesis route and the solvothermal method [27,28], but its processing is complex and needs HF as a capping agent to form the anatase phase. It would therefore be interesting to investigate how a simple method can be used for preparing a glaze containing Mg(II)-doped anatase that is stable in a medium-/high-temperature (>1000 °C) ceramic glaze [29] and has self-cleaning properties, as anatase TiO$_2$ has a nanometer size.

This study presents the simple synthetic procedure of producing Mg-doped TiO_2 anatase samples without surfactants or templates and evaluates the influence of the structure and $V_{water}/V_{ethanol}$ on their photocatalytic activity in decomposing rhodamine B (RhB). The self-cleaning activities of Mg-doped and undoped TiO_2 anatase glaze samples are evaluated by comparing their anti-pollution ability.

2. Experimental Section

2.1. Preparation of the Samples

The samples, with various deionized water and absolute ethanol contents, were prepared from tetrabutyl titanate (TBOT), $MgCl_2 \bullet 6H_2O$, and NaOH using the hydrothermal method. In a typical synthesis, firstly, solution A was made, which included $MgCl_2 \bullet 6H_2O$, deionized water, and absolute ethanol. Subsequently, solution B was made, which included TOBT and ethanol. Finally, suspension C was prepared by dripping solution B into system A. The molar ratio of $MgCl_2 \bullet 6H_2O$:TBOT:ethanol: water was 0.03:1:10:50. After 15 min, after adding suspension C into the reactor, it was heated at 180 °C for 36 h and then naturally cooled to room temperature. The final sample obtained was centrifuged and washed with deionized water and absolute ethanol. The photocatalytic properties of the samples were investigated by changing the molar ratio of water/ethanol ($V_{water}:V_{ethanol}$), keeping other experimental parameters unchanged. Figure 1 is the schematic diagram of Mg-doped TiO_2 sample preparation.

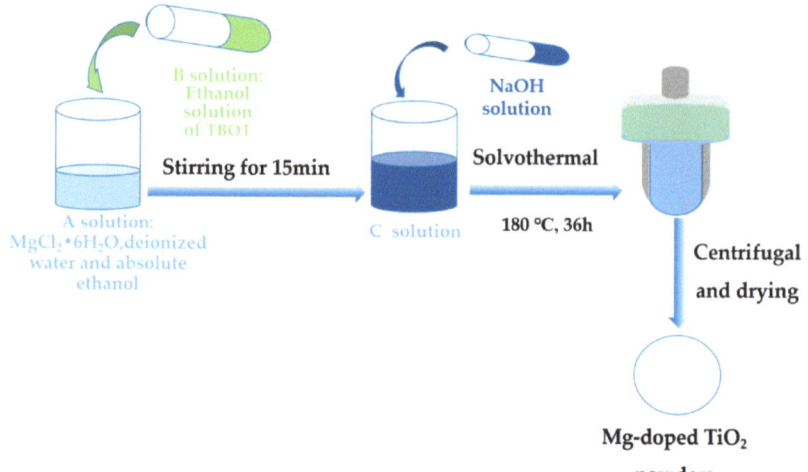

Figure 1. Schematic diagram of Mg-doped TiO_2 sample preparation.

The Mg-doped TiO_2 in the glaze sample was fabricated by sintering at 1180~1200 °C using raw powders, i.e., 95% of the as-prepared Kaolin clay was subjected to phase separation melting at 1500 °C for 4 h and 5% by adding 5% Mg-doped TiO_2 (V_{water}/V_{ethnol} of 12.5:2.5) photocatalysts, and the self-cleaning and hyper-hydrophilic properties of the fired glaze samples were characterized and tested, respectively. Figure 2 is the schematic diagram of the glaze firing processes.

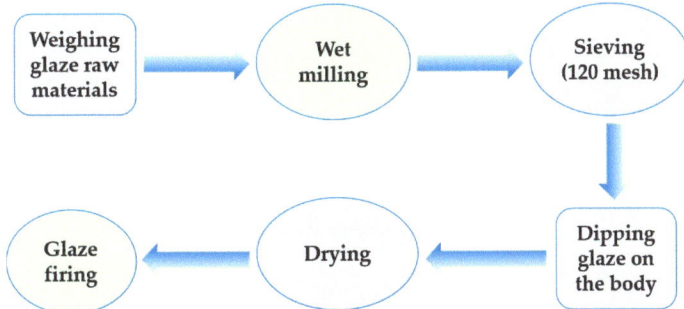

Figure 2. Schematic diagram of the glaze firing processes.

2.2. Characterization of the Samples

The crystalline phase was identified by X-ray diffractometer (XRD, D8 Advance Bruker AXS, Germany) using Cu Kα radiation. Compared with the standard pattern in the XRD standard database, including JCPDS (i.e., PDF cards), the phase composition of the sample was analyzed using Jade 6.0 software. Photocatalyst morphology was investigated by scanning electron microscopy (SEM, JSM-6700F, Japan) using a device equipped with an EDS system operating at an accelerating voltage of 5.0 kV or 15 kV (15 kV for EDS). The crystal surface of nanocrystals was evaluated by high-resolution microscopy. The microstructures of the samples were studied by transmission electron microscopy (TEM, FEI Tecnai G2 F-30, Holland) and high-resolution transmission electron microscopy (HRTEM, FEI Tecnai G2 F-30, Holland) at accelerating voltages of 160 kV and 200 kV, respectively. The valence states of the samples were characterized by X-ray photoelectron spectroscopy (XPS, ESCALAB Xi+, United States) using Al Kα radiation. The specific surface areas were determined by the Brunauer–Emmett–Teller method, and the pore size was determined by the Barrett–Joyner–Hallenda method. Nitrogen adsorption-desorption isotherms were collected on a Micromeritics TriStar ii 3020 analyzer at 77 K. The analysis of samples by UV-Vis diffuse reflectance spectroscopy was carried out. The hydrophilicity of the samples was tested by a contact angle meter (JGW-360D, China).

2.3. Photocatalytic Activity of the Samples

The photocatalytic activity of the TiO_2 was evaluated by bleaching the RhB with a concentration of 10^{-4} mol/L. The total volume of RhB was 50 mL, irradiated with 0.05 g of the photocatalyst and a 500 WXeon light with a cut-off filter of 420 nm. This was to prove that the RhB was exhibiting bleaching rather than adsorption after the dark experiment was carried out. Samples were taken out at 20 min intervals and analyzed with a spectrophotometer. The photocatalytic activity was characterized by the apparent first-order rate constant k, as in equation $k = \ln(A_0/A)$, where A was the absorbance of RhB at 553 nm after bleaching and A_0 was the absorbance of the initial RhB solution at 553 nm.

3. Results and Discussion

3.1. Structural and Morphology

The crystal phase of the samples was studied as shown in Figure 3. The obtained diffraction peak of the doped TiO_2 matched very well with the standard values (PDF-#21-1272) and the diffraction peaks at 2θ = 25.281(101), 37.800(004), 48.049(200), 53.890(105), and 62.688(204), illustrating that the samples were in the anatase phase. However, the obtained undoped TiO_2 was in a mixed phase of anatase and brookite. The cell volume was calculated by Fourier synthesis with the program SHELXS−97 [30]. When the solvent was water, the sample consisted of nanoparticles 10~20 nm in mean size, as determined by Nano Measurer 1.2 software using 10 nanoparticles. The average crystallite size of TiO_2 samples with different Mg-doped ions was calculated by XRD–Scherrer formula:

d = 0.91 λ/βcos θ, where d is the mean crystallite size, k is 0.9, λ is the wavelength of Cu Kα (i.e., λ = 0.15420 nm), β is the full width at half maximum intensity of the peak (FWHM) in radian, and θ is Bragg's diffraction angle [31]. The crystallite size and cell volume were calculated as shown Table 1. When increasing $V_{water}/V_{ethanol}$, there are differences in the diffraction peak intensity and minor shifts in the peak occur, which indicates a reduction in crystalline size and an increase in the volume of unit cells (Table 1). Since the ionic radius of Mg^{2+} (0.072 nm) is close to that of Ti^{4+} (0.061 nm), Mg^{2+} easily enters the TiO_2 lattice [32] and the lattice volume increases (Table 1), indicating that the formation of a crystal defect. Based on the experimental results, the formation of the crystal defect promotes the formation of the anatase phase, which is accordance with the reported literature [27,29]. Hence, after the addition of the magnesium source, a pure-anatase TiO_2 phase appears. The intensity of the (004) direction is significantly enhanced compared to undoped TiO_2. In addition, the FWHM of the (101) peak was calculated by using Lorentz fitting. According to the Scherrer formula, d = 0.91 λ/βcos θ, the crystallite size was calculated; it is shown in Table 1.

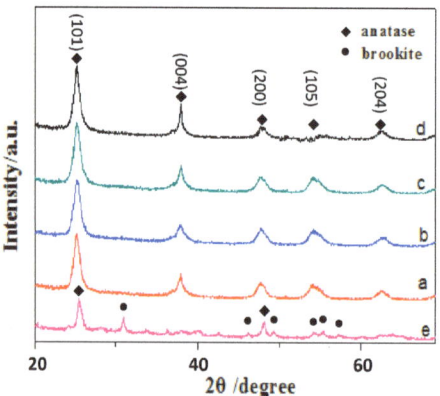

Figure 3. XRD patterns of the samples with different volume ratios of water: ethanol: (a) 15: 0, (b) 12.5:2.5, (c) 10:5, (d) 7.5:7.5, (e) undoped TiO_2 (12.5:2.5).

Table 1. Effect of different ratios of water: ethanol in the solvent on the crystal size, BET surface area, pore size, pore volume, and cell volume of Mg-doped TiO_2.

	V_{water}/V_{ethnol}	Crystal Size (nm)	BET (m²/g)	Pore Size (nm)	Pore Volume (cm³/g)	Cell Volume Å³
	15:0	13.6	152	13.8	0.415	136.458
Mg-doped TiO_2	12.5:2.5	13.2	148	12.5	0.402	136.315
	10:5	10.3	105	12.4	0.378	136.452
	7.5:7.5	8.1	101	12.0	0.350	136.689
Pure TiO_2	12.5:2.5	14.0	98	11.2	0.340	136.089

Figure 4 shows SEM images of the as-synthesized samples. When the solvent was water, the sample consisted of nanoparticles 5–10 nm in size. When the V_{water}/V_{ethnol} ratio was 12.5:2.5, agglomerated nanoparticles had a grape-like morphology (Figure 4b). With the increase in ethanol dosage, nanoparticles increased (Figure 4c,d). The experimental results show that the morphology of the samples was greatly affected by V_{water}/V_{ethnol}. Their morphology is determined by the relationship between crystal formation and growth. Moreover, crystal growth is influenced by the adsorption of certain crystalline facets into OH^-. This adsorption hinders the growth of these facets, resulting in different rates of crystalline growth. Ethanol is a typical polar solvent and amphiphilic molecule. It was vertically adsorbed on the hydrophilic surface of the TiO$_2$ particles, forming a two-amphiphilic bilayer, which limited the immersion of the water molecule in the hydrophilic side surface and the TiO$_2$ particles [33]. The rapid hydrolysis of TBOT promoted the rapid generation of TiO$_2$, which led to TiO$_2$ particle agglomeration with an increase in $V_{water}/V_{ethanol}$. Figure 5a,b show TEM and the corresponding SAED pattern (inset) and HRTEM images of the sample prepared at $V_{water}/V_{ethanol}$ = 12.5:2.5. From Figure 5a, it is observed that the aggregated particles in Figure 4b consist of nanoparticles. The major diffraction rings for the crystal surface at (101), (004), and (105) match well with XRD analysis. The d spacing is 0.325 nm (Figure 5b), and it matches well with the lattice spacing of anatase TiO$_2$ (101). Furthermore, the corresponding EDX spectrum shown in Figures 5c and S1 verifies the existence of Mg, Ti, and O ions. Other impurities were not detected in the EDX spectra.

Figure 4. SEM images of TiO$_2$ with different volume ratios of water: ethanol (**a**) 15:0, (**b**) 12.5:2.5, (**c**) 10:5, (**d**) 7.5:7.5.

As can be seen from Table 1 and Figure 3, the morphologies of the samples strongly depend on Mg-doped ions and $V_{water}/V_{ethanol}$. Because the current system contains ethanol, water, Mg-doped ions, and TBOT, we can reasonably assume that the formation of anatase TiO$_2$ is due to the dehydrating condensation between Ti(OH)$_6^{2-}$ and Mg-doped ions under solvothermal conditions [34]. Thus, due to the formation of a lower number of active OH$^-$ ions and a lower number of soluble species, Ti(OH)$_6^{2-}$ and TiO$_6$ octahedrons in one cluster may construct a chain via the corner-sharing of Ti(OH)$_6^{2-}$ growth units. Due to doped Mg ions entering the TiO$_2$ lattice, resulting in TiO$_6$ octahedron lattice distortion (Table 1) and an increase in the charge density of Ti and reduction in the electron density of oxygen, the preferred TiO$_6$ octahedron chain-shaped clusters further adsorb OH$^-$ soluble species into the (101) plane (Figure 5b) and anatase TiO$_2$ monomers form through a dehydrating condensation process. Therefore, these planes could be freely bonded by interactions between OH$^-$ and nuclei to obtain aggregated nanoparticles (Figure 4). The solubility of salt increases with the dielectric constant of the solvent [35], and the dielectric constant of water is bigger than that of ethanol. When V_{water}/V_{ethnol} decreases, that is, ethanol content increases, this could decrease the solubility of the precursor and increase the viscosity of the solution, thereby decreasing the diffusion ability of Ti(OH)$_6^{2-}$ ions and causing the crystal size of the TiO$_2$ sample to decrease (Table 1).

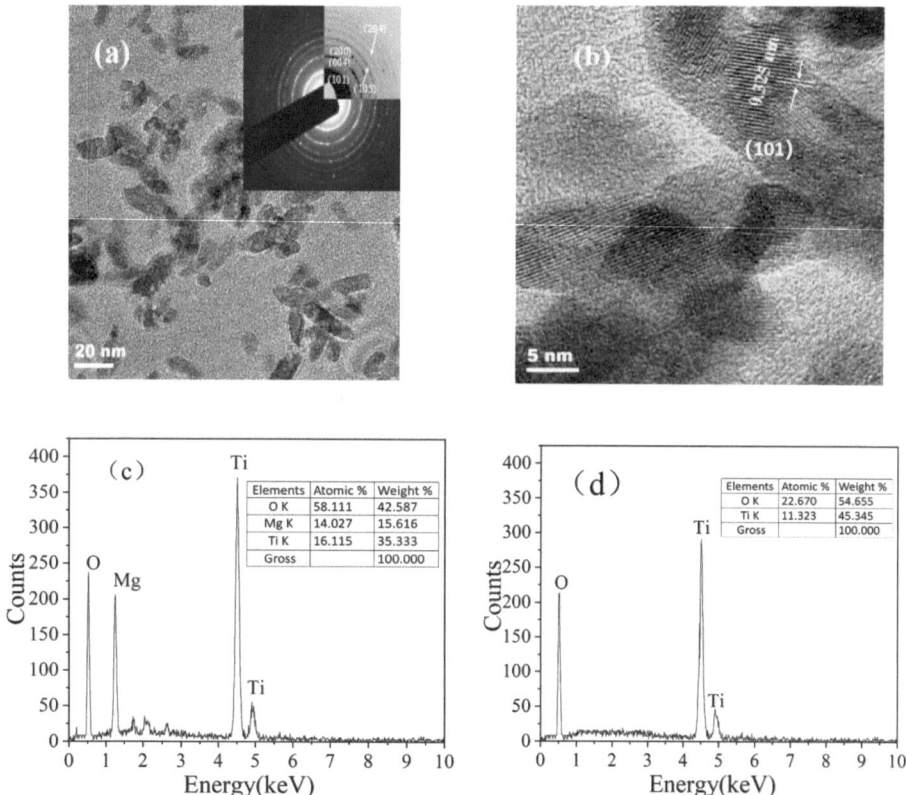

Figure 5. (a) TEM images and SAED pattern, (b) HRTEM images, (c,d) EDX spectra. (a–c) Mg-doped sample prepared using V_{water}/V_{ethnol} = 12.5:2.5, (d) pure TiO$_2$ sample prepared using V_{water}/V_{ethnol} = 12.5:2.5.

Figure 6 shows XPS spectra of pure and Mg-doped TiO$_2$ samples. Peaks located around 457 eV and 464 eV resulted from Ti 2p$_{3/2}$ and Ti 2p$_{1/2}$, respectively, corresponding to the oxidation state of Ti^{4+}. Meanwhile, due to the partial substitution of Tg^{4+} ions by Mg^{2+}, the binding energy of Ti decreases, thus increasing the charge density of Ti. The binding energy of O 1s in the pure TiO$_2$ sample is 529.8 eV, owing to the intrinsic binding energy of oxygen in TiO$_2$. The Mg-doped TiO$_2$ sample shows a shoulder peak near 532.3 eV in addition to the intrinsic binding energy of O 1s (shown in Figure 6b). This may be due to the addition of small amounts of Mg atoms, causing new oxygen vacancies [36]. Oxygen vacancies in TiO$_2$ are usually created in doped TiO$_2$ to maintain charge neutrality and improve the service life of the photocatalyst [37]. When oxygen vacancies are generated, a higher energy peak can be seen due to the decrease in the electron density of oxygen [37]. A peak at 49.93 eV was associated with Mg 2p, which is further verified by the incorporation of Mg^{2+} into the titanium dioxide lattice.

Figure 7 shows the typical FT-IR spectrum of undoped TiO$_2$ and Mg-doped TiO$_2$ samples with different V_{water}/V_{ethnol} ratios. All samples have absorption peaks at 3380 cm^{-1} and 1640 cm^{-1}, corresponding to O-H stretching vibration and bending vibration, respectively [38]. For the undoped TiO$_2$ sample, the bands at 1450 cm^{-1} and 1538 cm^{-1} are attributed to the H-O-H bending of the lattice water [39]. The band centered at 510 cm^{-1} is due to isolated tetrahedral TiO$_4$ stretching vibrations and only occurs in the pure TiO$_2$ sample [40]. As a result of Mg-doping, the bands at 1065 cm^{-1} and 458 cm^{-1} show the vibration of Ti-O-Mg [41]. With the increase in ethanol content, the intensities of the absorption peaks at 3380 cm^{-1} and 458 cm^{-1} increase, respectively. This indicates that Mg ions are doped into the lattice of TiO$_2$, and the HRTEM, TEM, and XRD results further confirmed this point.

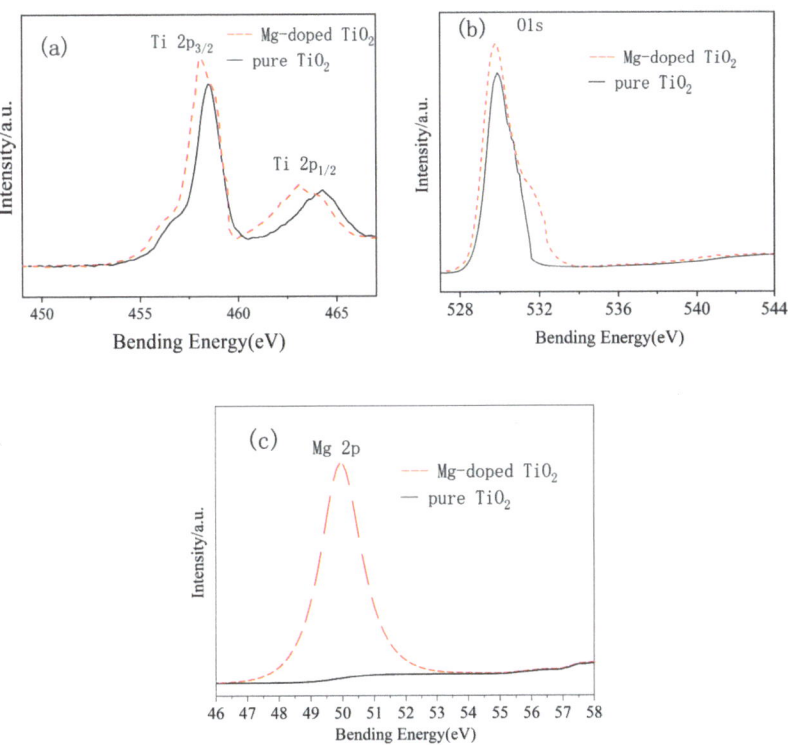

Figure 6. XPS spectra of the undoped TiO$_2$ and Mg-doped TiO$_2$ samples prepared using V_{water}/V_{ethnol} ratio of 15:0 (**a**) Ti, (**b**) O, (**c**) Mg.

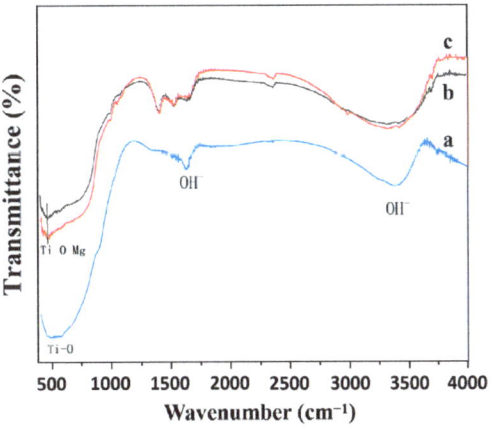

Figure 7. FT-IR spectrum of the typical samples: (**a**) undoped TiO$_2$, (**b**) Mg-doped TiO$_2$ samples prepared using V_{water}/V_{ethnol} ratio of 15:0, (**c**) Mg-doped TiO$_2$ samples prepared using V_{water}/V_{ethnol} ratio of 12.5:2.5.

3.2. BET Analysis

Figure 8 shows the BET analysis of the samples using nitrogen adsorption-desorption. For all samples, the isotherms are type IV, and clear hysteresis loops can be identified. With the increase in V_{water}/V_{ethnol}, the BET surface area of the Mg-doped TiO$_2$ samples decreases. However, the pore volume and porosity of the samples exhibit a prominent enhancement compared with the undoped TiO$_2$ sample, as shown in Table 1 and Figure 8. The BJH average pore diameters, calculated from the adsorption branch of the isotherms, are 11.205 nm, 12.560 nm, 12.365 nm, and 12.807 nm for pure TiO$_2$ and Mg-doped TiO$_2$ samples prepared with different V_{water}/V_{ethnol} ratios of 12.5:2.5, 10:5, and 7.5:7.5, respectively. The mesoporous structure is mainly due to the porous accumulation of nanoparticles [42]. The porosity increase is due to the crystal size reducing with the decrease in V_{water}/V_{ethnol}.

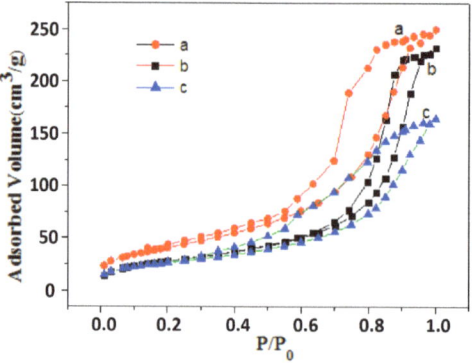

Figure 8. N$_2$ adsorption-desorption isotherm samples with different ratio of water: ethanol (a) doped TiO$_2$ 12.5:2.5, (b) doped TiO$_2$ 10:5, (c) undoped TiO$_2$ 12.5:2.5.

3.3. Optical Properties

Figure 9 shows the UV-Visible diffuse reflectance spectra of TiO$_2$. The absorption edge of doped TiO$_2$ had more of a blue shift than the undoped TiO$_2$. The Kulbeka–Munk formula, (E(ev) = hC/λ, h = 6.626 × 10^{-34} Js, C = 3.0 × 10^8 ms^{-1}), was used to acquire the exact band gap of TiO$_2$ from 3.26 eV to 3.13 eV, which can be attributed to the Mg^{2+}-doped TiO$_2$ in the framework. Since Mg^{2+} ions generated from oxygen vacancies are known to cause the photoexcitation of long-wavelength light, the UV-Vis absorption spectrum was inferred to verify the presence of Mg^{2+} in the TiO$_2$-doped sample.

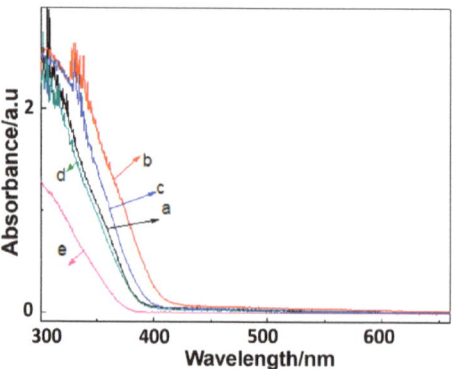

Figure 9. UV-Vis spectra of TiO$_2$ with different rate of water: ethanol (a) 15:0, (b) 12.5:2.5, (c) 10:5, (d) 7.5:7.5, (e) undoped TiO$_2$ 12.5:2.5.

Moreover, from the spectrum, the energy gap of the semiconductor nanoparticles is related to the particle size. The band gap increases as the particle size decreases, resulting in a phenomenon known as a "blue shift" in light absorption at a specific wavelength due to the quantum size effect [43]. With the increase in ethanol content, the absorption edge of the doped TiO_2 is blue-shifted, illustrating the particle size reduction. The results obtained are well-matched with the sizes of the crystals that were measured. The band gap energies of the prepared TiO_2 doped by adding 0 to 7.5 mL ethanol were found to be 3.17 ev, 3.03 ev, 3.13 ev, and 3.25 ev, respectively. From Figure 4, it is clear that the size of anatase nanoparticles increases with the increase in ethanol content. Optical absorption is highly dependent on the internal structure of the material [44]. Compared with pure TiO_2, the longer-wavelength region of Mg-doped TiO_2 samples implies that the only possible transition is from the oxygen vacancies causing a red shift of the absorption edge (Figure 6), which also implies that Mg^{2+} has been incorporated into the lattice of TiO_2 (Table 1). From Figure 6, it can be observed that compared with the pure TiO_2 sample, the Ti and O binding energy in Mg-doped TiO_2 samples has been shifted to a lower energy and a higher energy peak, because some Ti^{4+} ions are replaced by Mg^{2+} ions in order to increase the charge density of Ti and reduce the electron density of oxygen [45]. The new oxygen vacancies are created through the doping of small amounts of Mg atoms [46]. For the Mg-doped TiO_2 sample, the peak of 49.9 eV is ascribed to Mg 2p (Figure 6c), which is consistent with the value of Mg^{2+} [27,41]. These observations further verify the existence of Mg^{2+} in the Mg-doped TiO_2 sample, which is consistent with XRD (Figure 3), increased cell volume (Table 1), and FT-IR spectrum (Figure 7).

3.4. Photocatalytic Activity

Figure 10 shows the photocatalytic bleaching of RhB through the as-prepared sample under visible light. As shown in Figure 10, RhB concentration is unchanged, illustrating that RhB adsorbed on the TiO_2 surface had reached equilibrium in 30 min. Figure 10b shows kinetic curves of $\ln(C_0/C)$ versus irradiation time during RhB bleaching under visible light irradiation. It has been found that the apparent rate constants [47] for the reaction of RhB with Mg-doped TiO_2 samples ($V_{water}/V_{ethanol}$ = 15:0, 12.5:2.5, 10:5, 7.5:7.5) and Mg-undoped TiO_2 ($V_{water}/V_{ethanol}$ = 12.5:2.5) were 0.01704, 0.06335, 0.04153, 0.01668, and 0.00203 min^{-1}, respectively, which illustrates that the photocatalytic activity of the samples was effectively improved by Mg^{2+}-doping (due to pure anatase phase formation (Figure 3)). Moreover, the photocatalytic properties of Mg-doped TiO_2 can be further improved by changing the ratio of water to ethanol. The photocatalytic properties of the samples increased first and then decreased gradually with the increase in $V_{water}/V_{ethanol}$. When the $V_{water}/V_{ethanol}$ ratio was 12.5:2.5, Mg-doped TiO_2 had the maximum photocatalytic activity. In addition, by combining Table 1 with Figures 4 and 9, we can observe that the aggregated nanoparticles increase in size and thus E_g increases, which leads to the easy recombination of the electron and hole in the migration process, and therefore, the photocatalytic activity of the samples decreases with the increase in ethanol volume (i.e., $V_{water}/V_{ethanol}$ decreases). Although TiO_2 ($V_{water}/V_{ethanol}$ = 15:0) has a larger specific surface area and smaller crystal size (Table 1) compared with the Mg-doped samples, the sample had lower porosity and pore size, which caused the decrease in the sample of RhB adsorption. This clearly indicates that the adsorption of samples was determined by the surface area and characteristics of the pore. Obviously, Mg-doped TiO_2 samples exhibited better photocatalytic activities than pure TiO_2 samples. The narrowing of the band gap is a result of Mg doping into the TiO_2 lattice, which enables the trapping of the photo-induced electron and facilitates the separation of electron-hole pairs (Figure 11a).

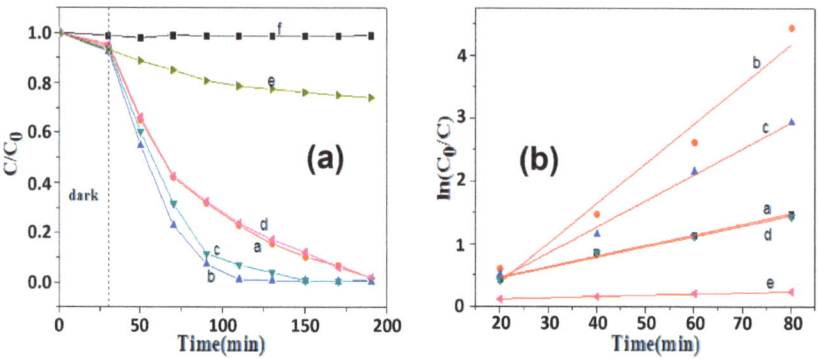

Figure 10. (a) Photocatalytic performance, (b) Kinetic curves of $\ln(C_0/C)$ as a function of irradiation time for RhB bleaching under visible light irradiation: (a) 15:0, (b) 12.5:2.5, (c) 10:5, (d) 7.5:7.5 of the Mg-doped samples, (e) Mg-undoped TiO_2 sample, (f) without photocatalyst.

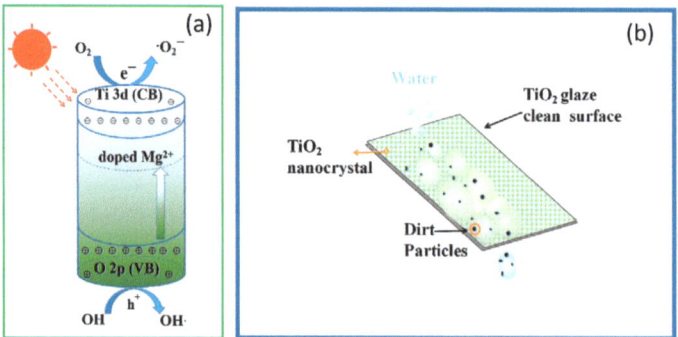

Figure 11. Schematic representation of photo processes using Mg-doped TiO_2 photocatalysts (a) and the self-cleaning process using Mg-doped TiO_2 in glaze sample (b), respectively.

3.5. Self-Cleaning Properties of Mg-Doped TiO_2 in Glaze Sample

It can be seen the wet angle of pure TiO_2 glaze samples is obviously higher than those of Mg-doped TiO_2 glaze samples (Figure 12). The super-hydrophilicity of Mg-doped TiO_2 glaze samples is attributed to several comprehensive factors. Based on the experimental results, Mg ions are helpful for the growth of the TiO_2 crystal grain, and thus separates the phase size in Mg-doped TiO_2 glaze more than pure TiO_2. This makes the Mg-doped TiO_2 glaze surface rougher than that of the pure TiO_2 glaze (Figure 12). A large surface roughness could improve the hydrophilicity, according to the Wenzel equation (1): $\cos\theta_r = r\cos\theta$, where r denotes the surface roughness of the glaze, $\cos\theta$ is the classical contact angle depicted by the Young equation, and θ_r is the measured real contact angle. Moreover, the partial substitution of Mg^{2+} ions for Ti sites increases the slight TiO_2 lattice distortion, which is available for a low initial contact angle and hydrophilicity [48]. From Figure 12, it can be seen that the contact angles of Mg-doped TiO_2 samples are smaller than that of the pure TiO_2 glaze sample in the dark condition, indicating that the greater roughness and lattice distortion are helpful for decreasing the contact angle. This could be because the incorporation of Mg makes the band gap of TiO_2 narrow, thus the visible light can excite pairs of electrons and holes (Figure 11a), just as in the case of ultraviolet irradiation for the pure TiO_2 glaze. Ti^{4+} ions could be united with the photo-induced electron and thus Ti^{3+} ions could be obtained. Ti^{3+} sites can be substituted by Mg^{2+} ions, which produces one excess positive charge. Those excess positive charges could capture the photo-induced electrons quickly, and thus photo-generated holes are available for combining more H₂O adsorbed on the glaze surface and react with water, producing hydroxyl radicals that are also available for maintaining the hydrophilicity of Mg-doped TiO$_2$ glaze samples [29]. Therefore, the super-hydrophilicity of Mg-doped TiO$_2$ glaze samples could be attributed to the visible-light-exciting photo-induced pairs of electrons and holes. For the sample with a V$_{water}$/V$_{ethanol}$ ratio of 10:5 and 7.5:7.5, the contact angles of water droplets on Mg-doped TiO$_2$ glaze samples increase slightly, which could be attributed to the decrease in the V$_{water}$/V$_{ethanol}$ ratio. However, when V$_{water}$/V$_{ethanol}$ is 10:5 and 7.5:7.5, the hydrophilicity of Mg-doped TiO$_2$ glaze samples decreases slightly, though it still has super-hydrophilicity. The hydroxy groups anchoring on the Mg-doped TiO$_2$ glaze surface have a significant impact on the hydrophilicity. The formation of hydroxy groups results in the dissociative adsorption of water molecules at oxygen vacancy sites on the Mg-doped TiO$_2$ glaze surface. The extra hydroxy groups and oxygen vacancies on the surface are produced by electron–hole pairs, which lead to the hydrophilicity of the Mg-doped TiO$_2$ glaze surface [39]. Because oxygen vacancy is produced by the doping of Mg in the TiO$_2$ crystal and the separation of electron–hole pairs is facilitated (Figure 11a), the Mg-doped TiO$_2$ glaze surface has more photo-induced wettability than the pure TiO$_2$ glaze surface.

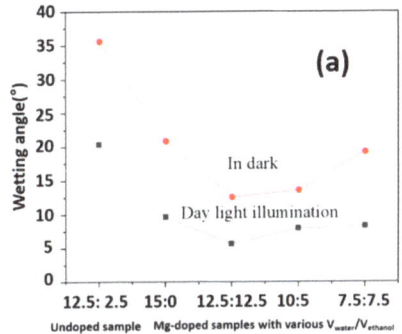

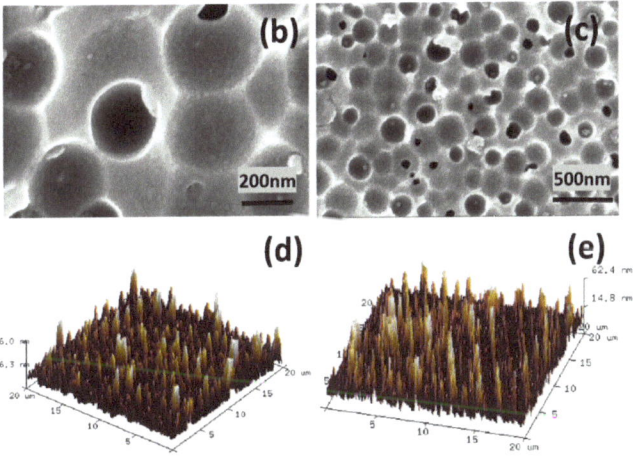

Figure 12. Wetting angle of Mg-undoped TiO$_2$ glaze sample prepared at V$_{water}$/V$_{ethanol}$ of 12.5:2.5, and Mg-doped TiO$_2$ glaze samples with various V$_{water}$/V$_{ethanol}$ ratios in dark and daylight illumination (**a**), respectively; SEM images of the Mg-doped TiO$_2$ glazes prepared using V$_{water}$/V$_{ethano}$ of 12.5:2.5 (**b**) and pure TiO$_2$ glaze prepared using V$_{water}$/V$_{ethano}$ of 12.5:2.5 (**c**); AFM surfaces of the Mg-doped TiO$_2$ glaze (**d**) and pure TiO$_2$ glaze sample (**e**) with V$_{water}$/V$_{ethano}$ of 12.5:2.5, respectively.

The self-cleaning performance was tested using a Japan Marker pen. The glaze surface was drawn on after drying for 1 h. After that, after placing a few drops of water on the glaze, we could observe whether the ink blots were floating. Table 2 shows that after firing at 1180~1200 °C, the water contact angle (5.623° vs. 15.23°) and stain resistance (the blot floats as a whole vs. not floating, as shown in Figure 13) of the sample fabricated were improved compared to commercial self-cleaning ceramic glazes [49]. The above results indicate the great potential application for enhancing the self-cleaning properties of glazes by introducing Mg-doped TiO_2.

Table 2. Performances of Mg-doped TiO_2 in the ceramic samples obtained in this study and from other literature studies.

Type	Firing Temperature (°C)	Water Contact Angle (°)		Stain Resistance	Ref.
		Before Use	Irradiation after Use		
Mg-doped TiO_2 in glaze sample	1180~1200	5.623	5.124	After dripping water droplets, the blot floats as a whole	This work
TiO_2 doped in glaze sample	1180~1200	12.26	13.56	Not floating	This work
The commercial self-cleaning ceramic products	1180~1200	21.23	28.96	Not floating	This work
C-PEG/TiO_2 coating	-	26	11	Blot cannot be completely removed	[50]
Commercial ceramic tiles with groove-like microstructure surfaces	-	164.75	-	Blot cannot be completely removed	[51]
Hybrid sol–gel coating and industrial application on polished porcelain stoneware tiles	-	-	-	With the help of cleaning agent, the stains can be removed from the surface	[52]

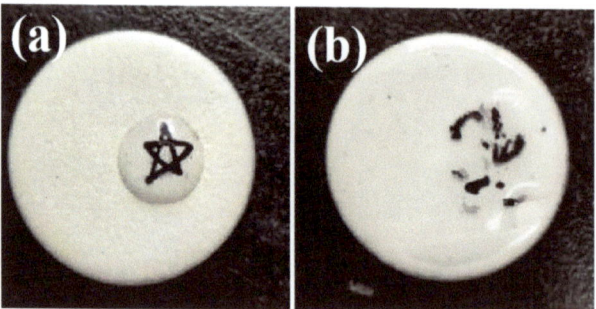

Figure 13. Photos of the as-prepared samples immediately meeting the water in this study (**a**) Mg-undoped TiO_2 in glaze sample, (**b**) Mg-doped TiO_2 in glaze sample.

4. Conclusions

In this paper, Mg-doped TiO_2 samples with various $V_{water}/V_{ethanol}$ ratios were successfully prepared through the solvothermal method at 180 °C for 36 h. The Mg-doped ($V_{water}/V_{ethanol}$ = 12.5:2.5) sample had higher surface area, porosity, optical performance, and photocatalytic activity than other samples. Undoped and Mg-doped TiO_2 glaze ceramic samples were prepared using a medium-/high-temperature solid-firing process. Mg-doped TiO_2 samples ($V_{water}/V_{ethanol}$ = 12.5:2.5) illustrated superior hydrophilicity properties, photocatalytic activity in terms of bleaching organic dye, and self-cleaning capability in ceramic glaze samples than other samples after visible light exposure. This study provides a preparation approach for the synthesis of TiO_2 while controlling crystal size and morphology, which can be utilized with solar energy for bleaching the contaminants in water and enhancing the self-cleaning properties of medium-/high-temperature glazes.

Supplementary Materials: The following supporting information can be downloaded at: https://www.mdpi.com/article/10.3390/inorganics11080341/s1, Figure S1: EDX spectroscopy mapping performed in the TEM microscope.

Author Contributions: Conceptualization, T.Z., Q.B. and W.D.; methodology, T.C. and Q.B.; software, P.L. and Y.L.; validation, X.G. and Q.B.; formal analysis, T.Z. and J.Z.; investigation, T.C. and Y.L.; resources, Q.B. and J.Z.; data curation, T.Z. and W.D.; writing—original draft preparation, T.Z. and W.D.; writing—review and editing, T.Z. and W.D.; visualization, P.L. and X.G.; supervision, X.G. and J.Z.; project administration, T.Z. and T.C.; funding acquisition, W.D. All authors have read and agreed to the published version of the manuscript.

Funding: We would like to express our gratitude for the financial support from Major Project of Natural Science Foundation of Jiangxi Province (20232ACB204017), Jingdezhen technology bureau (2021GYZD009-18 and 20224GY008-16) and Jiangxi Province Key R&D Program in China (No. 20202BBE53012), Graduate Innovation Fund Project of Jingdezhen Ceramic University (JYC202004).

Institutional Review Board Statement: Not applicable.

Informed Consent Statement: Not applicable.

Data Availability Statement: The raw data required to reproduce these findings cannot be shared at this time as the data also forms part of an ongoing study.

Conflicts of Interest: The authors declare no conflict of interest.

References

1. Mathew, S.; John, B.K.; Abraham, T.; Mathew, B. Metal-Doped Titanium Dioxide for Environmental Remediation, Hydrogen Evolution and Sensing: A Review. *Chemistryselect* **2021**, *6*, 12742–12751. [CrossRef]
2. Lettieri, S.; Pavone, M.; Nanostructures, B. Composites and Hybrid Photocatalysts. *Materials* **2022**, *15*, 1271. [CrossRef] [PubMed]
3. Jahdi, M.; Mishra, S.B.; Nxumalo, E.N.; Mhlanga, S.D.; Mishra, A.K. Synergistic effects of sodium fluoride (NaF) on the crystallinity and band gap of Fe-doped TiO_2 developed via microwave-assisted hydrothermal treatment. *Opt. Mater.* **2020**, *104*, 109844. [CrossRef]
4. Wang, Y.; Chen, Y.-X.; Barakat, T.; Wang, T.-M.; Krief, A.; Zeng, Y.-J.; Laboureur, M.; Fusaro, L.; Liao, H.-G.; Su, B.-L. Synergistic effects of carbon doping and coating of TiO_2 with exceptional photocurrent enhancement for high performance H_2 production from water splitting. *J. Energy Chem.* **2021**, *56*, 141–151. [CrossRef]
5. de Almeida, G.C.; Mohallem, N.D.; Viana, M.M. Ag/GO/TiO_2 nanocomposites: The role of the interfacial charge transfer for application in photocatalysis. *Nanotechnology* **2022**, *33*, 035710. [CrossRef]
6. Byrne, C.; Dervin, S.; Hermosilla, D.; Merayo, N.; Blanco, A.; Hinder, S.; Harb, M.; Dionysiou, D.D.; Pillai, S.C. Solar light assisted photocatalytic degradation of 1,4-dioxane using high temperature stable anatase W-TiO_2 nanocomposites. *Catal. Today* **2021**, *380*, 199–208. [CrossRef]
7. Tao, J.; Luttrell, T.; Batzill, M. A two-dimensional phase of TiO_2 with a reduced bandgap. *Nat. Chem.* **2011**, *3*, 296–300. [CrossRef]
8. Sivkov, A.; Vympina, Y.; Ivashutenko, A.; Rakhmatullin, I.; Shanenkova, Y.; Nikitin, D.; Shanenkov, I. Plasma dynamic synthesis of highly defective fine titanium dioxide with tunable phase composition. *Ceram. Int.* **2022**, *48*, 10862–10873. [CrossRef]
9. Buchalska, M.; Kobielusz, M.; Matuszek, A.; Pacia, M.; Wojtyła, S.; Macyk, W. On oxygen activation at rutile-and anatase-TiO_2. *ACS Catal.* **2015**, *5*, 7424–7431. [CrossRef]
10. Ji, Y.; Luo, Y. New mechanism for photocatalytic reduction of CO_2 on the anatase TiO_2 (101) surface: The essential role of oxygen vacancy. *J. Am. Chem. Soc.* **2016**, *138*, 15896–15902. [CrossRef]
11. Estrada-Flores, S.; Martinez-Luevanos, A.; Perez-Berumen, C.M.; Garcia-Cerda, L.A.; Flores-Guia, T.E. Relationship between morphology; porosity, and the photocatalytic activity of TiO_2 obtained by sol-gel method assisted with ionic and nonionic surfactants. *Bol. La Soc. Esp. Ceram. Y Vidr.* **2020**, *59*, 209–218. [CrossRef]
12. Byrne, C.; Rhatigan, S.; Hermosilla, D.; Merayo, N.; Blanco, A.; Michel, M.C.; Hinder, S.; Nolan, M.; Pillai, S.C. Modification of TiO_2 with hBN: High temperature anatase phase stabilisation and photocatalytic degradation of 1,4-dioxane. *J. Phys.-Mater.* **2020**, *3*, 015009. [CrossRef]
13. Yu, Z.H.; Zhu, S.L.; Zhang, L.H.; Watanabe, S. Mesoporous single crystal titanium oxide microparticles for enhanced visible light photodegradation. *Opt. Mater.* **2022**, *127*, 112297. [CrossRef]
14. Kumar, A.; Choudhary, P.; Krishnan, V. Selective and efficient aerobic oxidation of benzyl alcohols using plasmonic Au-TiO_2: Influence of phase transformation on photocatalytic activity. *Appl. Surf. Sci.* **2022**, *578*, 151953. [CrossRef]
15. Wu, J.M.; Xing, H. Facet-dependent decoration of TiO_2 mesocrystals on TiO_2 microcrystals for enhanced photoactivity. *Nanotechnology* **2020**, *31*, 025604. [CrossRef] [PubMed]
16. Song, C.G.; Won, J.; Jang, I.; Choi, H. Fabrication of high-efficiency anatase TiO_2 photocatalysts using electrospinning with ultra-violet treatment. *J. Am. Ceram. Soc.* **2021**, *104*, 4398–4407. [CrossRef]

17. George, S.; Pokhrel, S.; Ji, Z.; Henderson, B.L.; Xia, T.; Li, L.; Zink, J.I.; Nel, A.E.; Mädler, L. Role of Fe doping in tuning the band gap of TiO$_2$ for the photo-oxidation-induced cytotoxicity paradigm. *J. Am. Chem. Soc.* **2011**, *133*, 11270–11278. [CrossRef] [PubMed]
18. Galindo-Hernandez, F.; Gomez, R.; De la Torre, A.I.R.; Mantilla, A.; Lartundo-Rojas, L.; Martinez, A.M.M.; Cipagauta-Diaz, S. Structural changes and photocatalytic aspects into anatase network after doping with cerium: Comprehensive study via radial distribution functions, electron density maps and molecular hardness. *J. Photochem. Photobiol. A-Chem.* **2022**, *428*, 113855. [CrossRef]
19. Shymanovska, V.V.; Khalyavka, T.A.; Manuilov, E.V.; Gavrilko, T.A.; Aho, A.; Naumov, V.V.; Shcherban, N.D. Effect of surface doping of TiO$_2$ powders with Fe ions on the structural, optical and photocatalytic properties of anatase and rutile. *J. Phys. Chem. Solids* **2022**, *160*, 110308. [CrossRef]
20. Huang, J.X.; Li, D.G.; Li, R.B.; Chen, P.; Zhang, Q.X.; Liu, H.J.; Lv, W.Y.; Liu, G.G.; Feng, Y.P. One-step synthesis of phosphorus/oxygen co-doped g-C$_3$N$_4$/anatase TiO$_2$ Z-scheme photocatalyst for significantly enhanced visible-light photocatalysis degradation of enrofloxacin. *J. Hazard. Mater.* **2020**, *386*, 121634. [CrossRef]
21. Sudrajat, H.; Babel, S.; Hartuti, S.; Phanthuwongpakdee, J.; Laohhasurayotin, K.; Nguyen, T.K.; Tong, H.D. Origin of the overall water splitting activity over Rh/Cr$_2$O$_3$ @anatase TiO$_2$ following UV-pretreatment. *Int. J. Hydrog. Energy* **2021**, *46*, 31228–31238. [CrossRef]
22. Alshehri, A.; Narasimharao, K. PtOx-TiO$_2$ anatase nanomaterials for photocatalytic reformation of methanol to hydrogen: Effect of TiO$_2$ morphology. *J. Mater. Res. Technol.-JmrT* **2020**, *9*, 14907–14921. [CrossRef]
23. Zhang, Q.; Li, C.Y. Effects of water-to-methanol ratio on the structural, optical and photocatalytic properties of titanium dioxide thin films prepared by mist chemical vapor deposition. *Catal. Today* **2020**, *358*, 172–176. [CrossRef]
24. Alotaibi, A.M.; Williamson, B.A.D.; Sathasivam, S.; Kafizas, A.; Alqahtani, M.; Sotelo-Vazquez, C.; Buckeridge, J.; Wu, J.; Nair, S.P.; Scanlon, D.O.; et al. Enhanced Photocatalytic and Antibacterial Ability of Cu-Doped Anatase TiO$_2$ Thin Films: Theory and Experiment. *ACS Appl. Mater. Inter.* **2020**, *12*, 15348–15361. [CrossRef] [PubMed]
25. Wang, Q.Y.; Li, H.L.; Yu, X.L.; Jia, Y.; Chang, Y.; Gao, S.M. Morphology regulated Bi$_2$WO$_6$ nanoparticles on TiO$_2$ nanotubes by solvothermal Sb^{3+} doping as effective photocatalysts for wastewater treatment. *Electrochim. Acta* **2020**, *330*, 135167. [CrossRef]
26. Li, J.; Chu, B.X.; Xie, Z.; Deng, Y.Q.; Zhou, Y.M.; Dong, L.H.; Li, B.; Chen, Z.J. Mechanism and DFT study of degradation of organic pollutants on rare earth ions doped TiO$_2$ photocatalysts prepared by sol-hydrothermal synthesis. *Catal. Lett.* **2022**, *152*, 489–502. [CrossRef]
27. Nithya, N.; Gopi, S.; Bhoopathi, G. An Amalgam of Mg-Doped TiO$_2$ Nanoparticles Prepared by Sol-Gel Method for Effective Antimicrobial and Photocatalytic Activity. *J. Inorg. Organomet. Polym. Mater.* **2021**, *31*, 4594–4607. [CrossRef]
28. Barbierikova, Z.; Dvoranova, D.; Sofianou, M.V.; Trapalis, C.; Brezova, V. UV-induced reactions of Mg^{2+}-doped anatase nanocrystals with exposed {001} facets: An EPR study. *J. Catal.* **2015**, *331*, 39–48. [CrossRef]
29. Cao, T.H.; Liang, Y.Z.; Bao, Q.F.; Xu, C.L.; Bai, M.M.; Luo, T.; Gu, X.Y. A simple solvothermal preparation of Mg-doped anatase TiO$_2$ and its self-cleaning application, solar energy. *Sol. Energy* **2023**, *249*, 12–20. [CrossRef]
30. Sheldrick, G.M. Phase annealing in SHELX-90: Direct methods for larger structures. *Acta Crystallogr. Sect. A Found. Crystallogr.* **1990**, *46*, 467–473. [CrossRef]
31. Cullity, B.D. *Elements of X-ray Diffraction*, 2nd ed.; Addison-Wesley: London, UK, 1978.
32. Olowoyo, J.O.; Kumar, M.; Singhal, N.; Jain, S.L.; Babalola, J.O.; Vorontsov, A.V.; Kumar, U. Engineering and modeling the effect of Mg doping in TiO$_2$ for enhanced photocatalytic reduction of CO$_2$ to fuels. *Catal. Sci. Technol.* **2018**, *8*, 3686–3694. [CrossRef]
33. Dong, W.X.; Zhao, G.L.; Song, B.; Xu, G.; Zhou, J.; Han, G.R. Surfactant-free fabrication of CaTiO$_3$ butterfly-like dendrite via a simple one-step hydrothermal route. *Crystengcomm* **2012**, *14*, 6990–6997. [CrossRef]
34. Zhang, J.; Sun, P.; Jiang, P.; Guo, Z.; Liu, W.; Lu, Q.; Cao, W. The formation mechanism of TiO$_2$ polymorphs under hydrothermal conditions based on the structural evolution of [Ti(OH)$_h$(H$_2$O)$_{6-h}$]$^{4-h}$ monomers. *J. Mater. Chem. C* **2019**, *7*, 5764–5771. [CrossRef]
35. Kolker, A.; Pablo, J. Thermodynamic modeling of concentrated aqueous. *Ind. Eng. Chem. Res.* **1996**, *35*, 228–233. [CrossRef]
36. Wang, C.C.; Wang, K.W.; Perng, T.P. Electron field emission from Fe-doped TiO$_2$ nanotubes. *Appl. Phys. Lett.* **2010**, *96*, 143102. [CrossRef]
37. Wang, Q.; Li, H.J.; Zhang, R.X.; Liu, Z.Z.; Deng, H.Y.; Cen, W.G.; Yan, Y.G.; Chen, Y.G. Oxygen vacancies boosted fast Mg^{2+} migration in solids at room temperature. *Energy Storage Mater.* **2022**, *51*, 630–637. [CrossRef]
38. Ali, T.; Ahmed, A.; Siddique, M.N.; Alam, U.; Muneer, M.; Tripathi, P. Influence of Mg^{2+} ion on the optical and magnetic properties of TiO$_2$ nanostructures: A key role of oxygen vacancy. *Optik* **2020**, *223*, 165340. [CrossRef]
39. Dong, W.X.; Song, B.; Zhao, G.L.; Han, G.R. Controllable synthesis of CaTi$_2$O$_4$(OH)$_2$ nanoflakes by a facile template-free process and its properties. *Ceram. Int.* **2013**, *39*, 6795–6803. [CrossRef]
40. Nakamoto, K. *Infrared and Raman Spectra of Inorganic and Coordination Compounds, Part B: Applications in Coordination, Organometallic, and Bioinorganic Chemistry*; John Wiley & Sons: New York, NY, USA, 2009.
41. Shivaraju, H.P.; Yashas, S.R.; Harini, R. Application of Mg-doped TiO$_2$ coated buoyant clay hollow-spheres for photodegradation of organic pollutants in wastewater. *Mater. Today Proc.* **2020**, *27*, 1369–1374. [CrossRef]
42. Deng, R.H.; Liu, S.H.; Li, J.Y.; Liao, Y.G.; Tao, J.; Zhu, J.T. Mesoporous block copolymer nanoparticles with tailored structures by hydrogen-bonding-assisted self-assembly. *Adv. Mater.* **2012**, *24*, 1889–1893. [CrossRef]

43. Huang, W.; Cheng, H.; Feng, J.; Shi, Z.; Bai, D.; Li, L. Synthesis of highly water-dispersible N-doped anatase titania based on low temperature solvent-thermal method. *Arab. J. Chem.* **2018**, *11*, 871–879. [CrossRef]
44. Singh, M.K.; Mehata, M.S. Enhanced photoinduced catalytic activity of transition metal ions incorporated TiO_2 nanoparticles for degradation of organic dye: Absorption and photoluminescence spectroscopy. *Opt. Mater.* **2020**, *109*, 110309. [CrossRef]
45. Bally, A.R.; Korobeinikova, E.N.; Schmid, P.E.; Levy, F.; Bussy, F. Structural and electrical properties of Fe-doped thin films. *J. Phys. D Appl. Phys.* **1998**, *31*, 1149. [CrossRef]
46. Hoffmann, M.R.; Martin, S.T.; Choi, W.; Bahnemann, D.W. Environmental applications of semiconductor photocatalysis. *Chem. Rev.* **1995**, *95*, 69–96. [CrossRef]
47. Augugliaro, V.; Loddo, V.; Marcì, G.; Palmisano, L.; López-Muñoz, M.J. Photocatalytic oxidation of cyanides in aqueous titanium dioxide suspensions. *J. Catal.* **1997**, *166*, 272–283. [CrossRef]
48. Zubair, M.A.; Al Mamun, A.; McNamara, K.; Tofail, S.A.M.; Islam, F.; Lebedev, V.A. Amorphous interface oxide formed due to high amount of Sm doping (5–20 mol%) stabilizes finer size anatase and lowers indirect band gap. *Appl. Surf. Sci.* **2020**, *529*, 146967. [CrossRef]
49. Ferreira-Neto, E.P.; Ullah, S.; Martinez, V.P.; Yabarrena, J.M.C.; Simões, M.B.; Perissinotto, A.P.; Wender, H.; De Vicente, F.S.; Noeske, P.-L.M.; Ribeiro, S.J. Thermally stable $SiO_2@TiO_2$ core@shell nanoparticles for application in photocatalytic self-cleaning ceramic tiles. *Mater. Adv.* **2021**, *2*, 2085–2096. [CrossRef]
50. Kim, S.M.; In, I.; Park, S.Y. Study of photo-induced hydrophilicity and self-cleaning property of glass surfaces immobilized with TiO_2 nanoparticles using catechol chemistry. *Surf. Coat. Technol.* **2016**, *294*, 75–82. [CrossRef]
51. Li, K.; Yao, W.; Liu, Y.; Wang, Q.; Jiang, G.; Wu, Y.; Lu, L. Wetting and anti-fouling properties of groove-like microstructured surfaces for architectural ceramics. *Ceram. Int.* **2022**, *48*, 6497–6505. [CrossRef]
52. Akarsu, M.; Burunkaya, E.; Tunalı, A.; Selli, N.T.; Arpaç, E. Enhancement of hybrid sol–gel coating and industrial application on polished porcelain stoneware tiles and investigation of the performance. *Ceram. Int.* **2014**, *40*, 6533–6540. [CrossRef]

Disclaimer/Publisher's Note: The statements, opinions and data contained in all publications are solely those of the individual author(s) and contributor(s) and not of MDPI and/or the editor(s). MDPI and/or the editor(s) disclaim responsibility for any injury to people or property resulting from any ideas, methods, instructions or products referred to in the content.

Review

Research Progress of TiO₂ Modification and Photodegradation of Organic Pollutants

Tan Mao [1,2,*], Junyan Zha [1], Ying Hu [1], Qian Chen [1], Jiaming Zhang [1] and Xueke Luo [1,2]

[1] College of Mechanical and Material Engineering, North China University of Technology, Beijing 100144, China; erstazha@163.com (J.Z.); yhhhh2022@163.com (Y.H.); 15701973987@163.com (Q.C.); zhangjiaming4715@163.com (J.Z.); 18610547828@163.com (X.L.)

[2] College of Mechanical and Precision Instrument Engineering, Xi'an University of Technology, Xi'an 710000, China

* Correspondence: t_maoncut@163.com

Abstract: Titanium dioxide (TiO₂) photocatalysts, characterized by exceptional photocatalytic activity, high photoelectric conversion efficiency, and economic viability, have found widespread application in recent years for azo dye degradation. However, inherent constraints, such as the material's limited visible light absorption stemming from its bandgap and the swift recombination of charge carriers, have impeded its broader application potential. Encouragingly, these barriers can be mitigated through the modification of TiO₂. In this review, the common synthesis methods of TiO₂ are reviewed, and the research progress of TiO₂ modification technology at home and abroad is discussed in detail, including precious metal deposition, transition metal doping, rare earth metal doping, composite semiconductors, and composite polymers. These modification techniques effectively enhance the absorption capacity of TiO₂ in the visible region and reduce the recombination rate of carriers and electrons, thus significantly improving its photocatalytic performance. Finally, this paper looks forward to the future development direction of TiO₂ photocatalytic materials, including the exploration of new modified materials, in-depth mechanism research, and performance optimization in practical applications, to provide useful references for further research and application of TiO₂ photocatalytic materials.

Keywords: titanium dioxide; modification; photocatalysis; organic pollutants

Citation: Mao, T.; Zha, J.; Hu, Y.; Chen, Q.; Zhang, J.; Luo, X. Research Progress of TiO₂ Modification and Photodegradation of Organic Pollutants. *Inorganics* **2024**, *12*, 178. https://doi.org/10.3390/inorganics12070178

Academic Editors: Roberto Nisticò and Silvia Mostoni

Received: 3 May 2024
Revised: 20 June 2024
Accepted: 20 June 2024
Published: 26 June 2024

Copyright: © 2024 by the authors. Licensee MDPI, Basel, Switzerland. This article is an open access article distributed under the terms and conditions of the Creative Commons Attribution (CC BY) license (https://creativecommons.org/licenses/by/4.0/).

1. Introduction

In recent years, acid rain, eutrophication of water bodies, black and smelly water bodies, PM2.5 exceeding standards, and other environmental issues have occurred frequently, warning us that environmental protection cannot be delayed. Among them, the content of organic pollutants and pigments is high, which seriously affects the regional water quality [1,2]. Due to their highly stable structure and difficulty in handling, most of them are toxic to organisms, including direct lethality and carcinogenicity [3–5]. This highlights the urgent need for effective treatment of pollutants in water, and the current treatment methods for organic pollutants are mainly the combination of physical, biochemical, and chemical treatment methods [6].

Physical treatment methods include adsorption, extraction, radiation-based techniques, and membrane separation [7–9]. Biochemical treatment methods include aerobic treatment, anaerobic treatment [10,11], and so on. Chemical treatment methods include chemical oxidation, photocatalytic oxidation, electrochemical methods, chemical coagulation, and others [12–14]. Although the traditional physical and chemical treatment methods can remove azo dyes to a certain extent, there are often problems such as high energy consumption, long treatment cycles, and secondary pollution. Therefore, the development of efficient and environmentally friendly treatment of water pollutants has become the focus of current research.

As a new type of wastewater treatment technology being gradually developed based on photochemical oxidation [15], photocatalytic technology has garnered widespread favor among scholars due to its numerous advantages including good stability, high activity, non-toxicity, a wide range of applicability, and the absence of secondary pollution. Its ability to effectively degrade dye pollutants in water bodies is particularly noteworthy [16,17]. Photocatalysts are primarily composed of n-type semiconductor materials, including TiO_2, ZnO, CdS, and others. TiO_2 stands out as the most extensively studied and utilized n-type semiconductor material due to its exceptional chemical stability, non-toxicity, low cost, and strong photosensitivity [18,19]. Its application areas span water treatment, air purification, solar cell photosensitizers, self-cleaning materials, and medical applications [20]. In the degradation of organic pollutants in water, TiO_2 photocatalysis technology has shown unique advantages. TiO_2 photocatalyst can generate photogenerated electrons and holes under ultraviolet irradiation and then trigger a series of REDOX reactions. These reactions can break the chemical bonds of organic pollutant molecules and degrade them into small molecular compounds, ultimately achieving the purpose of removing the pollutants [21].

Although TiO_2 photocatalysis technology has many advantages in the degradation of pollutants in water, pure TiO_2, with a band gap of 3.0~3.2 eV [22], can only be excited by ultraviolet light with a wavelength less than 387 nm and photogenerated electron-hole pairs are easy to recombine, resulting in low photo quantum efficiency and limited photocatalytic activity [23]. To further improve the photocatalytic performance of TiO_2, researchers modified TiO_2 through doping [24], composite, surface modification [25], and other means to optimize the structure of TiO_2, thereby expanding its photoresponse range and reducing the carrier recombination rate [26,27]. Thus, its application effect in the photodegradation of organic pollutants was enhanced. These modification methods can significantly improve the efficiency and performance of TiO_2 photocatalysts in degrading pollutants and provide better technical support for practical applications. In addition, the study of TiO_2-modified photocatalysis technology has far-reaching significance [28]. It not only helps to solve the environmental problems of organic pollutants such as azo dyes but also promotes the application and development of photocatalysis technology in other fields [29]. Through in-depth research on the mechanism and methods of TiO_2 modification, theoretical guidance, and practical experience can be provided for the development of more efficient and environmentally friendly photocatalytic materials and contribute to environmental protection and sustainable development [30].

2. Mechanism and Kinetics of Photodegradation of Organic Pollutants by TiO_2 Photocatalyst

2.1. Mechanism of Photodegradation of Organic Pollutants by TiO_2 Photocatalyst

TiO_2 is a semiconductor, and its band structure is composed of a low-level valence band and a high-level conduction band [31]. The valence band is mainly composed of 2p orbitals of oxygen atoms, and the conduction band is mainly composed of 3d, 4s, and 4p orbitals of titanium atoms [32]. The band discontinuity of the semiconductor is bandgap between the valence band and the conduction band, and the energy difference between the conduction band and the valence band is called bandgap width [33].

TiO_2 mainly has three crystal structures, namely Anatase type, Rutile type, and Brookite type, of which rutile TiO_2 is the most stable crystal type, even at high temperatures will not be transformed and decomposed. The photocatalytic activity of anatase TiO_2 is higher than that of rutile TiO_2 because anatase TiO_2 has a lighter effective mass, smaller particle size, and longer life of photoexcited electrons and holes [34]. As a metastable phase, titanite is rarely found in nature, is difficult to synthesize, and has low practical application value, so anatase-type TiO_2 is often used to photocatalyze the degradation of organic pollutants in water. For anatase phase TiO_2, when it is irradiated by photons with energy greater than 3.2 eV (wavelength < 387.5 nm), the photoexcitation reaction will produce free electrons and holes:

$$TiO_2 + hv \rightarrow h^+ + e^- \quad (1)$$

Excited electrons react with oxygen in the air to produce superoxide radical anions, which combine with water molecules to form hydroxyl radicals:

$$e^- + O_2 \rightarrow \cdot O^{2-} \quad (2)$$

$$h^+ + H_2O \rightarrow \cdot OH + H^+ \quad (3)$$

Hydroxyl radicals and superoxide radicals with strong oxidation react with organic pollutants, destroying their molecular structure; they are easily broken by oxidation. After a series of reactions, organic pollutant molecules are decomposed into harmless small molecules or low-toxicity compounds, such as water and carbon dioxide. The process is as follows:

$$\text{Organic pollutant molecules} + \cdot OH + O^{2-} \rightarrow CO_2 + H_2O + \text{Other small molecules} \quad (4)$$

The photocatalytic mechanism of TiO_2 is depicted in Figure 1.

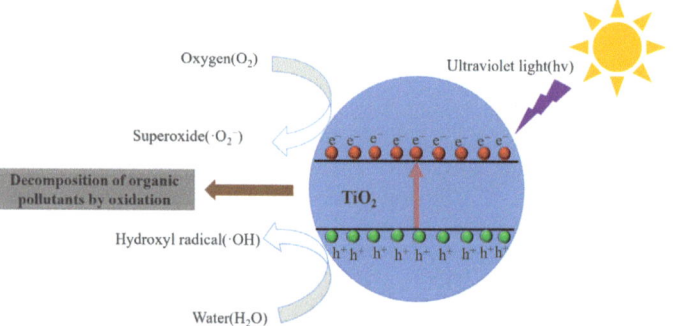

Figure 1. Photocatalytic mechanism of TiO_2.

2.2. Kinetics of Photodegradation of Organic Pollutants by TiO₂ Photocatalyst

The photocatalytic degradation of organic pollutants by TiO_2 is analyzed by the Langmuir–Hinshelwood model [35]; Arikal [36] et al. used chitosan as a carrier to immobilize TiO_2/MgO nanocomposites on chitosan beads. MO and azo red S (ARS) were used as model dye compounds. The experimental results showed that the degradation of the pollutants followed first-order kinetics, and the Langmuir–Hinshelwood model was suitable for describing the kinetics of the photocatalytic degradation of wastewater. This is simplified as the first-order kinetics formula: $\ln(C_0/C) = kt$, (C_0) the initial concentration of organic pollutants, (C) the concentration of organic pollutants after degradation, (k) the apparent reaction rate constant, and (t) the photodegradation time. The model takes into account the adsorption and reaction processes on the catalyst surface and can describe the kinetic behavior of photocatalytic degradation well. The kinetics of photocatalytic degradation can be described by the degradation rate constant. The degradation rate constant reflects the relationship between degradation rate and pollutant concentration.

The influencing factors of reaction kinetics are as follows:

- Solution pH value has an important effect on photocatalytic degradation. The photocatalytic activity of TiO_2 may vary under acidic and alkaline conditions.
- The amount of catalyst dosing will also affect the degradation rate. Less than or more than the optimal dosage will lead to a decrease in the degradation rate.
- Factors such as organic pollutant's initial concentration and light intensity also affect the kinetic process of photocatalytic degradation.

In summary, the process of photodegradation of organic pollutants using TiO$_2$ photocatalyst involves complex chemical reactions and kinetic behaviors. Through an in-depth understanding of its mechanism and kinetics, the photocatalytic degradation process can be optimized, and the degradation efficiency and environmental protection effect can be improved.

3. Synthesis Method of TiO$_2$ Photocatalyst

As a functional material, TiO$_2$ has high photocatalytic activity and, therefore, has an excellent advantage in the field of photocatalytic degradation. At present, the methods commonly used to synthesize TiO$_2$ include the sol-gel method, hydrothermal method, atomic layer deposition method, and microemulsion method. In terms of structure and morphology, a variety of morphologically controllable TiO$_2$ micro-nano meters, such as nanorods, nanotubes, nano-flowers, nano-hollow spheres, and mesoporous structures, were prepared through various experiments. As shown in Figure 2, these micro-nano TiO$_2$ materials with large specific surface areas usually have more excellent photocatalytic properties.

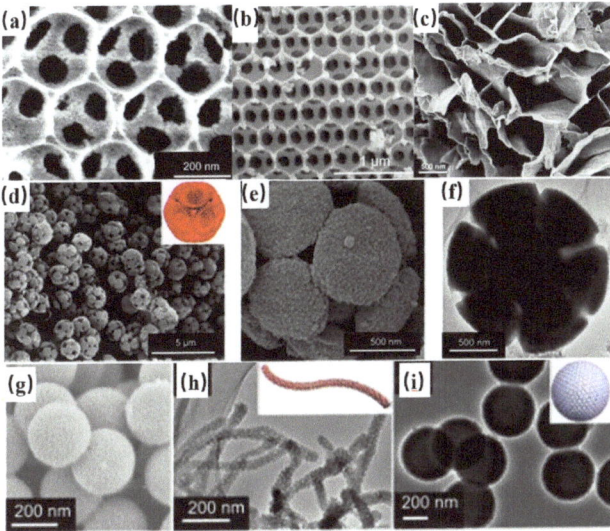

Figure 2. Different TiO$_2$ micro-nano structure diagram: (**a**) SEM image of ordered macroporous TiO$_2$; (**b**) Porous TiO$_2$ TEM image; (**c**) TEM image; TiO$_2$ flower-like superstructure self-assembled by limiting microemulsions. FESEM (**d**,**e**) and TEM (**f**) images of primary mesoporous TiO$_2$; (**g**) SiO$_2$ monolayer mesoporous TiO$_2$ core-shell structure; (**h**) Carbon nanotube-coated TiO$_2$ TEM and FESEM images; (**i**) carbon cluster TiO$_2$ TEM image [37].

3.1. Sol-Gel Method

Sol-gel is the most commonly used process method for preparing TiO$_2$ photocatalytic materials [38]. Inorganic salts and Alkoxides (including tetraethyl titanate, butyl titanate, isopropanol titanium, etc.) are placed in distilled water, mixed evenly at room temperature under liquid phase conditions, and a stable, transparent sol system is formed in solution through chemical reaction steps such as hydrolysis and condensation, as shown in Figure 3. After aging for some time, the colloidal gel slowly polymerizes to form a wet gel with a three-dimensional network structure, and the gel network is filled with solvent that loses fluidity. The wet gel is prepared by vacuum drying, high-temperature roasting, and curing to produce molecular and even nanostructured materials. Nanomaterials prepared by the sol-gel method have the advantages of high purity, uniformity, and controllable morphology and have been widely used in the preparation of TiO$_2$.

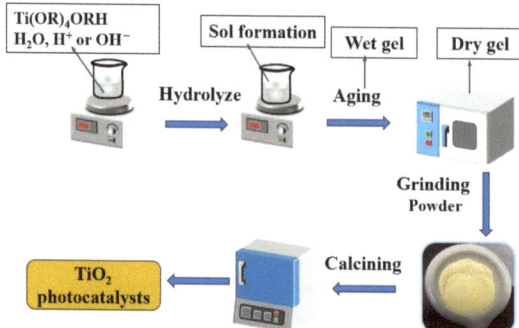

Figure 3. Schematic diagram of TiO$_2$ preparation process by sol-gel method.

Jamil [39], Huo [40], Fermeli [41], et al. prepared a composite photocatalyst by sol-gel method. The results show that the prepared TiO$_2$ composite photocatalyst has good dispersibility, a large specific surface area, and good photocatalytic activity. Dinkar Parashar [42] prepared photocatalysts by sol-gel method at different calcination temperatures (400–800 °C) and hydroalcohol ratios. The activity of TiO$_2$ prepared by the sol-gel method was higher than that of commercially available pure anatase TiO$_2$ nanoparticles due to the smaller average particle size. It was found that the ratio of water to alcohol in the preparation of TiO$_2$ catalysts had a significant effect on antibiotic removal. Namely, the removal-rate constants of metronidazole (MNZ), ciprofloxacin (CIP), and tetracycline (TET) were improved by a factor of 2.7, 3.3, and 1.6, which further indicates that the sol-gel prepared TiO$_2$ can effectively remove the harmful substance. Lalitha [43] et al. synthesized TiO$_2$/ZnO with quadrilateral and hexagonal structures by sol-gel method; the catalyst displayed 90% degradation within 40 min under UV light conditions.

3.2. Hydrothermal Synthesis

The hydrothermal synthesis method is a reaction occurring at high temperature and high pressure [44], which uses water or organic solvent as a medium through heating so that insoluble or insoluble substances are dissolved or recrystallized after washing and centrifugation to obtain the required nanoparticle. Rawat [45], Matakgane [46], et al. prepared composite photocatalysts by hydrothermal method. The results of photocatalytic performance show that it has good absorption ability to ultraviolet light and good transmittibility to visible light. Yang [47] et al. synthesized a novel 3D sea urchin-type titanium dioxide by the EDTA-Na2-assisted hydrothermal method (Figure 4). Benefiting from the conical structure prepared by the hydrothermal method and the rapid separation of photogenerated electron holes in the mixed crystal phase, the sub-micron-sized 3D sea urchin-type titanium dioxide can efficiently degrade 94.1% of the methicillin in the aqueous solution in 90 min, which is superior to that of the commercially available 25 nm-sized rutile titanium dioxide. Zhang [48] et al. modified TiO$_2$ with a carboxyl group and amino group by hydrothermal method, making the adsorption performance of functionalized TiO$_2$ better than P25 (unmodified commercial TiO$_2$). Moreover, the functionalized TiO$_2$ has good reusability for the removal of azo dye acid red G even after 5 adsorption–desorption cycles.

With the progress of scientific research, the hydrothermal method has attracted more and more attention from researchers, especially hydrothermal crystal growth, which has become the focus of research and will become the development object of the scientific research community. Hydrothermal preparation of nanomaterials equipment and conditions still need to be further explored and studied.

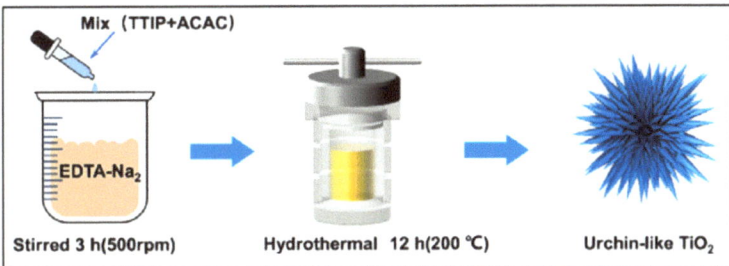

Figure 4. Preparation process of sea urchin-type titanium dioxide [47].

3.3. Atomic Layer Deposition Method

Atomiclayer deposition (ALD) technology, also known as Atomiclayer epitaxy (ALE) technology, is a chemical vapor deposition technology based on ordered surface autosaturation reactions. Generally speaking, it is a method of coating the material layer by layer on the substrate surface in the form of a single atomic film. Jialin [49], Abidi [50], et al. used the atomic deposition method to prepare TiO_2 and found that the prepared TiO_2 nanoparticles were still anatase-type at high temperatures, had particularly good dispersion, and had high photocatalytic activity after degrading dyes. Cao [51] et al. modified the ultra-thin Fe_2O_3 layer of industrial anatase TiO_2 powder by atomic layer deposition (ALD). The ultra-thin Fe_2O_3 coating with a small band gap of 2.20 eV can increase the visible light absorption of the TiO_2 carrier, and the degradation efficiency of TiO_2 powder coated with ALD Fe_2O_3 is the highest within 90 min, reaching 97.4%. Feng [52] et al. used atomic layer deposition (ALD) to deposit TiO_2 nanoparticles on carbon nanotube membranes to prepare hydrophilic electrodes, as shown in Figure 5. After 20 ALD cycles, the modified carbon nanotube membranes showed better electrosorption performance and reusability in the CDI process. The total Cr and Cr(VI) removal significantly increased to 92.1% and 93.3%, respectively. This work demonstrates that ALD is a highly controllable and simple method for the preparation of advanced CDI electrodes, broadening the application of metal oxide/carbon composites in electrochemical processes, especially in the field of photocatalytic degradation of organic pollutants.

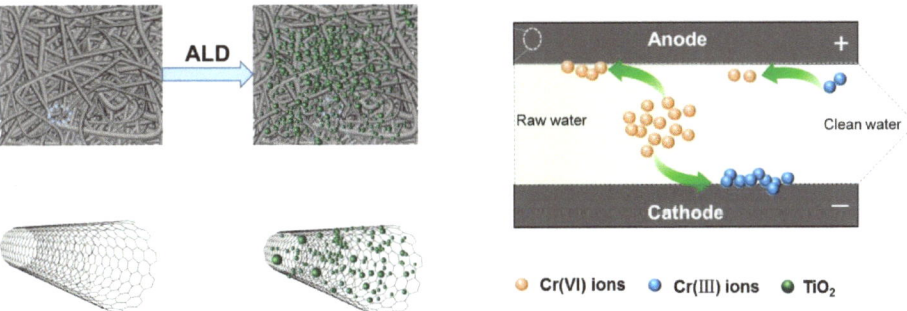

Figure 5. Atomic layer deposition (ALD) deposition of TiO_2 nanoparticles on carbon nanotube film [52].

Furthermore, Ostyn et al. [53] prepared TiO_2 film on graphene with ALD and oxidized the surface of graphite. The film has a high UV transmittance of 95%, which simplifies the experimental design. No chemical reagents are used to reduce the risk of contamination. TiO_2 films prepared by ALD react rapidly in graphite photooxidation, which is superior to traditional powder photocatalysts. Ke et al. [54] grew TiO_2 films by depositing atomic layers on silica support (SBA-15) and depositing precious metal Au on them and found that titanium dioxide films can successfully grow on mesoporous materials, increasing the specific surface area of the catalyst. Lys et al. [55] successfully combined laser-induced Si nano-fringes (SiNR), MXene, and TiO_2 to produce efficient ternary photocatalytic materials through ALD technology. Experiments show that the material has strong absorption capacity in the spectrum range, high stability, repeatability, and photocatalytic efficiency, which provides a new way for the photodegradation of azo dyes in wastewater treatment and shows broad application prospects.

3.4. Microemulsion Method

The microemulsion method is used for the preparation of nano or polymeric materials by adding surfactants and co-solvents and mixing the immiscible solution into a homogeneous phase [56]. Its advantage is that it can prepare nano-sized and highly dispersible materials with mild conditions and high-quality performance. Because of the large interfacial area and high surface activity of microemulsion, it can improve the reaction efficiency and yield, which is widely used in the preparation of nanomaterials, polymer materials, and catalysts.

Sun [57] et al. prepared $Fe_3O_4@TiO_2$ composite photocatalyst by microemulsion-solvothermal method. It was found that $Fe_3O_4@TiO_2$ could reduce the band gap, and the degradation rates of $Fe_3O_4@TiO_2$ and TiO_2 for acid red 73 were 93.56% and 74.47%, respectively. Yang [58] et al. successfully prepared TiO_2 nanofibers with porous and mixed-crystal structures by the microemulsion method, which can directly regulate the structure of titanium dioxide nanofibers by changing the microemulsion system. The titanium dioxide obtained by the microemulsion method shows a porous and mixed-crystal structure and excellent photocatalytic properties. It is simple to prepare and is of great importance for the application of the preparation and enhancement of the performance of the special-shape photocatalytic materials. The degradation rate of methylene blue solution was as high as 98% after 90 min, which was effective in the treatment of printing and dyeing wastewater.

However, the microemulsion method also has limitations that include the following: the product purity is not high, the reagent is difficult to completely remove, the precursor must be water-soluble or oil-soluble, and the preparation and control of the microemulsion is complicated, which requires experimental experience and technical skill.

3.5. Other Synthesis Methods

In addition to the above four most common methods for synthesizing TiO_2, there is also the Suzuki coupling reaction method [59], the electrostatic spinning method [60], the one-pot method [61], etc. Suzuki coupling reaction is a palladium-catalyzed cross-coupling reaction of aryl or alkenyl boronic acids or boronic esters with chlorine, bromine, iodine-substituted aromatic hydrocarbons or olefins. The one-pot method is a method in which the reactants are made to undergo successive multi-step reactions in a single reactor to improve the reaction efficiency. TiO_2 preparation by electrostatic spinning is a method of stretching titanium-containing solution into fibers using a high-voltage electric field, followed by heat treatment to obtain TiO_2 nanofibers. Djeda [62] et al. prepared layered double hydroxide (LDH)/TiO_2 nanocomposites with photocatalytic properties by immersion and direct coprecipitation methods and compared them with pure TiO_2 colloidal solutions. The degradation experiments showed that MgAl LDH/TiO_2 prepared by the coprecipitation method had the highest photodegradation efficiency for Orange II, which emphasized the importance of the preparation method of nanocomposites. Suitable synthesis methods can be selected according to different application requirements.

In summary, the most commonly used synthesis methods of TiO_2 photocatalysts, their reaction principles, and their advantages and disadvantages are shown in Table 1.

Table 1. Reaction Principle of TiO$_2$ Photocatalyst Common Preparation Methods and Their Advantages and Disadvantages.

Preparation Methods	Reaction Principle	Advantages	Disadvantages
sol-gel method	Inorganic salts and alcohol salts are hydrolyzed in distilled water, polymerized into a gel after hydrolysis, dried in a vacuum, and cured by high-temperature calcination.	Good dispersion; easy to control the reaction; simple process; low cost and cost-effective.	Long preparation time; many operating steps.
hydrothermal synthesis	In a closed system, using water as a solvent, the mixture reacts under certain temperature conditions.	Mild reaction conditions; high purity, good dispersion, crystalline form, controllable shape; environmentally friendly.	High equipment requirements; technically difficult and costly.
atomic layer deposition	A chemical vapor phase thin film deposition technique in which a substance is deposited on the surface of a substrate layer by layer in a single-atom-film format.	High accuracy; high atom utilization.	Expensive equipment; cumbersome process; difficult to promote industrialization.
microemulsion method	Mutually incompatible liquids form microreactors in the presence of surfactants for the preparation of nanomaterials.	Good dispersion of prepared samples; mild conditions; improved precursor reaction rate.	Poor purity; precursors may not be soluble; complex preparation.

Combined with the advantages and disadvantages of the four preparation methods—the sol-gel method, hydrothermal method, microemulsion method, and atomic deposition method—the improvement measures of TiO$_2$ preparation in the future can be focused on improving the preparation efficiency, reducing the cost, optimizing the performance, and meeting specific application needs. Through the above different preparation methods, two or more preparation methods can be combined according to the need to prepare the new, more stable, and higher photocatalytic efficiency of terpolymer or multi-component modified TiO$_2$ photocatalytic materials.

4. Modification Method of TiO$_2$ Photocatalyst

The direct preparation of TiO$_2$ sols can effectively solve the application challenges of TiO$_2$ particles, such as easy agglomeration and difficult loading [63]. However, TiO$_2$ nanocrystals in sol still face the problems of limited photoresponse range and high carrier complexation rate. To further enhance its photocatalytic performance, it is necessary to broaden the light absorption range of TiO$_2$ nanocrystalline sols and improve the separation efficiency of their photogenerated carriers by various modification methods.

4.1. Precious Metal-Doped Titanium Dioxide

Precious metals and TiO$_2$ have different Fermi energy levels, and the work function of the metal is higher than that of the semiconductor TiO$_2$; electrons usually tend to flow from the semiconductor to the metal when the two are in contact [64]. This cross-interface migration of electrons contributes to the efficient separation of electrons and holes in the semiconductor, as shown in Figure 6 Therefore, noble metals such as gold (Au), silver (Ag), platinum (Pt), and palladium (Pd) can significantly enhance the photocatalytic activity of TiO$_2$ materials [65]. In these applications, noble metal nanoparticles play a key role in trapping or transferring photogenerated electrons [66].

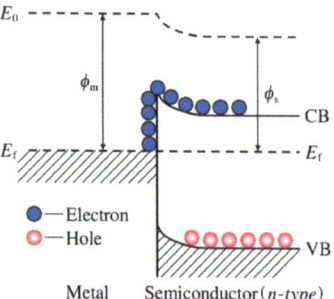

Figure 6. Schematic diagram of electron transfer in the noble metal deposition system [67].

The precious metal silver is relatively cheap and readily available [68]. Several studies, such as Chacon-Argaez [69], Borrego Pérez [70], etc., have found that the photocatalytic activity of composites is indeed significantly improved. Mohammed, W [71] et al. found that Schottky promoted electron transfer at Ag-TiO$_2$, transferred the absorption to the visible region, reduced the band gap of TiO$_2$, and inhibited electron-hole recombination, thus enhancing the photocatalytic activity and stability.

Other precious metal deposits, such as Wang et al. [72] promoted the interaction between Pt and TiO$_2$ by introducing oxygen vacancies through the carrier material, which facilitated the charge transfer from the carrier to Pt and exhibited excellent redox properties. Liu et al. [73] successfully prepared Au-TiO$_2$ nanoparticles by the hydrothermal synthesis method and found that the optical band gap value of the composites decreased due to the doping of Au or the formation of oxygen vacancies, etc., which induced a narrowing of the Au-TiO$_2$ band gap and increased visible light absorption, thus effectively degrading the MB.

The noble metal nanoparticles have a high specific surface area, good surface activity, and small particle size, which are characteristic of multiphase catalysis [74,75]. However, the catalytic activity may be limited by the limited ability of noble metals to absorb light at specific wavelengths [76,77]. In addition, the precious metal deposited TiO$_2$ is less efficient in utilizing visible light and requires higher energy of UV light for better catalytic effect. However, precious metals are expensive resources, and doping them into TiO$_2$ also increases the processing cost, which may be less economical in large-scale applications.

4.2. Transition Metal-Doped Titanium Dioxide

Interestingly, doping a moderate amount of transition metal ions in TiO$_2$, the electrons in the d- and f-orbitals can undergo a transition and enter into the TiO$_2$ structure, which reduces the bandgap and shifts the absorption edge to the visible region and reduces the rate of carrier complexation, thus improving the photocatalytic efficiency of TiO$_2$ [78]. At present, the transition metal doped ions are applied more: iron (Fe^{3+}), copper (Cu^{2+}), manganese (Mn^{2+}), vanadium (V^{4+}), zinc (Zn$^{2+)}$, etc [79].

4.2.1. Doping with a Single Transition Metal Elements

When TiO$_2$ is doped with a single transition metal ion, the addition of the transition metal elements will change the lattice structure of TiO$_2$ and thus its crystal crystallinity since they have multiple valences. This change affects the compounding process of photogenerated electrons and holes, thus improving the photocatalytic activity. Meshram et al. [80] microwave synthesized TiO$_2$-Al composite photocatalysts and found that the surface area and porosity of the composites were significantly reduced compared to their pristine components and were successfully used for the degradation of mixed azodyes such as methylene blue and rhodamine B. Zhu et al. [81] prepared iron-doped TiO$_2$ thin films, and concluded that the right amount of iron ions effectively inhibited the composite to enhance the photogeneration of the rate of electron holes and that the enhanced photoactivity of

Fe-doped TiO$_2$ samples was due to the generation of new elements to generate electronic states within the TiO$_2$ band gap. Mingmongkol et al. [82] prepared TiO$_2$-doped composite photocatalysts with transition metal Cu and found that Cu doping did not cause any difference in the particle size or specific surface area, and on the other hand, the surface of the TiO$_2$ material doped with high concentration of Cu charge transfer had a negative effect. Cu-doped TiO$_2$ showed much greater photocatalytic degradation of methylene blue compared to undoped TiO$_2$.

4.2.2. Co-Doping of Transition Metal Elements

In contrast, co-doping transition metals may produce more complex effects. Co-doping implies the simultaneous introduction of two or more different transition metal ions into TiO$_2$. Co-doping may further enhance the efficiency of photogenerated electron-hole separation through synergistic effects or broaden the light absorption range of TiO$_2$ by introducing new energy levels. Rao et al. [83] prepared Cu and Zn co-doped TiO$_2$ nano-photocatalysts and found that the addition of Cu and Zn to TiO$_2$ hindered the growth of nanoparticles, and there existed a more efficient electron-hole generation. Sukhadeve et al. [84] prepared Zn and Fe co-doped TiO$_2$ nanoparticles using a simple sol-gel process, and the absorption spectra of the prepared nanoparticles showed strong absorption in visible light. The synergistic effect produced by Zn and Fe blocked the photoinduced charge carriers and delayed the complexation probability, which greatly improved the RhB, MG, and MB mixture pollution degradation efficiency.

Transition metal elements have multiple valence, which can promote the chemical reaction in the electronic conduction of TiO$_2$, but it is worth noting that the doping amount should be controlled. Otherwise, it is not conducive to the improvement of the catalytic performance, and dopants are prone to agglomeration [85], such as the surface of the enriched or even the formation of new phases, so that the effective surface area of the semiconductor material is reduced, resulting in a decrease in the activity [86].

4.3. Rare Earth Metal-Doped Titanium Dioxide

Due to their unique electronic orbital structure [87], the doping of rare earth metal ions can adjust the energy level structure of TiO$_2$, enabling it to absorb more light within the visible light range and thus enhancing photocatalytic activity [88]. Additionally, rare earth metal doping can introduce additional energy levels, promoting the separation of photogenerated electrons and holes [89]. It can also improve the photostability of titanium dioxide, enhancing its long-term stability and extending the lifespan of photocatalytic materials [90]. This opens up new possibilities for its application in environmental purification, water treatment, energy conversion, and other fields [91]. Common rare earth metals used for doping include ytterbium, erbium, holmium, lan, cerium, yttrium, and europium [92].

4.3.1. Doping with a Single Rare Earth Metal Element

Single rare earth element doped TiO$_2$ is relatively easy to synthesize and can be more easily achieved in the TiO$_2$ lattice for control and regulation [93]. Ćurković et al. [94] prepared ce doped TiO$_2$ nanocomposites by sol-gel method and found that the absorption of the composites in the visible light band was higher than that of pure TiO$_2$ nanomaterials, and three cycles could be reused. Ikram Benammar et al. [95] used a hydrothermal-assisted sol-gel-gel method to prepare rare earth ytterbium and erbium-doped TiO$_2$, respectively, and the optical properties were improved after heat treatment of the powders and down-conversion of erbium doped nanoparticles was observed.

4.3.2. Co-Doping with Rare Earth Metal Elements

Single transition metal element doping may not be able to fully optimize the energy band structure of TiO$_2$ [96], while rare earth element co-doping can extend the light absorption range of TiO$_2$, enabling it to absorb wider wavelength bands of light [97], enhancing the light absorption capacity and photoelectric conversion efficiency of TiO_2, and improving the stability of the catalyst surface [98].

Pascariu [99] et al. used electrospinning calcination of normal TiO_2 doped with Sm^{3+} and Er^{3+} to demonstrate the stability and reusability of the catalyst in five repeated cycles of photodegradation of MB. Guetni et al. [100] designed a new co-doped TiO_2 with Nd-Sm and La/Y apatite nanoparticles, which showed the highest degradation rate of 96.49% for azo dye Orange Yellow G within 105 min. Ren et al. [101] used the classical sol-gel method to prepare TiO_2 nanoparticles co-doped with rare earth elements Ce and Er, as shown in Figure 7. The composite photocatalyst is a spherical particle with an uneven radius. It can be seen in Figure 7c that the polyhedral structure is mainly composed of hexagons and rhomboids, with an average size of 30 nm. The results show that Ce doping can reduce the band gap width of the composite and give it the ability to respond to visible light. At the same time, the upconversion luminescence characteristics of Er can convert near-infrared light in the solar spectrum into short-wave light that is more easily absorbed by TiO_2, thus enhancing the utilization rate of sunlight by the material. In addition, the doping of these two rare earth elements can effectively promote the separation and migration of photogenerated electrons and holes and optimize carrier utilization efficiency. Therefore, doped TiO_2 nanoparticles show better photocatalytic performance.

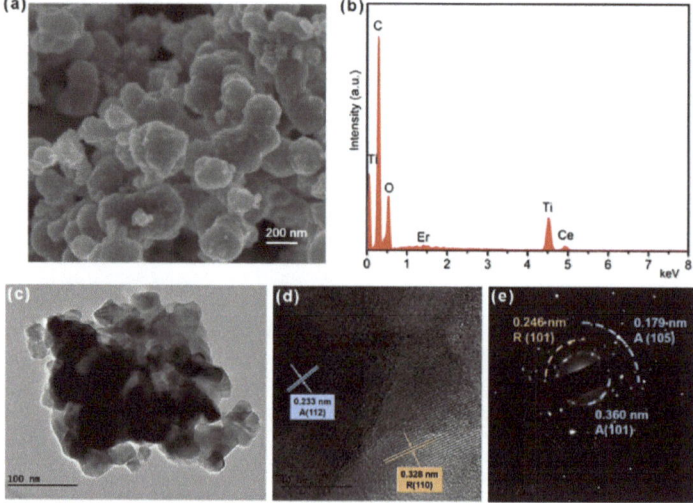

Figure 7. Characterizations of the $Er_{0.5}Ce_{0.2}Ti–O$: (**a**) SEM image; (**b**) EDS spectrum; (**c**) TEM image; (**d**) High-resolution TEM image; (**e**) selected area electron diffraction (SAED) pattern [101].

4.3.3. Co-Doping of Rare Earth Elements with Other Elements

Co-doping of rare earth elements with other elements has also attracted extensive research [102,103]. Li Jia et al. [104] synthesized lanthanum, graphene oxide (GO), and TiO_2 by sol-gel method to prepare an efficient photocatalytic material with visible light response. Lattice distortion occurred in the samples added with the rare earth element lanthanum, which increased the surface area of the photocatalyst and significantly improved the photocatalytic activity of TiO_2. The degradation efficiency of acid red B reached 95.6% after 5 h of simulated sunlight irradiation.

Co-doping of rare earth elements with other elements increases the separation efficiency of photogenerated electron-hole pairs and the number of surface active sites and improves the utilization efficiency of visible and infrared light. However, the co-doping of rare earth elements with other elements involves the synthesis and control of multi-element systems, so the preparation process and conditions are more complicated [105,106].

4.4. Compound Semiconductors Based on Titanium Dioxide

Semiconductor compounding is a method used to enhance photocatalytic efficiency by combining different types of semiconductor materials to form heterojunctions, as exemplified in Figure 8, which depicts a p-n heterojunction formed by TiO_2 and ZnO. The semiconductor composite effectively improves photocatalytic efficiency due to the following advantages:

1. By manipulating the size of the modified particles, the spectral absorption range and bandgap of the semiconductor material can be effectively tuned utilizing the quantum size effect [107,108].
2. Surface modification of TiO_2 plays a role in improving the photostability of the semiconductor materials [109].
3. Given that light absorption in semiconductors mainly occurs at the band-edge, the semiconductor composite facilitates more efficient harvesting of sunlight [110].

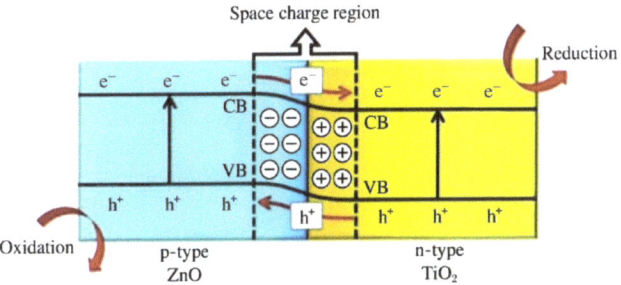

Figure 8. Schematic diagram of p-n junction formation between TiO_2 and ZnO [110].

4.4.1. Titanium Dioxide Composites with Common Semiconductors

Wongburapachart [111] et al. prepared $TiO_2/NiO\text{-}TiO_2$ bilayer film photocatalyst (BLF) for photocatalytic degradation measurement using acid orange 7 (AO7) solution under light and dark conditions. It was found that compared with ordinary TiO_2, the photocatalytic activity of the prepared sample was increased by about 8 times after 48 h of AO7 degradation. Bai [112] et al. synthesized TiO_2/ZnO composites by the sol-gel method and the hydrothermal method and found that they have obvious heterostructures, which can reduce the band gap width and improve the light absorption intensity. The photocatalytic mechanism is shown in Figure 9. Ratanathavorn Wittawat et al. [113] prepared TiO_2/ZnO composite spherical particles. TiO_2/ZnO composite also showed much higher catalytic activity compared to a single component. In addition, Wang et al. [114] also prepared $ZnO\text{-}TiO_2$ materials with different composite ratios and found that the specific surface area, pore volume, and pore diameter of $ZnO\text{-}TiO_2$ composites were significantly larger than those of TiO_2 and the $ZnO\text{-}TiO_2$ composites were more surface acidic. The energy band structure facilitates the efficient separation of electrons and holes, and the catalytic reduction activity and selectivity are stronger. It is further shown that TiO_2 composite with ZnO can inhibit TiO_2 crystal transition and particle growth, and the UV absorption ability is enhanced [115]. Abumousa [116] et al. prepared ternary $TiO_2/Y_2O_3@g\text{-}C_3N_4$ nanocomposites by a simple sonochemical method, which showed excellent photocatalytic properties in the degradation of Congo red dye, Malachite green, and other dyes in aqueous solutions in a short time. These binary semiconductor composites and ternary composites are based on TiO_2 to form heterostructures, and the interface effect and energy band migration in the heterostructures can promote the transfer of photogenerated electrons and holes, thereby improving the rate and efficiency of the photocatalytic reaction.

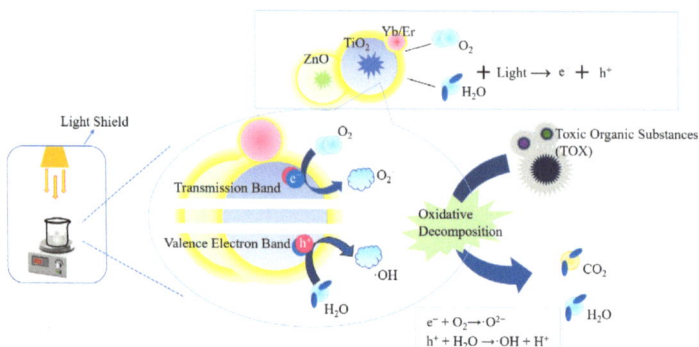

Figure 9. Proposed schematic diagram of the TiO_2/ZnO photocatalyst.

TiO_2 is composited with other semiconductors to form a composite semiconductor material. This new composite material broadens the excitation wavelengths available to the catalyst, effectively regulates the performance of individual materials, and generates numerous novel photochemical and photophysical properties. In recent years, numerous studies have been conducted on binary semiconductor composites, such as TiO_2/ZnO, TiO_2/CdS, TiO_2/WO_3, and ZnO/ZnS, among others. The photocatalytic performance of these compound semiconductors surpasses that of a single semiconductor.

4.4.2. Titanium Dioxide Hybrid Materials with Graphene

In recent years, the modification of graphene composite TiO_2 has become the focus of many researchers. Graphene with a two-dimensional structure has a large number of two-dimensional conjugated structures, so people combine graphene with TiO_2 to enhance the photocatalytic performance of TiO_2, which has been extensively studied in this area. Heltina [117], Wang [118], and others prepared TiO_2 with different graphene (GO) composite ratios compared to the pure TiO_2 and the composite TiO_2/GO catalysts. The smaller grain size and higher adsorbed oxygen/lattice oxygen ratio exhibited superior photocatalytic performance. In addition, GO acts as a capture center for photogenerated electrons and transfers electrons to the target reactants, thus inhibiting the recombination rate of photogenerated electron-hole pairs and increasing the photocatalytic rate of TiO_2.

However, graphene tends to aggregate in solution to form clusters, which can lead to difficulties in controlling the homogeneity and dispersion of graphene in the composites [119], and the preparation of composites of titanium dioxide and graphene usually requires special synthesis techniques and equipment, which may lead to high preparation costs. This makes these composites difficult to commercialize on a large scale for some applications.

4.5. TiO_2 Composite Polymers

Combining modified TiO_2 with polymers can endow nanocomposite materials with new properties. Additionally, polymers can enhance the adsorption and photocatalytic performance of nano-TiO_2, facilitating its separation and recovery [120].

Maeda et al. [121] first employed a phosphorus coupling agent to modify the surface of the modified TiO_2 particles, aiming to enhance their compatibility with specific organic monomers (Figure 10). TiO_2@PMMA hybrids were successfully obtained through in situ polymerization. It can be observed that as the in situ polymerization time elongates, the PMMA polymer chains growing on the surface of TiO_2 particles lead to an increase in the distance between the nanoparticles, thereby enhancing the transparency of the PMMA composites. This method maintains the flexibility, film forming, and electrical conductivity of the polymer. In addition, the composite may also exhibit better photocatalytic activity and antimicrobial properties.

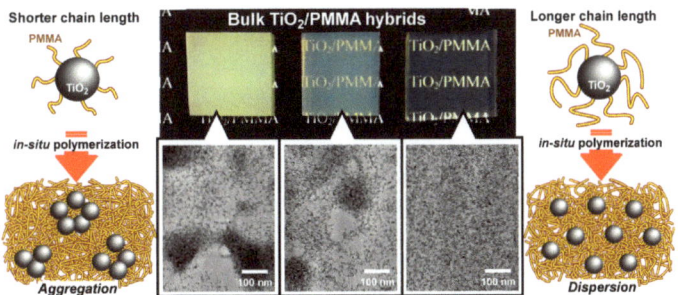

Figure 10. Appearance and TEM image of TiO$_2$/PMMA hybrid plate at different polymerization times [121].

Interestingly, Shi et al. [122] first coated stearic acid on the surface of TiO$_2$ particles by impregnation method and then further modified it with paraffin wax to prepare TiO$_2$ with Janus structure, which enhances the surface charge separation efficiency and adsorption capacity of organic matter (Figure 11). In addition, the modified Janus-like TiO$_2$ can be used as an additive to stabilize Pickering emulsion, which in turn enhances its efficiency in degrading oil-phase pollutants in high-concentration kerosene and nitrobenzene wastewater.

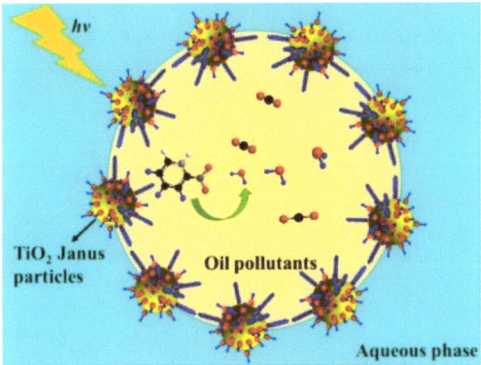

Figure 11. Photocatalytic mechanism diagram of modified TiO$_2$/polymer composites [122].

Nair [123] et al. fixed PANI-TiO$_2$ nanocomposites in polystyrene cubes to form PANi-TiO$_2$@polystyrene cubes for photocatalytic degradation of acid yellow 17 (AY 17) dyes under visible light. And the photocatalytic activity was significantly improved. It provides a way to prepare more excellent new photocatalysts. Tran et al. [124] formulated different concentrations of acrylic acid (AA) mixed with isopropanol to modify the nano-TiO$_2$ and then uniformly coated it on the surface of polyvinylidene fluoride (PVDF) film. At high temperatures, AA reacts with PVDF, and nano-TiO$_2$ is fixed on the surface of the film to maintain its flexibility and film formation, which can improve the photocatalytic activity of the composite. Neves et al. [125] used polydimethylsiloxane-modified TiO$_2$ nanoparticles to increase the surface roughness, enhance their hydrophobic strength and photocatalytic activity, and provide ideas for achieving complete degradation of polymers.

5. Summary and Outlook

In recent years, significant progress has been made in the field of TiO$_2$ modified photodegradation of organic pollutants. Through a series of modification methods, the photocatalytic performance of TiO$_2$ has been significantly improved, providing a more efficient and reliable solution for the photodegradation process of organic pollutants. The following is a summary of the research progress in this field in recent years, including prospects of future research development.

5.1. Summary of Research Progress

- By doping precious metals, transition metals, rare earth metals, or non-metallic elements, TiO_2 can change the band structure, broaden its light absorption range, and improve the separation efficiency of photoelectron-hole pairs, which can significantly improve the photocatalytic activity of TiO_2.
- TiO_2 was mixed with other semiconductor materials to form a composite photocatalytic material with a heterogeneous structure. This modification method can make use of the synergistic effect between different materials to improve the utilization efficiency of photogenerated electron-hole pairs and enhance photocatalytic performance. For example, the composite of TiO_2 with SiO_2, ZnO, and other materials can form heterogeneous structures and improve photocatalytic efficiency.
- By introducing functional material polymer on the surface of TiO_2, the surface properties of TiO_2 are improved, and the photocatalytic performance is improved. Surface modification can increase the active sites on the surface of TiO_2 and promote the separation and migration of photogenerated electrons and holes.

The five modification methods mentioned in this paper, the dopants commonly used, the specific surface area, the band gap energy, and degradation effects of the modified TiO_2 photocatalyst are shown in Table 2.

Table 2. Characterization and degradation effect of modified TiO_2 photocatalyst.

Modification Methods of TiO_2	Doping Agents	The Surface Area (Before~After/m^2 g^{-1})	Bandgap Energy (Before~After/eV)	Catalytic Effect	Reference
Precious Metal-Doped TiO_2	Ag	——	3.15~2.31	The degradation rate of methylene blue under visible light was 93%.	[126]
	Au	17.8~28.7	3.15~2.9	Methyl orange is completely dissolved within 90 min.	[127]
	Pt	42~68	3.24~2.92	The degradation rate of Dichlo-Rophenoxyacid (2,4-D) was 99%.	[128]
Transition Metal-Doped TiO_2	Fe	——	3.22~3.20	The removal rate of pollutants reached 97% within 240 min.	[129]
	Mn	50~93.35	3.20~2.21	The degradation rate of pollutants increased from 48.17% to 60.12%.	[130]
	Cu	43~46	3.08~2.78	The reduction rate of organic carbon within 6 h is 75%.	[131]
Rare Earth Metal-Doped TiO_2	La	——	3.16~3.12	The degradation rate of p-azo dye orange-yellow G was 96.49% in 105 min under UV-VIS spectral radiation.	[100]
	Er	——	3.15~2.69	The degradation rate of methylene blue was 80% under visible light.	[132]
	Eu	——	3.43~3.40	The degradation rate of Congo red reached 97%.	[133]
TiO_2 Composites with Common Semiconductors	ZnO	50.05~107.98	3.26~2.76	Under sunlight irradiation, when pH is 5.8, the degradation efficiency of the dye is the highest, which is 92%.	[134]
	SiO_2	217~256	3.22~3.22	At 300 W Xenon lamp irradiation for 60 min, the degradation efficiency of TC is 96%.	[135]
	$BiVO_4$	60.6~95.3	3.2~3.03	The degradation rate of formaldehyde reached 97.1%.	[136]
	WO_3	95~117	3.0~2.6	Under no light conditions, the degradation rate of pollutants reached 22%.	[137]

Table 2. Cont.

Modification Methods of TiO_2	Doping Agents	The Surface Area (Before~After/m^2 g^{-1})	Bandgap Energy (Before~After/eV)	Catalytic Effect	Reference
TiO_2 composite polymers	Triformyl chlorine-melamine polymer (TMP)	13~17	3.78~2.82	It can degrade 96.1% RhB.	[138]
	Polyaniline titanium Dioxide quantum Dots (PAN-TiQD)	— —	2.95~2.82	The degradation rate of Dianix blue dye reached 91%.	[139]
	Polydopamine (PDA)	— —	3.22~3.15	The photocatalytic CO_2 reduction yield of CH_4 by the composite was up to 1.50 μmol/g·h, which was 5 times that of pure TiO_2.	[140]

In summary, the modified TiO_2 photocatalyst significantly improves its photocatalytic performance in UV and visible light by increasing the specific surface area and reducing the band gap energy. The larger specific surface area increases the catalyst's active site and improves the contact efficiency with pollutant molecules, while the reduced band gap energy enables the catalyst to absorb a wider spectrum, including visible light, thereby broadening the photoresponse range and enhancing the utilization of sunlight. In addition, the specific modification method can further promote the separation and migration of photogenerated electrons and holes and further improve photocatalytic efficiency. These improvements make the modified TiO_2 photocatalyst show a broader application potential in the field of environmental governance and new energy development.

The photocatalytic performance of TiO_2 was improved to varying degrees. However, due to problems such as low efficiency, insufficient stability, and secondary pollution, TiO_2 modification methods cannot be widely used. The advantages and disadvantages of TiO_2 modification methods are shown in Table 3.

Table 3. Advantages and disadvantages of TiO_2 modification methods.

Modification Methods	Advantages	Disadvantages
TiO_2-doped noble metals	High specific surface area; good surface activity; good stability; characteristics of multiphase catalysis.	Inefficient use of visible light; Expensive precious metals.
TiO_2-doped transition metal	The presence of polyvalent transition metals promotes chemical reactions, modulates the electronic structure of TiO_2 to improve its photocatalytic properties, and extends the light absorption range.	Prone to focusing; not environmentally friendly.
TiO_2-doped rare earth metals	Good stability; high catalytic activity; expanding the range of TiO_2 light absorption and promoting photocatalytic reactions.	Complicated operating procedures; not easy to recycle.
TiO_2 compound semiconductors	Formation of heterojunctions to expand the light absorption range of the material and improve photocatalytic efficiency. Reduction in electron-hole complex reaction.	The complexity of design and preparation.
TiO_2 composite polymers	More environmentally friendly; Mechanical properties will be improved; Can be repeated many times.	The dispersion is not good; May degrade the polymer matrix.

5.2. Future Research Direction and Development Trend

Although TiO_2 modified photodegradation of organic pollutants has made remarkable progress, there are still some key problems and shortcomings. First, the stability of the modification effect is still a challenge, which can lead to performance fluctuations in practical applications. Secondly, the high preparation cost limits the possibility of large-scale applications. Therefore, future research needs to further explore and optimize modification methods to improve their stability and reduce costs.

In addition, the importance of mechanism research cannot be ignored. Strengthening the mechanism study can not only provide solid theoretical support for the optimization of modification methods but also help us to understand more deeply the reaction mechanism of TiO_2 photocatalytic degradation of organic pollutants. This will lay the foundation for the design of more efficient and stable TiO_2 photocatalytic materials.

In practical applications, the effect of TiO_2 photocatalytic degradation of organic pollutants may be affected by a variety of factors, including water quality, light conditions, dye type, and concentration. Therefore, future research also needs to focus on how to optimize the performance of TiO_2 photocatalytic degradation of organic pollutants in various real-world environments.

In conclusion, optimizing the preparation method and process of TiO_2 composite photocatalytic materials has become an important direction of future research on the photodegradation of organic pollutants. By introducing visible light absorbing inorganic substances or modifying TiO_2, we can prepare multi-functional composite materials that can expand the range of light absorption and, at the same time, play the role of photocatalysis, adsorption, and catalysis, thus significantly improving the degradation efficiency of organic matter. With continuous research and exploration, we look forward to developing more efficient, stable, and environmentally friendly TiO_2 photocatalytic materials, making important contributions to environmental protection and sustainable development.

Author Contributions: Writing—original draft preparation, T.M. and J.Z. (Junyan Zha); writing—review and editing, Y.H., Q.C., J.Z. (Jiaming Zhang) and X.L. All authors have read and agreed to the published version of the manuscript.

Funding: This research was funded by the National Key Research and Development Program of China (No. 2021YFD1600402), the Key Research and Development Plan of Shaanxi Province (No. 2020GXLH-Z-031), the Beijing University Students' Innovation and Entrepreneurship Training Program (No. 10805136024XN139-78), the Project of Postgraduate Education and Teaching Reform Research at North China University of Technology (No. YJS2023JG17), and the Undergraduate Innovation and Entrepreneurship Training Project of North China University of Technology (No. 10805136024XN139-50).

Data Availability Statement: No new data were created or analyzed in this study. Data sharing is not applicable to this article.

Conflicts of Interest: The authors declare no conflicts of interest.

References

1. Song, Y.; Wang, L.; Qiang, X.; Gu, W.; Ma, Z.; Wang, G. An Overview of Biological Mechanisms and Strategies for Treating Wastewater from Printing and Dyeing Processes. *J. Water Process Eng.* **2023**, *55*, 104242. [CrossRef]
2. Ewuzie, U.; Saliu, O.D.; Dulta, K.; Ogunniyi, S.; Bajeh, A.O.; Iwuozor, K.O.; Ighalo, J.O. A Review on Treatment Technologies for Printing and Dyeing Wastewater (PDW). *J. Water Process Eng.* **2022**, *50*, 103273. [CrossRef]
3. Dong, H.; Guo, T.; Zhang, W.; Ying, H.; Wang, P.; Wang, Y.; Chen, Y. Biochemical Characterization of A Novel Azoreductase from *Streptomyces* sp.: Application in Eco-Friendly Decolorization of Azo Dye Wastewater. *Int. J. Biol. Macromol.* **2019**, *140*, 1037–1046. [CrossRef] [PubMed]
4. Hu, Z.; Guan, D.; Sun, Z.; Zhang, Z.; Shan, Y.; Wu, Y.; Gong, C.; Ren, X. Osmotic Cleaning of Typical Inorganic and Organic Foulants on Reverse Osmosis Membrane for Textile Printing and Dyeing Wastewater Treatment. *Chemosphere* **2023**, *336*, 139162. [CrossRef] [PubMed]
5. Jing, X.; Yuan, J.; Cai, D.; Li, B.; Hu, D.; Li, J. Concentrating And Recycling of High-Concentration Printing and Dyeing Wastewater by A Disc Tube Reverse Osmosis-Fenton Oxidation/Low-Temperature Crystallization Process. *Sep. Purif. Technol.* **2021**, *266*, 118583. [CrossRef]
6. Deng, D.; Lamssali, M.; Aryal, N.; Ofori-Boadu, A.; Jha, M.K.; Samuel, R.E. Textiles Wastewater Treatment Technology: A Review. *Water Environ. Res. Res. Publ. Water Environ. Fed.* **2020**, *92*, 1805–1810. [CrossRef] [PubMed]
7. Katheresan, V.; Kansedo, J.; Lau, S.Y. Efficiency of Various Recent Wastewater Dye Removal Methods: A Review. *J. Environ. Chem. Eng.* **2018**, *6*, 4676–4697. [CrossRef]

8. Li, W.; Mu, B.; Yang, Y. Feasibility Of Industrial-Scale Treatment of Dye Wastewater Via Bio-Adsorption Technology. *Bioresour. Technol.* **2019**, *277*, 157–170. [CrossRef] [PubMed]
9. Zhao, L.; Huang, L.; Zheng, Z.; Wei, J.; Qiu, Z.; Zeng, D. Enhanced Degradation Performance of Fe(75)B(12.5)Si(12.5) Amorphous Alloys On Azo Dye. *Environ. Sci. Pollut. Res. Int.* **2023**, *30*, 34428–34439. [CrossRef] [PubMed]
10. González-Martínez, S.; Piña-Mondragón, S.; González-Barceló, Ó. Treatment of The Azo Dye Direct Blue 2 in A Biological Aerated Filter under Anaerobic/Aerobic Conditions. *Water Sci. Technol. J. Int. Assoc. Water Pollut. Res.* **2010**, *61*, 789–796. [CrossRef] [PubMed]
11. Araújo, S.; Damianovic, M.; Foresti, E.; Florencio, L.; Kato, M.T.; Gavazza, S. Biological Treatment of Real Textile Wastewater Containing Sulphate, Salinity, and Surfactant Through An Anaerobic-Aerobic System. *Water Sci. Technol. J. Int. Assoc. Water Pollut. Res.* **2022**, *85*, 2882–2898. [CrossRef] [PubMed]
12. Liu, X.; Chen, Z.; Du, W.; Liu, P.; Zhang, L.; Shi, F. Treatment of Wastewater Containing Methyl Orange Dye By Fluidized Three Dimensional Electrochemical Oxidation Process Integrated with Chemical Oxidation And Adsorption. *J. Environ. Manag.* **2022**, *311*, 114775. [CrossRef] [PubMed]
13. Ramalingam, G.; Perumal, N.; Priya, A.K.; Rajendran, S. A Review of Graphene-Based Semiconductors for Photocatalytic Degradation of Pollutants in Wastewater. *Chemosphere* **2022**, *300*, 134391. [CrossRef] [PubMed]
14. Silva, T.E.M.; Moreira, A.J.; Nobrega, E.T.D.; Alencar, R.G.; Rabello, P.T.; Blaskievicz, S.F.; Marques, G.N.; Mascaro, L.H.; Paris, E.C.; Lemos, S.G.; et al. Hierarchical Structure of 3D Zno Experimentally Designed to Achieve High Performance in The Sertraline Photocatalysis in Natural Waters. *Chem. Eng. J.* **2023**, *475*, 146235. [CrossRef]
15. Syahidatul Insyirah Mohd Foad, N.; Abu Bakar, F. Synthesis of TiO_2 Photocatalyst with Tunable Optical Properties and Exposed Facet for Textile Wastewater Treatment. *Results Opt.* **2023**, *13*, 100545.
16. Thuan, D.V.; Chu, T.T.H.; Thanh, H.D.T.; Le, M.V.; Ngo, H.L.; Le, C.L.; Thi, H.P. Adsorption and Photodegradation of Micropollutant in Wastewater by Photocatalyst TiO_2/Rice Husk Biochar. *Environ. Res.* **2023**, *236*, 116789. [CrossRef] [PubMed]
17. Santhosh, A.M.; Yogendra, K.; Madhusudhana, N.; Mahadevan, K.M.; Veena, S.R. Efficient Photodegradation of Victoria Blue B and Acridine Orange Dyes by Nickel Oxide Nanoparticles. *Mater. Today Proc.* **2023**, *92*, 1616–1622. [CrossRef]
18. Chen, D.; Cheng, Y.; Zhou, N.; Chen, P.; Wang, Y.; Li, K.; Huo, S.; Cheng, P.; Peng, P.; Zhang, R.; et al. Photocatalytic Degradation of Organic Pollutants Using TiO_2-Based Photocatalysts. *J. Clean. Prod.* **2020**, *268*, 121725. [CrossRef]
19. Saeed, M.; Muneer, M.; Haq, A.U.; Akram, N. Photocatalysis: An Effective Tool for Photodegradation of Dyes—A Review. *Environ. Sci. Pollut. Res.* **2022**, *29*, 293–311. [CrossRef] [PubMed]
20. Ul Haq, I.; Ahmad, W.; Ahmad, I.; Yaseen, M. Photocatalytic Oxidative Degradation of Hydrocarbon Pollutants In Refinery Wastewater Using TiO_2 As Catalyst. *Water Environ. Res. Res. Publ. Water Environ. Fed.* **2020**, *92*, 2086–2094. [CrossRef] [PubMed]
21. Armaković, S.J.; Savanović, M.M.; Šiljegović, M.V.; Kisić, M.; Šćepanović, M.; Grujić-Brojčin, M.; Simić, N.; Gavanski, L.; Armaković, S. Self-Cleaning and Charge Transport Properties of Foils Coated with Acrylic Paint Containing TiO_2 Nanoparticles. *Inorganics* **2024**, *12*, 35. [CrossRef]
22. Ravelli, D.; Dondi, D.; Fagnoni, M.; Albini, A. Titanium Dioxide Photocatalysis: An Assessment of The Environmental Compatibility for The Case of The Functionalization of Heterocyclics. *Appl. Catal. B Environ.* **2010**, *99*, 442–447. [CrossRef]
23. Li, M.; Liu, H.; Song, Y.; Li, Z. TiO_2 Homojunction with Au Nanoparticles Decorating as An Efficient and Stable Electrocatalyst for Hydrogen Evolution Reaction. *Mater. Charact.* **2019**, *151*, 286–291. [CrossRef]
24. El Mragui, A.; Zegaoui, O.; Daou, I. Synthesis, Characterization and Photocatalytic Properties under Visible Light of Doped and Co-Doped TiO_2-Based Nanoparticles. *Mater. Today Proc.* **2019**, *13*, 857–865. [CrossRef]
25. El-Kholy, R.A.; Isawi, H.; Zaghlool, E.; Soliman, E.A.; Khalil, M.M.H.; Said, M.M.; El-Aassar, A.M. Preparation and Characterization of Rare Earth Element Nanoparticles for Enhanced Photocatalytic Degradation. *Environ. Sci. Pollut. Res. Int.* **2023**, *30*, 69514–69532. [CrossRef] [PubMed]
26. Kumar, S.G.; Rao, K.S.R.K. Comparison Of Modification Strategies towards Enhanced Charge Carrier Separation and Photocatalytic Degradation Activity of Metal Oxide Semiconductors (TiO_2, WO_3 And Zno). *Appl. Surf. Sci.* **2017**, *391*, 124–148. [CrossRef]
27. Srinithi, S.; Balakumar, V.; Chen, T.-W.; Chen, S.-M.; Akilarasan, M.; Lou, B.-S.; Yu, J. In-Situ Fabrication of TiO_2-MWCNT Composite for An Efficient Electron Transfer Photocatalytic Rhodamine B Dye Degradation under UV–Visible Light. *Diam. Relat. Mater.* **2023**, *138*, 110245. [CrossRef]
28. Makota, O.; Dutková, E.; Briančin, J.; Bednarčik, J.; Lisnichuk, M.; Yevchuk, I.; Melnyk, I. Advanced Photodegradation of Azo Dye Methyl Orange Using H_2O_2-Activated Fe_3O_4@SiO_2@ZnO Composite under UV Treatment. *Molecules* **2024**, *29*, 1190. [CrossRef] [PubMed]
29. Rubesh Ashok Kumar, S.; Vasvini Mary, D.; Suganya Josephine, G.A. Design of solar-light-driven agglomerated cluster-like transition/rare-earth metal oxide-supported carbon-based nanomaterial for the degradation of azo dye. *Chem. Phys. Impact* **2024**, *8*, 100563.
30. Kumar, S.R.A.; Mary, D.V.; Josephine, G.A.S.; Sivasamy, A. Hydrothermally synthesized WO_3:CeO_2 supported gC_3N_4 nanolayers for rapid photocatalytic degradation of azo dye under natural sunlight. *Inorg. Chem. Commun.* **2024**, *164*, 112366. [CrossRef]
31. Luo, L.; Shen, J.; Jin, B. Construction of $Zn_{0.5}Cd_{0.5}S$/$Bi_4O_5Br_2$ Heterojunction for Enhanced Photocatalytic Degradation of Tetracycline Hydrochloride. *Inorganics* **2024**, *12*, 127. [CrossRef]

32. Nawaz, R.; Hanafiah, M.M.; Ali, M.; Anjum, M.; Baki, Z.A.; Mekkey, S.D.; Ullah, S.; Khurshid, S.; Ullah, H.; Arshad, U. Review of the performance and energy requirements of metals modified TiO_2 materials based photocatalysis for phenolic compounds degradation: A case of agro-industrial effluent. *J. Environ. Chem. Eng.* **2024**, *12*, 112766. [CrossRef]
33. Zhao, H.; Dong, J.; Xie, Y.; Meng, L.; Shen, S.; Chen, J.G.; Hu, D.; Yang, G. Construction of thin-shell TiO_2 vesicles inspired by the shell-deposition of diatoms for chlorophyll-sensitized photocatalyst. *Solid State Sci.* **2024**, *152*, 107520. [CrossRef]
34. Eddy, D.R.; Permana, M.D.; Sakti, L.K.; Sheha, G.A.N.; Solihudin; Hidayat, S.; Takei, T.; Kumada, N.; Rahayu, I. Heterophase Polymorph of TiO_2 (Anatase, Rutile, Brookite, TiO_2 (B)) for Efficient Photocatalyst: Fabrication and Activity. *Nanomaterials* **2023**, *13*, 704. [CrossRef] [PubMed]
35. Razdan, N.K.; Bhan, A. Catalytic site ensembles: A context to reexamine the Langmuir-Hinshelwood kinetic description. *J. Catal.* **2021**, *404*, 726–744. [CrossRef]
36. Arikal, D.; Kallingal, A. Photocatalytic degradation of azo and anthraquinone dye using TiO_2/MgO nanocomposite immobilized chitosan hydrogels. *Environ. Technol.* **2021**, *42*, 2278–2291. [CrossRef] [PubMed]
37. Zhang, W.; Tian, Y.; He, H.; Xu, L.; Li, W.; Zhao, D. Recent advances in the synthesis of hierarchically mesoporous TiO_2 materials for energy and environmental applications. *Natl. Sci. Rev.* **2020**, *7*, 1702–1725. [CrossRef] [PubMed]
38. Naas, L.A.; Bouaouina, B.; Bensouici, F.; Mokeddem, K.; Abaidia, S.E. Effect of TiN thin films deposited by oblique angle sputter deposition on sol-gel coated TiO_2 layers for photocatalytic applications. *Thin Solid Film.* **2024**, *793*, 140275. [CrossRef]
39. Jamil, A.; Sawaira, T.; Ali, A. Ce-TiO_2 nanoparticles with surface-confined Ce^{3+}/Ce^{4+} redox pairs for rapid sunlight-driven elimination of organic contaminants from water. *Environ. Nanotechnol. Monit. Manag.* **2024**, *21*, 100946. [CrossRef]
40. Huo, J.; Xiao, Y.; Yang, H. Ultrasmall TiO_2/C nanoparticles with oxygen vacancy-enriched as an anode material for advanced Li-ion hybrid capacitors. *J. Energy Storage* **2024**, *89*, 111586. [CrossRef]
41. Fermeli, P.N.; Lagopati, N.; Gatou, A.M. Biocompatible PANI-Encapsulated Chemically Modified Nano-TiO_2 Particles for Visible-Light Photocatalytic Applications. *Nanomaterials* **2024**, *14*, 642. [CrossRef] [PubMed]
42. Parashar, D.; Achari, G.; Kumar, M. Multi-antibiotics removal under UV-A light using sol-gel prepared TiO_2: Central composite design, effect of persulfate addition and degradation pathway study. *Chemosphere* **2023**, *341*, 140025. [CrossRef] [PubMed]
43. Gnanasekaran, L.; Priya, A.K.; Ghfar, A.A.; Sekar, K.; Santhamoorthy, M.; Arthi, M.; Soto-Moscoso, M. The influence of heterostructured TiO_2/ZnO nanomaterials for the removal of azo dye pollutant. *Chemosphere* **2022**, *308*, 136161. [CrossRef] [PubMed]
44. Ioannidou, T.; Anagnostopoulou, M.; Vasiliadou, I.A.; Marchal, C.; Alexandridou, E.O.; Keller, V.; Christoforidis, K.C. Mixed phase anatase nanosheets/brookite nanorods TiO_2 photocatalysts for enhanced gas phase CO_2 photoreduction and H_2 production. *J. Environ. Chem. Eng.* **2024**, *12*, 111644. [CrossRef]
45. Rawat, J.; Sharma, H.; Dwivedi, C. Microwave-assisted synthesis of carbon quantum dots and their integration with TiO_2 nanotubes for enhanced photocatalytic degradation. *Diam. Relat. Mater.* **2024**, *144*, 111050. [CrossRef]
46. Matakgane, M.; Mokoena, T.; Kroon, R. Robust upconversion luminescence of Ho^{3+}/Yb^{3+} co-doped TiO_2 nanophosphors manifested by crystallinity. *J. Mol. Struct.* **2024**, *1305*, 137747. [CrossRef]
47. Yang, W.; Jia, N.; Xu, W.; Bu, Q. Fabrication of novel 3D urchin-like mixed-phase TiO_2 for efficient degradation of trimethoprim. *Mater. Lett.* **2023**, *351*, 135001. [CrossRef]
48. Zhang, W.; Zhao, X.; Zhang, L.; Zhu, J.; Li, S.; Hu, P.; Feng, J.; Yan, W. Insight into the effect of surface carboxyl and amino groups on the adsorption of titanium dioxide for acid red G. *Front. Chem. Sci. Eng.* **2021**, *15*, 1147–1157. [CrossRef]
49. Jialin, L.; Zhonghao, W.; Furui, C. Self-assembly, structure and catalytic activity of Ni_3 on TiO_2: A triple-atom catalyst for hydrogen evolution. *Appl. Surf. Sci.* **2024**, *643*, 158719.
50. Abidi, M.; Assadi, A.; Bouzaza, A. Photocatalytic indoor/outdoor air treatment and bacterial inactivation on CuO/TiO_2 prepared by HiPIMS on polyester cloth under low intensity visible light. *Appl. Catal. B Environ.* **2019**, *259*, 118074. [CrossRef]
51. Cao, Y.-Q.; Zi, T.-Q.; Zhao, X.-R.; Liu, C.; Ren, Q.; Fang, J.-B.; Li, W.-M.; Li, A.-D. Enhanced visible light photocatalytic activity of Fe2O3 modified TiO2 prepared by atomic layer deposition. *Sci. Rep.* **2020**, *10*, 13437. [CrossRef]
52. Feng, J.; Xiong, S.; Ren, L.; Wang, Y. Atomic layer deposition of TiO_2 on carbon-nanotubes membrane for capacitive deionization removal of chromium from water. *Chin. J. Chem. Eng.* **2022**, *45*, 15–21. [CrossRef]
53. Ostyn, N.R.; Sree, S.P.; Li, J.; Feng, J.Y.; Roeffaers, M.B.; De Feyter, S.; Dendooven, J.; Detavernier, C.; Martens, J.A. Covalent graphite modification by low-temperature photocatalytic oxidation using a titanium dioxide thin film prepared by atomic layer deposition. *Catal. Sci. Technol.* **2021**, *11*, 6724–6731. [CrossRef]
54. Ke, W.; Qin, X.; Vazquez, Y.; Lee, I.; Zaera, F. Direct characterization of interface sites in Au/TiO_2 catalysts prepared using atomic layer deposition. *Chem Catal.* **2024**, *4*, 100977. [CrossRef]
55. Lys, A.; Gnilitskyi, I.; Coy, E.; Jancelewicz, M.; Gogotsi, O.; Iatsunskyi, I. Highly regular laser-induced periodic silicon surface modified by MXene and ALD TiO_2 for organic pollutants degradation. *Appl. Surf. Sci.* **2023**, *640*, 15833. [CrossRef]
56. Domenico, R.; Francesca, D.A.; Irene, B.; Paola, B.M.; Luca, D.P. Easy way to produce iron-doped titania nanoparticles via the solid-state method and investigation their photocatalytic activity. *J. Mater. Res.* **2023**, *38*, 1282–1292.
57. Sun, L.; Zhou, Q.; Mao, J.; Ouyang, X.; Yuan, Z.; Song, X.; Gong, W.; Mei, S.; Xu, W. Study on Photocatalytic Degradation of Acid Red 73 by Fe_3O_4@TiO_2 Exposed (001) Facets. *Appl. Sci.* **2022**, *12*, 3574. [CrossRef]
58. Yang, J.X.; Chang, W.; Huang, L.; Qin, J.; Li, Y.F. Preparation and properties of titanium dioxide nanofibers by microemulsion electrospinning. *Chem. Ind. Eng.* **2022**, *39*, 21–28.

59. Sun, L.; Wang, G.; Bi, S.; Li, H.; Zhan, H.; Liu, W. Synergistic modulation of oxygen vacancies and heterojunction structure in Pd@CN@TiO$_2$ promotes efficient photocatalytic Suzuki–Miyaura C–C coupling reactions. *J. Appl. Organomet. Chem.* **2024**, *38*, e7353. [CrossRef]
60. Hongying, L.; Bicheng, Z.; Jian, S.; Haiming, G.; Jiaguo, Y.; Liuyang, Z. Photocatalytic hydrogen production from seawater by TiO$_2$/RuO$_2$ hybrid nanofiber with enhanced light absorption. *J. Colloid Interface Sci.* **2024**, *654*, 1010–1019.
61. Andrea, B.S.; Roberto, F.; Armeli, I.M.T.; Javier, L.-T.F.; José, U.F.; Salvatore, S. H2 production through glycerol photoreforming using one-pot prepared TiO$_2$-rGO-Au photocatalysts. *Mol. Catal.* **2023**, *547*, 113346.
62. Djeda, R.; Mailhot, G.; Prevot, V. Porous Layered Double Hydroxide/TiO$_2$ Photocatalysts for the Photocatalytic Degradation of Orange II. *ChemEngineering* **2020**, *4*, 39. [CrossRef]
63. Zhao, Y.; Ge, H.; Kondo, Y.; Kuwahara, Y.; Mori, K.; Yamashita, H. Photosynthesis of hydrogen peroxide in a two-phase system by hydrophobic Au nanoparticle-deposited plasmonic TiO$_2$ catalysts. *Catal. Today* **2024**, *431*, 114558. [CrossRef]
64. Li, S.; Hu, M.; Chen, X.; Sui, J.; Jin, L.; Geng, Y.; Jiang, J.; Liu, A. The Performance And Functionalization of Modified Cementitious Materials Via Nano Titanium-Dioxide: A Review. *Case Stud. Constr. Mater.* **2023**, *19*, E02414. [CrossRef]
65. Khitous, A.; Noel, L.; Molinaro, C.; Vidal, L.; Grée, S.; Soppera, O. Sol-Gel TiO$_2$ Thin Film on Au Nanoparticles for Heterogeneous Plasmonic Photocatalysis. *ACS Appl. Mater. Interfaces* **2024**, *16*, 10856–10866. [CrossRef] [PubMed]
66. Chen, J.; Qiu, F.; Xu, W.; Cao, S.; Zhu, H. Recent Progress In Enhancing Photocatalytic Efficiency of TiO$_2$-Based Materials. *Appl. Catal. A-Gen.* **2015**, *495*, 131–140. [CrossRef]
67. Hou, W.; Hung, W.H.; Pavaskar, P.; Goeppert, A.; Aykol, M.; Cronin, S.B.J.A.C. Photocatalytic Conversion of CO$_2$ to Hydrocarbon Fuels Via Plasmon-Enhanced Absorption and Metallic Interband Transitions. *ACS Catal.* **2011**, *1*, 929–936. [CrossRef]
68. Bhatti, M.A.; Shah, A.A.; Almaani, K.F.; Tahira, A.; Chandio, A.D.; Willander, M.; Nur, O.; Mugheri, A.Q.; Bhatti, A.L.; Waryani, B.; et al. TiO$_2$/ZnO Nanocomposite Material For Efficient Degradation Of Methylene Blue. *J. Nanosci. Nanotechnol.* **2021**, *21*, 2511–2519. [CrossRef] [PubMed]
69. Chacon-Argaez, U.; Cedeño-Caero, L.; Cadena-Nava, R.D.; Ramirez-Acosta, K.; Moyado, S.F.; Sánchez-López, P.; Alonso Núñez, G. Photocatalytic Activity and Biocide Properties of Ag-TiO$_2$ Composites on Cotton Fabrics. *Materials* **2023**, *16*, 4513. [CrossRef] [PubMed]
70. Borrego PÉRez, J.A.; Morales, E.R.; Paraguay Delgado, F.; Meza Avendaño, C.A.; Alonso Guzman, E.M.; Mathews, N.R. Ag Nanoparticle Dispersed TiO$_2$ Thin Films by Single Step Sol-Gel Process: Evaluation of The Physical Properties and Photocatalytic Degradation. *Vacuum* **2023**, *215*, 112276. [CrossRef]
71. Majeed, A.; Ibrahim, A.H.; Al-Rawi, S.S.; Iqbal, M.A.; Kashif, M.; Yousif, M.; Abidin, Z.U.; Ali, S.; Arbaz, M.; Hussain, S.A. Green Organo-Photooxidative Method for the Degradation of Methylene Blue Dye. *ACS Omega* **2024**, *9*, 12069–12083. [CrossRef] [PubMed]
72. Wang, Z.; Jin, X.; Chen, F.; Kuang, X.; Min, J.; Duan, H.; Li, J.; Chen, J. Oxygen Vacancy Induced Interaction between Pt and TiO$_2$ to Improve The Oxygen Reduction Performance. *J. Colloid Interface Sci.* **2023**, *650*, 901–912. [CrossRef] [PubMed]
73. Liu, L.; He, A. Optical And Electrochemical Study On The Performance Of Au@ TiO$_2$ Core-Shell Heterostructured Nanoparticles as Photocatalyst for Photodegradation of Methylene Blue under Solar-Light Irradiation. *Int. J. Electrochem. Sci.* **2022**, *17*, 220636. [CrossRef]
74. Zuo, J.; Ma, X.; Tan, C.; Xia, Z.; Zhang, Y.; Yu, S.; Li, Y.; Li, Y.; Li, J. Preparation of Au-RGO/TiO$_2$ Nanotubes and Study on The Photocatalytic Degradation of Ciprofloxacin. *Anal. Methods Adv. Methods Appl.* **2023**, *15*, 519–528. [CrossRef] [PubMed]
75. Zheng, F.; Martins, P.M.; Queirós, J.M.; Tavares, C.J.; Vilas-Vilela, J.L.; Lanceros-Méndez, S.; Reguera, J. Size Effect in Hybrid TiO$_2$:Au Nanostars for Photocatalytic Water Remediation Applications. *Int. J. Mol. Sci.* **2022**, *23*, 13741. [CrossRef] [PubMed]
76. Bhuskute, B.D.; Ali-Löytty, H.; Honkanen, M.; Salminen, T.; Valden, M. Influence of The Photodeposition Sequence on The Photocatalytic Activity of Plasmonic Ag-Au/TiO$_2$ Nanocomposites. *Nanoscale Adv.* **2022**, *4*, 4335–4343. [CrossRef] [PubMed]
77. Huerta-Aguilar, C.A.; Palos-Barba, V.; Thangarasu, P.; Koodali, R.T. Visible Light Driven Photo-Degradation of Congo Red by TiO$_2$-ZnO/Ag: DFT Approach on Synergetic Effect on Band Gap Energy. *Chemosphere* **2018**, *213*, 481–497. [CrossRef] [PubMed]
78. Stojadinović, S.; Radić, N.; Vasilić, R.; Tadić, N.; Tsanev, A. Photocatalytic Degradation of Methyl Orange in The Presence of Transition Metals (Mn, Ni, Co) Modified TiO$_2$ Coatings Formed by Plasma Electrolytic Oxidation. *Solid State Sci.* **2022**, *129*, 106896. [CrossRef]
79. Belošević-Čavor, J.; Koteski, V.; Umićević, A.; Ivanovski, V. Effect of 5d Transition Metals Doping on The Photocatalytic Properties of Rutile TiO$_2$. *Comput. Mater. Sci.* **2018**, *151*, 328–337. [CrossRef]
80. Meshram, A.A.; Sontakke, S.M. Rapid Degradation of Metamitron And Highly Complex Mixture of Pollutants Using MIL-53(Al) Integrated Combustion Synthesized TiO$_2$. *Adv. Powder Technol.* **2021**, *32*, 3125–3135. [CrossRef]
81. Zhu, Y. Enhanced Photoelectrochemical Properties of Fe-TiO$_2$ Nanotube Films: A Combined Experimental and Theoretical Study. *Int. J. Electrochem. Sci.* **2021**, *16*, 210662. [CrossRef]
82. Mingmongkol, Y.; Trinh, D.T.T.; Phuinthiang, P.; Channei, D.; Ratananikom, K.; Nakaruk, A.; Khanitchaidecha, W. Enhanced Photocatalytic and Photokilling Activities of Cu-Doped TiO$_2$ Nanoparticles. *Nanomater* **2022**, *12*, 1198. [CrossRef] [PubMed]
83. Rao, T.N.; Babji, P.; Parvatamma, B.; Naidu, T.M. Decontamination of Pesticide Residues in Water Samples Using Copper and Zinc Co-Doped Titania Nanocatalyst. *Environ. Eng. Manag. J.* **2020**, *19*, 721–731. [CrossRef]
84. Sukhadeve, G.K.; Gedam, R.S. Visible Light Assisted Photocatalytic Degradation of Mixture of Reactive Ternary Dye Solution by Zn–Fe Co-Doped TiO$_2$ Nanoparticles. *Chemosphere* **2023**, *341*, 139990. [CrossRef]

85. Liu, X. Preparation of Rare Earth Doped Modified TiO$_2$ Nanocrystalline Materials and Their Photocatalytic Properties. Master's Thesis, Changji College, Xinjiang, China, 2022.
86. Shi, Z.M.; Jin, L.N. Influence of La^{3+}/Ce^{3+}-Doping on Phase Transformation and Crystal Growth in TiO$_2$-15wt% Zno Gels. *J. Non-Cryst. Solids* **2009**, *355*, 213–220. [CrossRef]
87. Ricci, P.C.; Carbonaro, C.M.; Geddo Lehmann, A.; Congiu, F.; Puxeddu, B.; Cappelletti, G.; Spadavecchia, F. Structure and Photoluminescence of TiO$_2$ Nanocrystals Doped and Co-Doped with N and Rare Earths (Y^{3+}, Pr^{3+}). *J. Alloys Compd.* **2013**, *561*, 109–113. [CrossRef]
88. Hassan, M.S.; Amna, T.; Yang, O.B.; Kim, H.-C.; Khil, M.-S. TiO$_2$ Nanofibers Doped with Rare Earth Elements and Their Photocatalytic Activity. *Ceram. Int.* **2012**, *38*, 5925–5930. [CrossRef]
89. Wen, M.; Yang, H.; Tong, L.; Yuan, L. Study on Degradation of Gold-Cyanide Wastewater by Rare-Earth La Doped TiO$_2$ -MnO$_2$ Composite Photocatalyst. *Mol. Catal.* **2023**, *550*, 113506. [CrossRef]
90. Reszczyńska, J.; Grzyb, T.; Sobczak, J.W.; Lisowski, W.; Gazda, M.; Ohtani, B.; Zaleska, A. Visible Light Activity of Rare Earth Metal Doped (Er^{3+}, Yb^{3+} Or Er^{3+}/Yb^{3+}) Titania Photocatalysts. *Appl. Catal. B Environ.* **2015**, *163*, 40–49. [CrossRef]
91. Tang, X.; Xue, Q.; Qi, X.; Cheng, C.; Yang, M.; Yang, T.; Chen, F.; Qiu, F.; Quan, X. DFT and Experimental Study on Visible-Light Driven Photocatalysis of Rare-Earth-Doped TiO$_2$. *Vacuum* **2022**, *200*, 110972. [CrossRef]
92. Zhou, F.; Yan, C.; Sun, Q.; Komarneni, S. TiO$_2$/Sepiolite Nanocomposites Doped with Rare Earth Ions: Preparation, Characterization and Visible Light Photocatalytic Activity. *Microporous Mesoporous Mater.* **2019**, *274*, 25–32. [CrossRef]
93. Kobwittaya, K.; Oishi, Y.; Torikai, T.; Yada, M.; Watari, T.; Luitel, H.N. Bright Red Upconversion Luminescence from Er^{3+} and Yb^{3+} Co-Doped Zno-TiO$_2$ Composite Phosphor Powder. *Ceram. Int.* **2017**, *43*, 13505–13515. [CrossRef]
94. Ćurković, L.; Briševac, D.; Ljubas, D.; Mandić, V.; Gabelica, I. Synthesis, Characterization, and Photocatalytic Properties of Sol-Gel Ce-TiO$_2$ Films. *Processes* **2024**, *12*, 1144. [CrossRef]
95. Benammar, I.; Salhi, R.; Deschanvres, J.-L.; Maalej, R. The Effect Of Rare Earth (Er, Yb) Element Doping On The Crystallization Of Y$_2$Ti$_2$O$_7$ Pyrochlore Nanoparticles Developed by Hydrothermal-Assisted-Sol-Gel Method. *J. Solid State Chem.* **2023**, *320*, 123856. [CrossRef]
96. Arévalo-Pérez, J.C.; Cruz-Romero, D.D.L.; Cordero-García, A.; Lobato-García, C.E.; Aguilar-Elguezabal, A.; Torres-Torres, J.G. Photodegradation of 17 A-Methyltestosterone Using TiO$_2$-Gd^{3+} and TiO$_2$-Sm^{3+} Photocatalysts and Simulated Solar Radiation As An Activation Source. *Chemosphere* **2020**, *249*, 126497. [CrossRef] [PubMed]
97. Kuwik, M.; Kowalska, K.; Pisarska, J.; Kochanowicz, M.; Żmojda, J.; Dorosz, J.; Dorosz, D.; Pisarski, W.A. Influence of TiO$_2$ Concentration on Near-Infrared Emission of Germanate Glasses Doped with Tm^{3+} and Tm^{3+}/Ho^{3+} Ions. *Ceram. Int.* **2023**, *49*, 41090–41097. [CrossRef]
98. Alzahrani, J.S.; Alrowaili, Z.A.; Eke, C.; Olarinoye, I.O.; Al-Buriahi, M.S. Characterization and Applications of Highly Optical Transparency Tellurite Glasses Doped with Er^{3+}, Tm^{3+}, And Nd^{3+}. *Optik* **2023**, *281*, 170825. [CrossRef]
99. Pascariu, P.; Cojocaru, C.; Homocianu, M.; Samoila, P. Tuning of Sm^{3+} and Er^{3+}-doped TiO$_2$ nanofibers for enhancement of the photocatalytic performance: Optimization of the photodegradation conditions. *J. Environ. Manag.* **2022**, *316*, 115317. [CrossRef] [PubMed]
100. Guetni, I.; Belaiche, M.; Ferdi, C.A.; Oulhakem, O.; Alaoui, K.B.; Zaoui, F.; Bahije, L. Novel modified nanophotocatalysts of TiO$_2$ nanoparticles and TiO$_2$/Alginate beads with lanthanides [La, Sm, Y] to degrade the Azo dye Orange G under UV-VIS radiation. *Mater. Sci. Semicond. Process.* **2024**, *174*, 108193. [CrossRef]
101. Ren, Y.; Han, Y.; Li, Z.; Liu, X.; Zhu, S.; Liang, Y.; Yeung, K.W.K.; Wu, S. Ce and Er Co-doped TiO$_2$ for rapid bacteria-killing using visible light. *Bioact. Mater.* **2020**, *5*, 201–209. [CrossRef] [PubMed]
102. Seshaiah, K.V.; Dileep, R.K.; Easwaramoorthi, R.; Ganapathy, V.; Kumar, R.S.S. Deciphering the role of (Er^{3+}/Nd^{3+}) co-doping effect on TiO$_2$ as an improved electron transport layer in perovskite solar cells. *Sol. Energy* **2023**, *262*, 111801.
103. Guo, Y.T.; Wang, H.Q. Research on Rare Earth Lanthanum-Doped Nanotitanium Dioxide Composite Fresh Packaging Films. *Mater. Guide* **2018**, *32*, 4357–4362.
104. Li, J.; Zhang, X.; Tan, Z.; Wei, X.; Zhou, X. Effect of Lanthanum and Graphene Oxide Doping on The Photocatalytic Performance of TiO$_2$. *Chem. New Mater.* **2023**, *51*, 159–163+168.
105. Alanazi, H.E.; Emran, K.M. Nd-Gd–Platinum Doped TiO$_2$ Nanotube Arrays Catalyst for Water Splitting in Alkaline Medium. *Int. J. Electrochem. Sci.* **2023**, *18*, 100112. [CrossRef]
106. Mei, Q.F.; Zhang, F.; Wang, N.; Lu, W.; Su, X.; Wang, W.; Wu, R. Titanium Dioxide-Based Z-Type Heterojunction Photocatalysts. *J. Inorg. Chem.* **2019**, *35*, 1321–1339.
107. Qi, K.; Liu, N.; Cui, N.; Song, J.; Ma, Y.; Chen, Q. Design of Modified Titanium Dioxide Photocatalyst. *J. Shenyang Norm. Univ. (Nat. Sci. Ed.)* **2021**, *39*, 103–108.
108. Hou, J.; Wang, Y.; Zhou, J.; Lu, Y.; Liu, J.; Lv, X. Photocatalytic Degradation Of Methylene Blue Using A ZnO/TiO$_2$ Heterojunction Nanomesh Electrode. *Surf. Interfaces* **2021**, *22*, 100889. [CrossRef]
109. Wang, D. Preparation and Photocatalytic Properties of TiO$_2$/ZnO Micro- and Nanomaterials and Core-Shell Structures. Ph.D. Thesis, University of Chinese Academy of Sciences (Changchun Institute of Optical Precision Machinery and Physics, Chinese Academy of Sciences), Beijing, China, 2018.
110. Yu, J.; Wang, S.; Low, J.; Xiao, W. Enhanced Photocatalytic Performance of Direct Z-Scheme G-C$_3$N$_4$-TiO$_2$ Photocatalysts for The Decomposition of Formaldehyde in Air. *Phys. Chem. Chem. Phys. PCCP* **2013**, *15*, 16883–16890. [CrossRef] [PubMed]

111. Wongburapachart, C.; Pornaroontham, P.; Kim, K.; Rangsunvigit, P. Photocatalytic Degradation of Acid Orange 7 by NiO-TiO$_2$/TiO$_2$ Bilayer Film Photo-Chargeable Catalysts. *Coatings* **2023**, *13*, 141. [CrossRef]
112. Bai, N.; Liu, X.; Li, Z.; Ke, X.; Zhang, K.; Wu, Q. High-efficiency TiO$_2$/ZnO nanocomposites photocatalysts by sol–gel and hydrothermal methods. *J. Sol-Gel Sci. Technol.* **2021**, *99*, 92–100. [CrossRef]
113. Wittawat, R.; Rittipun, R.; Jarasfah, M.; Nattaporn, B. Synthesis of ZnO/TiO$_2$ Spherical Particles for Blue Light Screening by Ultrasonic Spray Pyrolysis. *Mater. Today Commun.* **2020**, *24*, 101126.
114. Wang, S.; Li, X.; Li, D. Microwave Assisted Synthesis Of ZnO-TiO$_2$ and Its Visible Light Catalytic Denitrification Activity. *J. Fuel Chem. Technol.* **2023**, *51*, 589–598. [CrossRef]
115. Ali, M.M.; Haque, M.J.; Kabir, M.H.; Kaiyum, M.A.; Rahman, M.S. Nano Synthesis Of ZnO–TiO$_2$ Composites by Sol-Gel Method and Evaluation of Their Antibacterial, Optical and Photocatalytic Activities. *Results Mater.* **2021**, *11*, 100199. [CrossRef]
116. Abumousa, R.A.; Bououdina, M.; Ben Aissa, M.A.; Khezami, L.; Modwi, A. Efficient photocatalytic degradation of Congo red and other dyes by ternary TiO$_2$/Y$_2$O$_3$@g-C$_3$N$_4$ nanohybrid. *J. Mater. Sci. Mater. Electron.* **2024**, *35*, 486. [CrossRef]
117. Heltina, D.; Imamatul Mastura, D.; Amri, A.; Peratenta Sembiring, M. Komalasari, Comparison of Synthesis Methods on TiO$_2$-Graphene Composites for Photodegradation of Compound Waste. *Mater. Today Proc.* **2023**, *87*, 293–298. [CrossRef]
118. Wang, S.; Li, X.; Wu, J. Preparation of TiO$_2$/Graphene Oxide and Their Photocatalytic Properties at Room Temperature. *J. Fuel Chem. Technol.* **2022**, *50*, 1307–1316. [CrossRef]
119. Kusiak-Nejman, E.; Wanag, A.; Kapica-Kozar, J.; Kowalczyk, Ł.; Zgrzebnicki, M.; Tryba, B.; Przepiórski, J.; Morawski, A.W. Methylene Blue Decomposition on TiO$_2$/Reduced Graphene Oxide Hybrid Photocatalysts Obtained by A Two-Step Hydrothermal and Calcination Synthesis. *Catal. Today* **2020**, *357*, 630–637. [CrossRef]
120. Zhou, M.; Jiang, Y.; Xie, Y.; Xie, D.L.; Mei, Y. Research progress on the preparation and modification of titanium dioxide nanoparticles and their application in polymer matrix composites. *J. Compos. Mater.* **2022**, *39*, 2089–2105.
121. Maeda, S.; Fujita, M.; Idota, N.; Matsukawa, K.; Sugahara, Y. Preparation of Transparent Bulk TiO$_2$/Pmma Hybrids with Improved Refractive Indices Via An In Situ Polymerization Process Using TiO$_2$ Nanoparticles Bearing Pmma Chains Grown by Surface-Initiated Atom Transfer Radical Polymerization. *ACS Appl. Mater. Interfaces* **2016**, *8*, 34762–34769. [CrossRef] [PubMed]
122. Shi, Y.; Zhang, Q.; Liu, Y. Preparation of amphiphilic TiO$_2$ Janus particles with highly enhanced photocatalytic activity. *Chin. J. Catal.* **2019**, *40*, 786–794. [CrossRef]
123. Nair, V.R.; Shetty Kodialbail, V. Floating bed reactor for visible light induced photocatalytic degradation of Acid Yellow 17 using polyaniline-TiO$_2$ nanocomposites immobilized on polystyrene cubes. *Environ. Sci. Pollut. Res.* **2020**, *27*, 14441–14453. [CrossRef] [PubMed]
124. Chien, H.; Chen, X.; Tsai, W. Poly(Methyl Methacrylate)/Titanium Dioxide (Pmma/TiO$_2$) Nanocomposite with Shark-Skin Structure for Preventing Biofilm Formation. *Mater. Lett.* **2021**, *285*, 129098. [CrossRef]
125. Neves, J.C.; Mohallem, N.D.; Viana, M. Polydimethylsiloxanes-Modified TiO$_2$ Coatings: The Role of Structural, Morphological and Optical Characteristics in A Self-Cleaning Surface. *Ceram. Int.* **2020**, *46 Pt B*, 11606–11616. [CrossRef]
126. Abdul, M.; Shahid, R.M.; Ghulam, M.M.; Tehreem, A.; Imran, S.; Maryum, R.; Aqsa, N. Fabrication of Ag–TiO$_2$ nanocomposite employing dielectric barrier discharge plasma for photodegradation of methylene blue. *Phys. B Condens. Matter* **2023**, *665*, 414995.
127. Dong, S.; Tebbutt, G.T.; Millar, R.; Grobert, N.; Maciejewska, B.M. Hierarchical porosity design enables highly recyclable and efficient Au/TiO$_2$ composite fibers for photodegradation of organic pollutants. *Mater. Des.* **2023**, *234*, 112318. [CrossRef]
128. Reguero-Márquez, G.A.; Lunagómez-Rocha, M.A.; Cervantes-Uribe, A.; Del Angel, G.; Rangel, I.; Torres-Torres, J.G.; González, F.; Godavarthi, S.; Arevalo-Perez, J.C.; de Los Monteros, A.E.; et al. Photodegradation of 2,4-D (dichlorophenoxyacetic acid) with Rh/TiO2; comparative study with other noble metals (Ru, Pt, and Au). *RSC Adv.* **2022**, *12*, 25711–25721. [CrossRef] [PubMed]
129. Solmaz, F.; Aziz, H.; Rafael, L. Preparation of TiO$_2$/Fe-MOF n–n heterojunction photocatalysts for visible-light degradation of tetracycline hydrochloride. *Chemosphere* **2023**, *336*, 139101.
130. Muhammad, I.; Rab, N.; Akbar, K.J.; Habib, U.; Tahir, H.; Stanislaw, L.; Saifur, R.; Jerzy, J.; Alsaiar, M.A.; Asif, K.M.K.; et al. Synthesis and Characterization of Manganese-Modified Black TiO$_2$ Nanoparticles and Their Performance Evaluation for the Photodegradation of Phenolic Compounds from Wastewater. *Materials* **2021**, *14*, 7422. [CrossRef]
131. Eleni, E.; Zoi, C.; Athanasios, T.; Kyriakos, P.; Pavlina, T.; Daniel, S.; Anastasia, K.; Lambropoulou, D.A. Photocatalytic degradation of a mixture of eight antibiotics using Cu-modified TiO$_2$ photocatalysts: Kinetics, mineralization, antimicrobial activity elimination and disinfection. *J. Environ. Chem. Eng.* **2021**, *9*, 105295.
132. Cho, H.; Joo, H.; Kim, H.; Kim, J.-E.; Kang, K.-S.; Yoon, J. Improved photoelectrochemical properties of TiO$_2$ nanotubes doped with Er and effects on hydrogen production from water splitting. *Chemosphere* **2021**, *267*, 129289. [CrossRef] [PubMed]
133. Rajeev, Y.N.; Magdalane, C.M.; Hepsibha, S.; Ramalingam, G.; Kumar, B.A.; Kumar, L.B.; Sangaraju, S. Europium decorated hierarchical TiO$_2$ heterojunction nanostructure with enhanced UV light photocatalytic activity for degradation of toxic industrial effluent. *Inorg. Chem. Commun.* **2023**, *157*, 111339. [CrossRef]
134. Hu, L.; Xing, M.; He, X.; Yang, K.; Zhu, J.; Wang, J.; He, J.; Shi, J. Photocatalytic degradation of tetracycline hydrochloride by ZnO/TiO$_2$ composite photocatalyst. *J. Mater. Sci. Mater. Electron.* **2023**, *34*, 2273. [CrossRef]
135. Huang, F.; Hao, H.; Sheng, W.; Lang, X. Dye-TiO$_2$/SiO$_2$ assembly photocatalysis for blue light-initiated selective aerobic oxidation of organic sulfides. *Chem. Eng. J.* **2021**, *423*, 129419. [CrossRef]

136. Yang, J.; Huang, Q.; Sun, Y.; An, G.; Li, X.; Mao, J.; Wei, C.; Yang, B.; Li, D.; Tao, T. Photocatalytic oxidation of formaldehyde under visible light using BiVO$_4$-TiO$_2$ synthesized via ultrasonic blending. *Environ. Sci. Pollut. Res. Int.* **2024**, *31*, 30085–30098. [CrossRef] [PubMed]
137. Paula, L.F.; Hofer, M.; Lacerda, V.P.B. Unraveling the photocatalytic properties of TiO$_2$/WO$_3$ mixed oxides. *Photochem. Photobiol. Sci.* **2019**, *18*, 2469–2483. [CrossRef] [PubMed]
138. Zhang, J.; Wang, C.; Shi, X.; Feng, Q.; Shen, T.; Wang, S. Modulation of the Structure of the Conjugated Polymer TMP and the Effect of Its Structure on the Catalytic Performance of TMP–TiO$_2$ under Visible Light: Catalyst Preparation, Performance and Mechanism. *Materials* **2023**, *16*, 1563. [CrossRef] [PubMed]
139. Abu-Melha, S. Distinguishable photocatalytic activity of nano polyaniline with quantum dots metal oxide as photocatalysts for photodegradation of Dianix blue dye and different industrial pollutants. *Polyhedron* **2024**, *252*, 116781. [CrossRef]
140. Meng, A.; Cheng, B.; Tan, H.; Fan, J.; Su, C.; Yu, J. TiO$_2$/polydopamine S-scheme heterojunction photocatalyst with enhanced CO$_2$-reduction selectivity. *Appl. Catal. B Environ.* **2021**, *289*, 120039. [CrossRef]

Disclaimer/Publisher's Note: The statements, opinions and data contained in all publications are solely those of the individual author(s) and contributor(s) and not of MDPI and/or the editor(s). MDPI and/or the editor(s) disclaim responsibility for any injury to people or property resulting from any ideas, methods, instructions or products referred to in the content.

Review

Morphological Dependence of Metal Oxide Photocatalysts for Dye Degradation

Ahmed H. Naggar [1,*], Abdelaal S. A. Ahmed [2], Tarek A. Seaf El-Nasr [3], N. F. Alotaibi [3], Kwok Feng Chong [4] and Gomaa A. M. Ali [2,5,6,*]

1. Department of Chemistry, College of Science and Arts, Jouf University, Al Qurayyat 75911, Saudi Arabia
2. Chemistry Department, Faculty of Science, Al-Azhar University, Assiut 71524, Egypt; abdelaalsaiyd@azhar.edu.eg
3. Department of Chemistry, College of Science and Arts, Jouf University, Sakaka 2014, Saudi Arabia; taahmed@ju.edu.sa (T.A.S.E.-N.); nfotaibi@ju.edu.sa (N.F.A.)
4. Faculty of Industrial Sciences & Technology, Universiti Malaysia Pahang Al-Sultan Abdullah, Gambang, Kuantan 26300, Malaysia; ckfeng@umpsa.edu.my
5. Faculty of Advanced Basic Science, Galala University, Suez 43511, Egypt
6. Faculty of Industry and Energy Technology, New Assiut Technological University, Assiut 71684, Egypt
* Correspondence: ahayoub@ju.edu.sa (A.H.N.); gomaasanad@azhar.edu.eg (G.A.M.A.)

Citation: Naggar, A.H.; Ahmed, A.S.A.; El-Nasr, T.A.S.; Alotaibi, N.F.; Chong, K.F.; Ali, G.A.M. Morphological Dependence of Metal Oxide Photocatalysts for Dye Degradation. *Inorganics* 2023, 11, 484. https://doi.org/10.3390/inorganics11120484

Academic Editor: Roberto Nisticò

Received: 12 October 2023
Revised: 11 December 2023
Accepted: 11 December 2023
Published: 18 December 2023

Copyright: © 2023 by the authors. Licensee MDPI, Basel, Switzerland. This article is an open access article distributed under the terms and conditions of the Creative Commons Attribution (CC BY) license (https://creativecommons.org/licenses/by/4.0/).

Abstract: There is no doubt that organic dyes currently play an indispensable role in our daily life; they are used in products such as furniture, textiles, and leather accessories. However, the main problems related to the widespread use of these dyes are their toxicity and non-biodegradable nature, which mainly are responsible for various environmental risks and threaten human life. Therefore, the elimination of these toxic materials from aqueous media is highly recommended to save freshwater resources, as well as our health and environment. Heterogeneous photocatalysis is a potential technique for dye degradation, in which a photocatalyst is used to absorb light (UV or visible) and produce electron–hole pairs that enable the reaction participants to undergo chemical changes. In the past, various metal oxides have been successfully applied as promising photocatalysts for the degradation of dyes and various organic pollutants due to their wide bandgap, optical, and electronic properties, in addition to their low cost, high abundance, and chemical stability in aqueous solutions. Various parameters play critical roles in the total performance of the photocatalyst during the photocatalytic degradation of dyes, including morphology, which is a critical factor in the overall degradation process. In our article, the recent progress on the morphological dependence of photocatalysts will be reviewed.

Keywords: organic dyes; wastewater treatment; photocatalyst; photocatalytic degradation; metal oxides; morphology

1. Introduction

Due to the growth of societies and the speeding up of industry over the last few decades, environmental pollution is considered the biggest challenge facing our societies [1]. Almost all activities of humans to produce commodities and services lead to the creation of environmental contaminants. These pollutants cause harm to the health of people, plants, animals, and microbes when they are discharged into the air, water, and soil [2]. Since humans depend on the creation and enhancement of commodities and services to survive on earth, these practices cannot be completely abandoned [2]. Major contributors toward aquatic pollution include industrial dyes, which are the greatest class of organic pollutants [3,4]. Due to their intricate chemical compositions, the majority of synthetic dyes are poisonous and extremely durable. Currently, synthetic dyes as organic compounds or mixtures are widely utilized in the leather, pharmaceutical, food, cosmetic, color photography, paper printing, textile dyeing, and textile dyeing industries [5–8].

The last two decades have seen a dramatic rise in public understanding of the toxic and cancer-causing effects of many pollutants that were previously not thought to be dangerous [2]. Some man-made chemicals can linger in the environment for a very long time without degrading, in contrast to naturally occurring compounds that break down immediately. These substances, which include pesticides, organochlorines, polychlorinated biphenyls, synthetic polymers, and synthetic dyes, are regarded as the main environmental contaminants [9]. Most dyes are organic, multidimensional compounds with the property of adhering to various surfaces, including fabrics, leathers, and others. The paper, textile, dyestuff, and distilling industries release highly colored wastewaters [10]. By increasing the demand for dyes, a larger amount of water is polluted [11,12]. As reported by Couto et al., in traditional dying processes, each 1 kg of textile materials required about 100 L of water to obtain the final product [5]. As per O'Neill et al., not every type of dye adheres to fabric, and their discharge in wastewater varies. Basic dyes may lose up to 2% while reactive dyes can lose up to 50%, resulting in the contamination of surface and ground waters in the dyeing industries [13].

It has been reported that each year about 280,000 tons of textile dyes are released into industrial wastewater by the textile industry [14]. In general, it is thought that throughout the industrial processes, roughly 12% of the total generated synthetic dyes, including methyl orange (MO), methyl red (MR), methylene blue (MB), rhodamine B (RhB), remazol brilliant blue (RB), congo red (CR), and many others, are lost [3,15]. Additionally, 15–20% of global dye production is wasted during the dying process and released into water without additional treatment, resulting in significant environmental damage [16]. The presence of color in dye effluents serves as a clear indicator that water is contaminated since it is easier to see, and the discharge of these strongly colored effluents can directly impact the receiving waterways [17]. Even a very small concentration (1 mg L^{-1}) of synthetic dyes in water can generate color and create an unfavorable concentration for ingestion, and also can significantly harm the environment and pose substantial health risks [18]. When utilizing the untreated dyeing effluents in agriculture purposes, both the environment and human health are badly affected [19]. In light of the harmful and cancer-causing properties associated with dyes, and considering that numerous dyes are known carcinogens, significant recent endeavors have been directed towards regulating the release of dyes into the environment [20,21]. The remediation of dye wastewaters can be accomplished using various techniques. A variety of procedures are involved, comprising biological and microbiological methods, along with physicochemical techniques such as adsorption, chemical oxidation, precipitation, coagulation, filtration, electrolysis, and photodegradation [22]. Based on its suitability, each strategy can be used to target a certain class of toxins and offers advantages of its own. Despite their usefulness, physicochemical techniques often suffer from several limitations, including high costs, low efficiency, limited flexibility, susceptibility to interference from other wastewater constituents, and the challenge of managing the waste generated [18].

Therefore, it is crucial to develop affordable, effective, and environmentally friendly methods to reduce the amount of dye in wastewater [5]. Advanced oxidation processes (AOPs) can be used for treating most industrial effluents. Among the numerous AOPs that are recognized, photocatalytic degradation has emerged as a promising method for destroying organic substances [23,24]. The photodegradation process of dyes involves oxidizing large dye molecules into smaller ones such as water, carbon dioxide, and other byproducts. Compared to other AOPs, photocatalytic degradation is more successful since semiconductors are less expensive and can easily mineralize a variety of organic molecules. The materials used as effective photocatalysts should have an appropriate energy band gap, appropriate morphology, high surface area, long term stability, and considerable recycling ability [25,26]. Due to the high surface-to-volume ratio of nanomaterials, more surface area is available for redox reactions. In recent years, metal oxides have garnered significant interest in environmental remediation due to their ability to generate charge carriers upon activation by sufficient energy [27]. Numerous metal oxide nanoparticles have been employed as photocatalysts in the past, including zinc oxide (ZnO), titanium dioxide (TiO_2), copper oxide (CuO), nickel oxide (NiO), and tungsten oxide (WO_3). This is a result of their potential optical, chemical, and physical capabilities, including their distinct electronic structures, abilities to absorb light, and capacities for charge transport [28]. The composition, size, doping, and shape of metal oxides are only a few of the variables that might influence their photocatalytic activity. These factors are all crucial for photocatalytic activity [29–32]. This study mainly reviews the recent progress on the morphologically dependent photocatalytic activity of the metal oxides toward degradation of organic dyes from aqueous solutions. From the metal oxides, TiO_2, ZnO, CuO, NiO, and WO_3, with various morphologies, will be presented. To provide a big picture, our article starts with an introduction that outlines the dye classifications and their environmental problems. Additionally, this article outlines the principle of the photocatalytic process and the different photocatalytic reaction mechanisms.

In addition, this review presents a comprehensive study of the recent progress made in the field of the morphological dependence of photocatalysts for the photocatalytic degradation of organic dyes. It discusses the latest developments in the synthesis and characterization of metal oxide photocatalysts with tailored morphologies. Furthermore, it analyzes the underlying mechanisms governing morphology-induced effects and provides insights into the prospects and challenges in this research area.

According to publication records during the last decade, there were about 77,214 publications on photocatalysis. The research output on photocatalysis has continued growing with time, with 4155 and 10,550 publications in 2013 and 2023, respectively, as shown in Figure 1. Among the metal oxides, most publications investigated TiO_2, ZnO, CuO, NiO, and WO_3 materials.

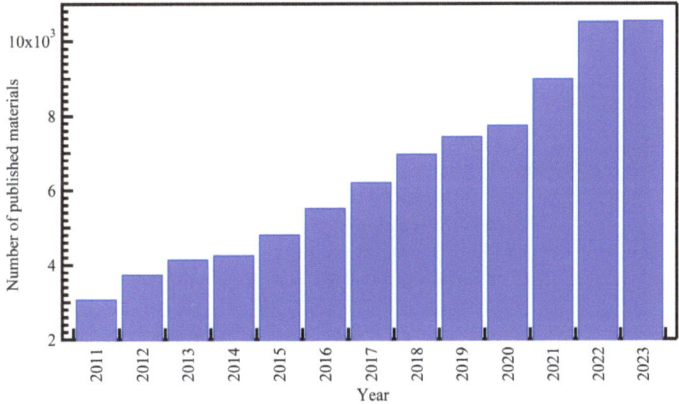

Figure 1. Distribution of the cumulative number of publications on photocatalysis per year from the Scopus database (November 2023).

2. Principles of Photocatalysis

In the last decades, the use of photocatalysis technology as a quick, affordable, and efficient way to get rid of the majority of poisonous dyes has received a lot of interest [33–35]. Photocatalysis is a field that utilizes catalysts to enhance the speed of some chemical reactions. A photocatalyst is a material capable of absorbing light, generating electron–hole pairs that enable the participants in a reaction to undergo chemical transformations [36]. The primary factor behind photocatalytic activity (PCA) is the catalyst's capability to produce electron–hole pairs, which leads to the creation of free radicals such as hydroxyl radicals (•OH). These radicals can then undergo secondary reactions. There are two primary categories of photocatalytic reactions: homogeneous photocatalysis and heterogeneous photocatalysis. The ability of any specific photocatalytic technique to work properly depends on several characteristics, including bandgap, shape, and high surface area [37].

Since dyes are colored, it is simple to track how they change color throughout the experiment. The dye sample is exposed to UV light for a predetermined amount of time in the photocatalytic process. The variations in solution color are then measured in terms of decreasing absorbance using a spectrophotometer. The degree of discoloration (τ) is determined from the dye absorption before and after irradiation by the following relation (Equation (1)) [38].

$$\tau = 1 - \frac{A_i}{A_o} * 100 \tag{1}$$

The equation includes A_o, which denotes the absorbance of the dye solution before irradiation, and A_i, which represents the absorbance of the dye solution after irradiation.

Heterogeneous photocatalysis is an environmentally friendly, low-cost method of decontaminating organic materials. Organic contaminants can be effectively reduced through heterogeneous photocatalysis in both the atmosphere and water. The mechanism of the photocatalytic degradation of organic pollutants has been previously discussed by several groups [39,40]. Usually, sunlight in the presence of a semiconductor photocatalyst is utilized to speed up the removal of environmental contaminants and the obliteration of extremely harmful compounds [41]. The process of heterogeneous photocatalysis comprises five sequential steps: (i) transferring the reactants from the liquid phase to the catalyst surface; (ii) adsorbing the reactants onto the catalyst surface; (iii) enabling the reaction to occur in the adsorbed phase; (iv) desorbing the final product; and (v) removing the final products from the liquid phase [42].

The following steps are stated as the mechanism of dye and other organic compound degradation via photocatalysis: When the catalyst is exposed to UV light, electrons move from the valence band (VB) to the conduction band (CB), resulting in the formation of an electron–hole pair (Equation (2)) [43,44].

$$\text{catalyst} + h\nu \rightarrow e^-_{CB} + h^+_{VB} \tag{2}$$

Here, e^-_{CB} and h^+_{VB}, respectively, represent the electrons in the CB and the holes in the VB.

The produced excited substances (excitons) may travel to the catalyst surface and engage in a redox reaction with other species already there. Usually, h^+_{VB} may readily form •OH radicals through the reaction with surface-bound H_2O molecules (Equation (3)), whereas e^-_{CB} can produce superoxide radical anions of oxygen through the reaction with O_2 (Equation (4)) [42].

$$H_2O + h^+_{VB} \rightarrow \bullet OH + H^+ \tag{3}$$

$$O_2 + e^-_{CB} \rightarrow O_2^{\bullet -} \tag{4}$$

The combination of the electron and the hole created in the first step is prevented by this reaction. The •OH and O_2 that are created in the following equations react with the dye to create additional species, which results in the dye discoloration.

$$O_2^- + H^+ \rightarrow \bullet OH_2 \tag{5}$$

$$H_2O_2 \rightarrow 2\bullet OH \tag{6}$$

$$\bullet OH + \text{dye} \rightarrow \text{dye}_{oxi} \tag{7}$$

$$\text{dye} + e^-_{CB} \rightarrow \text{Reductant products} \tag{8}$$

Due to the presence of dissolved oxygen and water molecules, all above reactions are possible in photocatalysis. In Figure 2, the pathways of oxidative species formation in photocatalytic research are presented schematically.

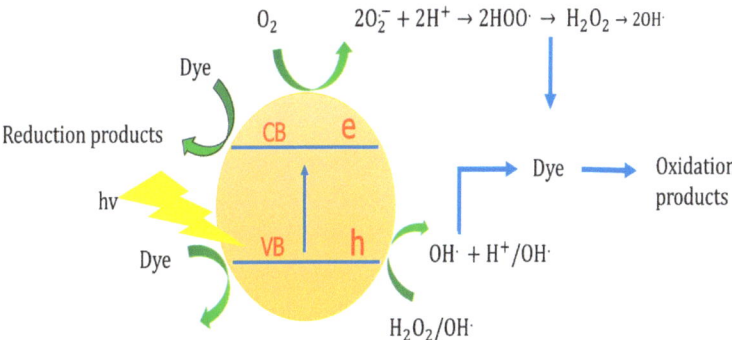

Figure 2. The mechanism of the oxidative species in heterogeneous photocatalysis. Adopted with permission from Ref. [42] Copyright 2017, Royal Society of Chemistry.

3. Mechanisms of Photocatalyzed Dye Degradation

There are two main pathways for photocatalytic reactions, as described below.

3.1. Direct Photocatalytic Pathway

There are two proposed mechanisms describing heterogeneous photocatalysis.

3.1.1. The Langmuir–Hinshelwood Model

This model is used to describe the mechanism of solid catalytic reactions [45]. As reported in the published review by Uyen N. P. Tran et al. [46], typically, the L–H mechanism model consists of four sequential steps: (1) molecules are absorbed on a catalytic surface, (2) adsorbed molecules dissociate, (3) products are produced by reactions between dissociated molecules, and (4) products are liberated to the medium. Usually, this model is used to study the heterogeneous catalytic degradation of organic pollutants in wastewater [47,48].

Based on step (1), the adsorption and desorption rate can be expressed as Equations (9) and (10).

$$r_a = k_a(1-\theta)C_P \tag{9}$$

$$r_d = k_d\theta \tag{10}$$

Here, k_a and k_d are the adsorption and desorption rate constants, θ is a fraction of the coverage site, and C_p is the concentration of pollutant "p".

At equilibrium, $r_a = r_d$, which results in Equation (11) [49].

$$\theta = \frac{kC_p}{1+kC_p} \tag{11}$$

k = the equilibrium constant, $k = \frac{k_a}{k_d}$

Usually, the photocatalytic degradation of an organic pollutant (P) occurs after adsorption; thus, the degradation rate is proportional to θ, as shown in Equation (12) [45].

$$r_{deg} = k_{deg}\theta = \frac{k_{deg}kC_P}{1+kC_P} \tag{12}$$

Here, k_{deg} is the degradation rate constant.

Based on the rate law of chemical reaction, the rate of the degradation can also be expressed by the following relation:

$$r_{deg} = -\frac{dC_P}{d_t} \tag{13}$$

By combining Equations (12) and (13), L–H kinetic model can be obtained.

$$-\frac{dC_P}{dt} = \frac{k_{deg}kC_P}{1+kC_P} \tag{14}$$

By integrating Equation (14) from $C_P = C_{P,0}$ at $t = 0$ to C_P at the interval time, t:

$$\frac{1}{k_{deg}k}\ln\frac{C_P}{C_{P,0}} + \frac{1}{k_{deg}}(C_P - C_{P,0}) = -t \tag{15}$$

$$\frac{-t}{C_P - C_{P,0}} = \frac{1}{k_{deg}} + \frac{1}{k_{deg}k}\ln\left(\frac{C_P}{C_{P,0}}\right)/(C_P - C_{P,0}) \tag{16}$$

By linearly plotting the above equation, the obtained intercept and slope are $\frac{1}{k_{deg}}$ and $\frac{1}{k_{deg}k}$, respectively.

According to previous studies, the kinetic of the photocatalytic degradation is fitted well with the L–H model. For instance, Cao et al. showed that the kinetics of the photodegradation of gaseous benzene by nitrogen-doped TiO_2 (N-TiO_2) under visible light irradiation agreed with the L–H model [50]. Another study by M. Klumpp et al. [51] confirmed that the experimental data of the degradation of rhodamine B (RhB) by TiO_2 thin film also agreed with L–H model. Furthermore, Yang et al. [52] reported that the catalytic photodegradation of norfloxacin and enrofloxacin under visible light irradiation with bismuth tungstate (Bi_2WO_6) synthesized by combining ultrasonic solvothermal treatment and high-temperature calcination adopted the L–H kinetic model with a high correlation coefficient ($R^2 > 0.95$).

3.1.2. The Eley–Rideal Model

In this method, the holes are first photo-created, then the free charged carriers are trapped by surface flaws. The dye is then chemically altered by the active centers (AS) to produce an adduct species such as (S-dye)$^+$, which can then degrade further to produce products or recombine with electrons. The subsequent reactions depict the reaction structure [34,53]:

$$\text{Catalyst} + h\nu \rightarrow e^- + h^+ \text{(photogeneration of charges)} \tag{17}$$

$$AS + h^+ \rightarrow AS^+ \text{(Trapping holes by AS)} \tag{18}$$

$$AS^+ + e^- \rightarrow AS \text{(Physical decay of AS)} \tag{19}$$

$$AS^+ + \text{dye} \rightarrow (AS - \text{dye})^+ \text{(Adduct; chemisorption)} \tag{20}$$

$$(AS - \text{dye})^+ + e^- \rightarrow AS + \text{products} \tag{21}$$

3.2. Indirect Photocatalytic Pathway

On the surface of the catalyst, electron–hole pairs are photogenerated during this process. After that, water molecules trap the holes and create H^+ and $^\bullet OH$ radicals. The $^\bullet OH$ radicals can attack the dye to produce intermediates and finished products, or they can interact with one another to form H_2O_2. The superoxide radical, which can start a chain reaction that produces HO_2 and H_2O_2, can also trap the electron by binding to molecules of oxygen.

The organic molecule is oxidized as a result of all these radicals being produced, creating intermediates and finished products [54]. The following equations serve as examples of the mechanism [54]:

$$TiO_2 + h\nu \rightarrow e^- + h^+ \tag{22}$$

$$h^+ + H_2O(\text{ads.}) \rightarrow \bullet OH(\text{ads.}) + H^+(\text{ads.}) \tag{23}$$

$$O_2 + e^- \rightarrow O_2^{-\cdot}(\text{ads.}) \quad (24)$$

$$O_2^{-\cdot}(\text{ads.}) + H^+ \leftrightarrow HO_2(\text{ads.}) \quad (25)$$

$$HO_2(\text{ads.}) \rightarrow H_2O_2(\text{ads.}) + O_2 \quad (26)$$

$$H_2O_2(\text{ads.}) \rightarrow 2HO\bullet(\text{ads.}) \quad (27)$$

$$\bullet OH + \text{dye} \rightarrow \text{intermediate} \rightarrow CO_2 + H_2O \quad (28)$$

The photodegradation efficiency and oxidation rates of a photocatalytic system are heavily dependent on various operational parameters. These parameters play a crucial role in regulating the degradation of organic molecules and include the effects of the dye concentration, catalyst dose, solution pH, light intensity, and exposure period. A comprehensive discussion of all these factors can be found in the literature [47,54].

4. Morphological Dependence of Metal Oxide Photocatalysts

Due to their capacity to produce charge carriers when stimulated with effective energy, metal oxides are generally of great interest in both environmental remediation and electronics. Numerous metal oxide nanoparticles have been utilized as photocatalysts and adhere to the fundamental principles of photocatalytic activity [28,55,56]. The first step in the photocatalytic reaction is the absorption of light, which separates charges and creates (h^+) positive holes that can oxidize substrates [57]. The metal oxide is activated by exposure to UV light, visible light, or a combination of both. Consequently, photoexcited electrons are raised from the valence band to the conduction band, forming an electron–hole pair (e^-/h^+). This generated pair (e^-/h^+) has the capability to either reduce or oxidize a compound adsorbed on the photocatalyst's surface. These excitons initiate the oxidation or reduction of substrates and reactants on the surface of photocatalysts. The photocatalytic efficacy of the metal oxide originates from two distinct mechanisms: (i) the generation of $\bullet OH$ radicals through the oxidation of OH^- anions and (ii) the production of O_2^- radicals through the reduction of O_2. Both these radicals and anions can subsequently interact with pollutants, resulting in their degradation or conversion into less harmful byproducts. These produced radicals and anions cause contaminants to degrade and change into byproducts with low risks [58]. In the following sections, the recent progress on the morphological dependence of TiO_2, ZnO, CuO, NiO, and WO_3 in the degradation of organic dyes in aqueous media will be presented.

4.1. Titanium Dioxide Photocatalysts

Since Fujishima and Honda's great discovery of water splitting in 1972 [59], great attention has been paid to utilizing the photocatalytic properties of some materials to convert solar energy into chemical energy for the oxidation or reduction of materials to produce useful materials such as hydrogen [60] and hydrocarbons [61], as well as to remove pollutants and bacteria from air, water, wall surfaces, and other environments [62,63].

TiO_2 has four polymorphs in nature, namely, tetragonal anatase, orthorhombic brookite, tetragonal rutile, and monoclinic. Two extra high-pressure forms were also produced using the rutile, TiO_2, and hollandite structures. Rutile, which is the most stable form of TiO_2, is the primary source of this compound. Conversely, anatase and brookite are metastable and transform into rutile during calcination [64]. TiO_2 is categorized as an n-type semiconductor and has varying energy bandgaps depending on its crystalline form. The energy bandgap is approximately 3.2 eV for anatase, 3.0 eV for rutile, and 3.2 eV for brookite [42]. Among the many different photocatalysts, TiO_2 has received the most attention and has been utilized in the most applications. This is due to its potent oxidizing properties [65–68], its ability to degrade organic pollutants [69], its superhydrophilicity [70], its chemical stability, long durability, nontoxicity, low cost, and transparency to visible light [62,71], as well as its good anti-corrosion performance, high mechanical strength, low density, and a competitive price.

TiO$_2$ exhibits photocatalytic properties by producing photogenerated charge carriers upon absorption of UV light which correspond to the energy band gap of TiO$_2$. These photogenerated holes in the valence band (VB) move to the TiO$_2$ surface and interact with adsorbed water molecules, resulting in the formation of hydroxyl radicals (•OH). Nearby organic molecules on the catalyst's surface are oxidized by the photogenerated holes and the •OH radicals. Meanwhile, superoxide radical anions (O$_2$•$^-$) are typically produced when molecular oxygen in the air reacts with electrons in reduction processes in the CB.

It has been reported that using pristine TiO$_2$ is not recommended because of its low electron transfer rate, light absorption, and electron–hole pair recombination. Thus, TiO$_2$ is produced in various nanostructures to improve electron transport and reduce electron–hole recombination in order to weaken these restrictions [72]. Several factors can significantly impact the effectiveness of photocatalytic systems, such as the size, specific surface area, pore volume, pore shape, crystalline phase, and exposed surface facets of the photocatalyst. Morphological factors can also affect photocatalytic performance as well as the properties of TiO$_2$ materials. Therefore, recently, there has been a lot of interest in the preparation of TiO$_2$ nano- or micro-structures with various morphologies [73,74], and a variety of TiO$_2$ nanostructural materials have been created, including spheres [74,75], nanorods [76], fibers [77], tubes [78], sheets [79], and interconnected architectures [80]. Due to their interconnected structure, three-dimensional (3D) monoliths may have high carrier mobility and be used in environmental decontamination as opposed to two-dimensional (2D) nanosheets, which have flat surfaces and good stickiness [69,81,82]. In the following section, the photocatalytic properties of TiO$_2$ with various morphologies are used for the breakdown of organic pigments.

In the last decade, Zhen et al. [83], simply treated amorphous anodic TiO$_2$ nanotubes (TiO$_2$ NTs) in situ hydrothermally at 70 °C to prepare anatase porous TiO$_2$ nanowires (TiO$_2$ NWs) (Figure 3a–f). The estimated BET surface area of the prepared TiO$_2$ NWs was 267.56 m^2 g^{-1}, nearly four times that of the utilized amorphous anodic TiO$_2$ NTs. The photocatalytic capability of porous TiO$_2$ NWs towards MB and Rhodamine 6G (Rh6G) were studied. Compared to TiO$_2$ NTs or Degussa P25, the porous TiO$_2$ NWs had superior photocatalytic activity (Figure 3g,h). The porous design and the substantial specific surface area are responsible for the increased photocatalytic activity.

Furthermore, anatase TiO$_2$ NWs were prepared by Lou et al. [84] using a facile one-pot solvothermal approach (Figure 4). Due to the high productivity and yield of the solvothermal reaction, a significant amount of what appears to be white mud is produced afterward. The choice of the DMF/HAc volumetric ratio in the solvent system used for synthesis determines whether the TiO$_2$ NWs assemble into hierarchical architectures or remain in freestanding form. The 1D nanostructure with outstanding photocatalytic activity for RhB degradation was perfectly preserved in both the synthesized and the annealed TiO$_2$ NWs.

Recent studies on the morphology of TiO$_2$ nanostructures in the photocatalytic degradation of several organic dyes (MB, MV, MO) have been published by Zhang et al. [85]. The balanced angle deposition technique (GLAD) was used to create nanorod, nanohelic, and nanozigzag TiO$_2$ nanofilms (Figure 5a). It can be observed that the morphology displayed significantly influences the performance of the photocatalytic degradation under UV–Vis light irradiation (Figure 5b,c). This is mostly explained by the variation in the produced nanostructures' specific surface area and pore volume. TiO$_2$ nanozigzag films demonstrate superior degradation capabilities compared to nanohelics and nanorods due to their extensive surface area, increased porosity, distribution of active sites at varying pore lengths, and the existence of oxygen vacancies.

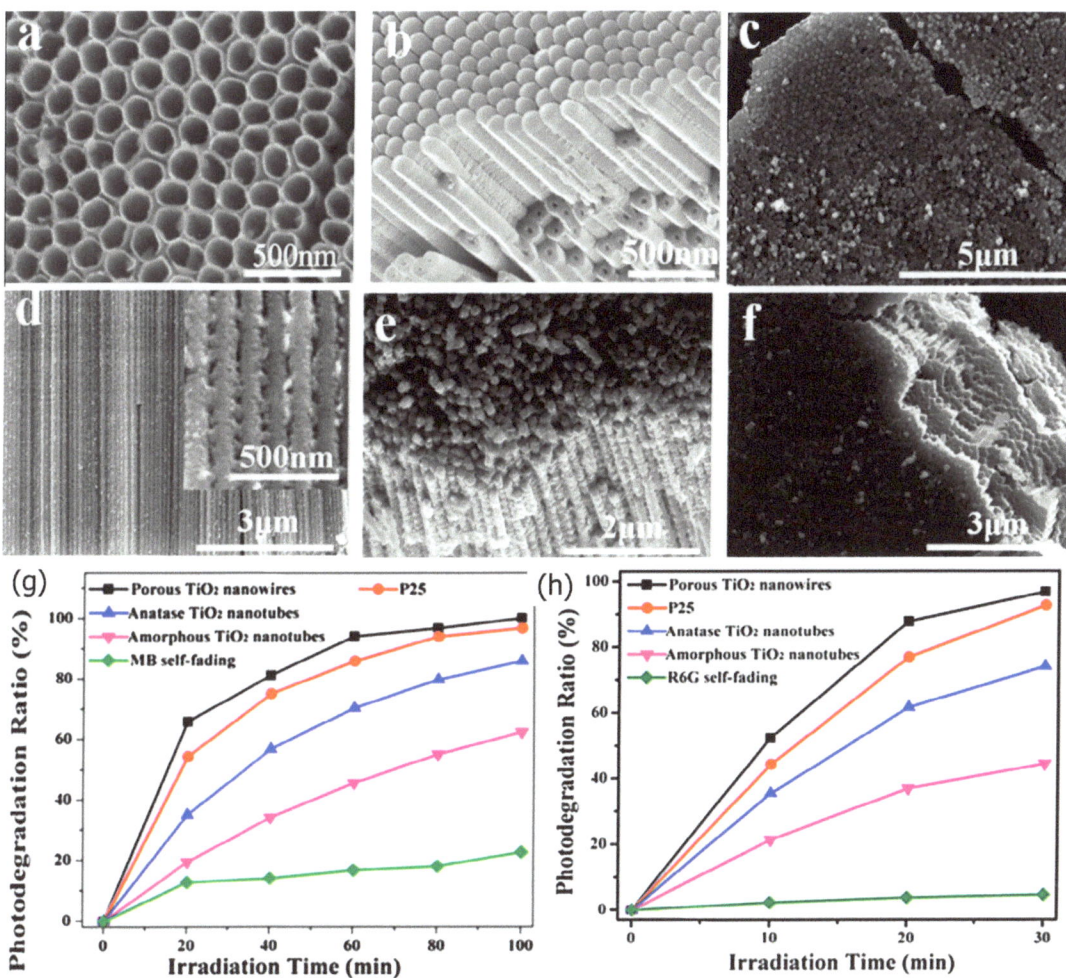

Figure 3. (**a**) SEM image of the anodic TiO$_2$ NTs, (**b**) cross-section and surface view image of the bottom side of the anodic TiO$_2$ NTs, (**c**) the porous TiO$_2$ NWs, (**d**) cross-section image of the porous TiO$_2$ NWs, (**e**) high and (**f**) low magnification cross-sections and surface views of the porous TiO$_2$ NWs. The photocatalytic degradation performance of (**g**) MB and (**h**) R6G with time of irradiation. Adopted with permission from Ref. [83] Copyright 2013, Elsevier.

By carefully hydrolyzing titanium tetrachloride (TiCl$_4$) with a homemade glass apparatus, Zidki et al. [86] prepared TiO$_2$ nanoparticles. The photoactive anatase and brookite phases make up the TiO$_2$ nanoparticles, as indicated by XRD and TEM/HRTEM image studies (Figure 6a,b). The photodegradation of MB and CR under UV–Vis light was used to illustrate the photocatalytic activity of TiO$_2$ nanoparticles. The degradation of MB showed that, in contrast to the CR degradation, the photocatalytic degradation of MB on TiO$_2$ nanoparticles is more effective in an alkaline environment (Figure 6c,d).

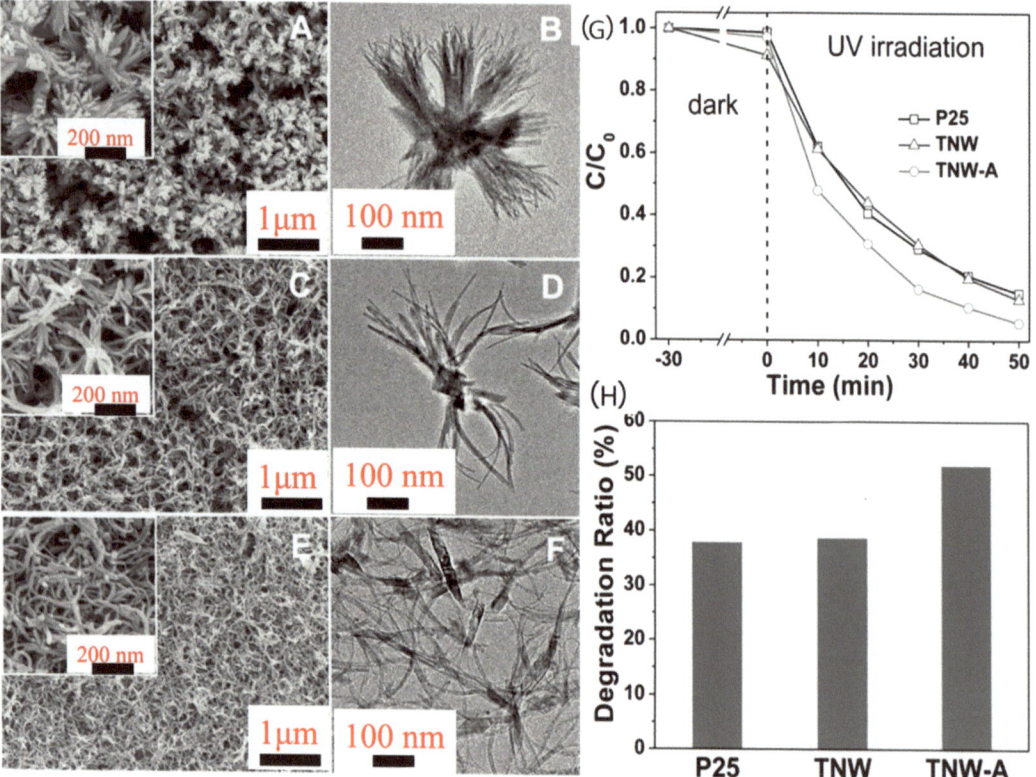

Figure 4. Morphological analysis (SEM and TEM images) of TiO_2 NWs prepared with different volumetric ratios of DMF/HAc: (**A,B**) 4/6, (**C,D**) 5/5, and (**E,F**) 6/4. (**G**) Photocatalytic degradation of RhB and (**H**) degradation ratio after 10 min irradiation over TNW, TNW − A, and P25. Adopted with permission from Ref. [84] Copyright 2012, Wiley.

Habibi and Jamshidi [87] conducted a study in which they synthesized TiO_2 in various shapes using cellulose nanofibers (CNFs) as a template and a sol-gel technique. Three different forms of TiO_2 were produced: hydrogel, aerogel, and alcogel (as shown in Figure 7). The hydrogel-produced TiO_2 nanofibers were porous and strongly entangled, while the aerogel-produced TiO_2 had a sheet-like structure. The alcogel-produced TiO_2 had a structure similar to a hydrogel, but water was replaced with isopropanol. By using the Stober method with ammonia, the sol-gel process was modified by the researchers to produce TiO_2 nanowhiskers and nanosheets. The morphology of the nanowhiskers transformed into nanosheets as the ammonia level was raised, as confirmed by FESEM images. Moreover, the specific surface area of the TiO_2 samples was found to have increased. The photocatalytic efficiency of the prepared samples was examined using methylene blue (MB) as a model pollutant. All samples exhibited high photocatalytic efficiency under UV light, with over 98% degradation of MB in 2 h. Additionally, TiO_2 nanowhiskers displayed higher photodegradation efficiency under visible light compared to TiO_2 NPs and nanosheets.

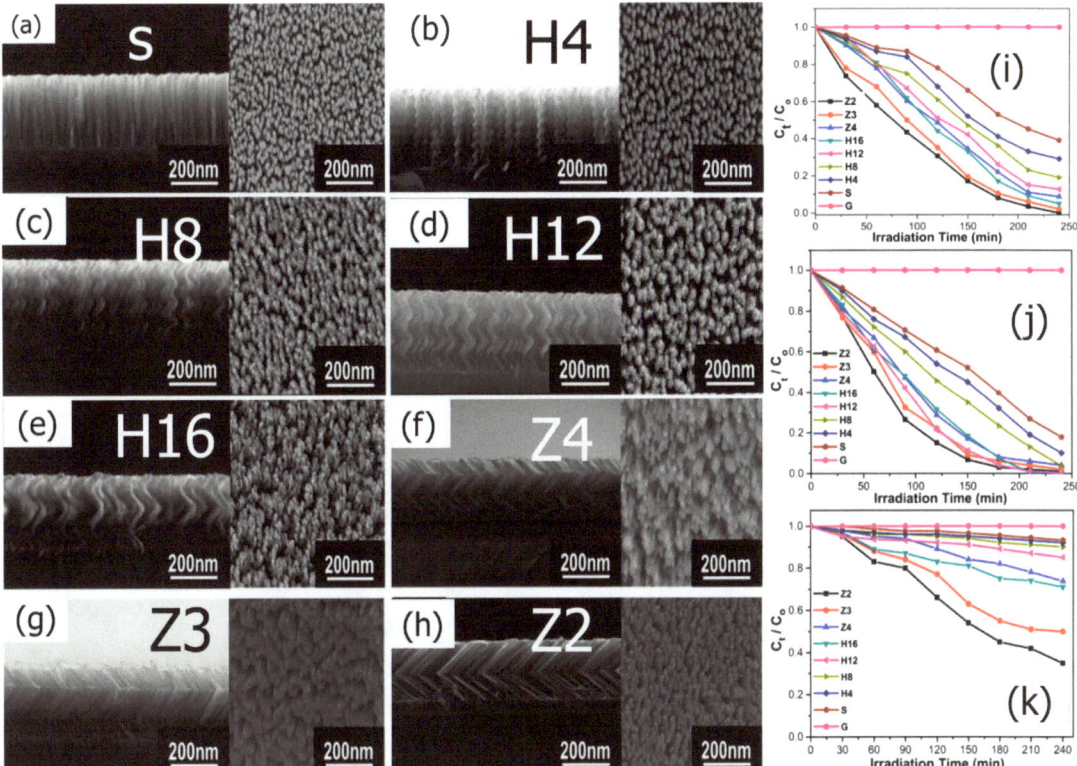

Figure 5. (**a**) SEM images (cross-sectional and top view) of TiO$_2$ nanostructures (**a**) Standing nanorod (**b**) Nanohelix-H4, (**c**) Nanohelix-H8, (**d**) Nanohelix-H12, (**e**) Nanohelix-H16, (**f**) Zigzag-Z4, (**g**) Zigzag-Z3, and (**h**) Zigzag-Z2. Pseudo-first-order kinetics (after exposure to UV–Vis light for 240 min for decolorization) of (**i**) MB, (**j**) MV, and (**k**) MO. Adopted with permission from Ref. [85] Copyright 2017, Elsevier.

TiO$_2$ NTs have received a lot of interest in recent years because of their one-dimensional ion exchange direction, increased surface area, and light absorption [88]. TiO$_2$ NTs can typically be made using one of three methods: hydrothermal, template, or electrochemical anodization. The most effective of these techniques, electrochemical anodization, is the least expensive and produces remarkably ordered nanotubes [89]. Additionally, the anodization technique stands out for its ability to control the morphology of nanotubes by simply changing the process parameters. In contrast to nanotubes grown over foils, Subramanian et al. [90] found that anodized TiO$_2$ NTs over titanium wires (TWs) significantly improved the photocatalytic degradation of MO.

In the presence of nanotubes grown on titanium wires, photocatalytic degradation rises from 20% to about 40%. Additionally, MO degradation in the presence of Pt-loaded TiO$_2$ nanotubes over foils matches the degradation in the presence of TiO$_2$ nanotubes on wires. This increased photoactivity is attributed to nanotubes formed in a radially outward orientation along a titanium backbone, which effectively absorb light that is reflected and refracted. Further enhancement was achieved by Rojviroon et al. [78]. In this study, electrochemical anodization was used to create TiO$_2$ nanotubes using thin titanium sheets at voltages of 20, 30, 40, and 50 V. The obtained TiO$_2$ nanotubes are shown in Figure 8a. The characterization investigation revealed that the inner diameter and depth of TiO$_2$ NTs rose with increasing anodization voltages, but their wall thickness dropped. It seems that the TiO$_2$ NTs anodized at 50 V have shown promising photocatalytic activities towards the decolorization of indigo carmine (IC) and reactive black 5 (RB5) dyes. The decolorization efficiencies were measured to be 74.14% and 65.71%, respectively, under UV irradiation for 180 min and with an initial dye concentration of 4 µM. These results suggest that TiO$_2$ NTs anodized at 50 V can be a potential photocatalytic material for dye wastewater treatment.

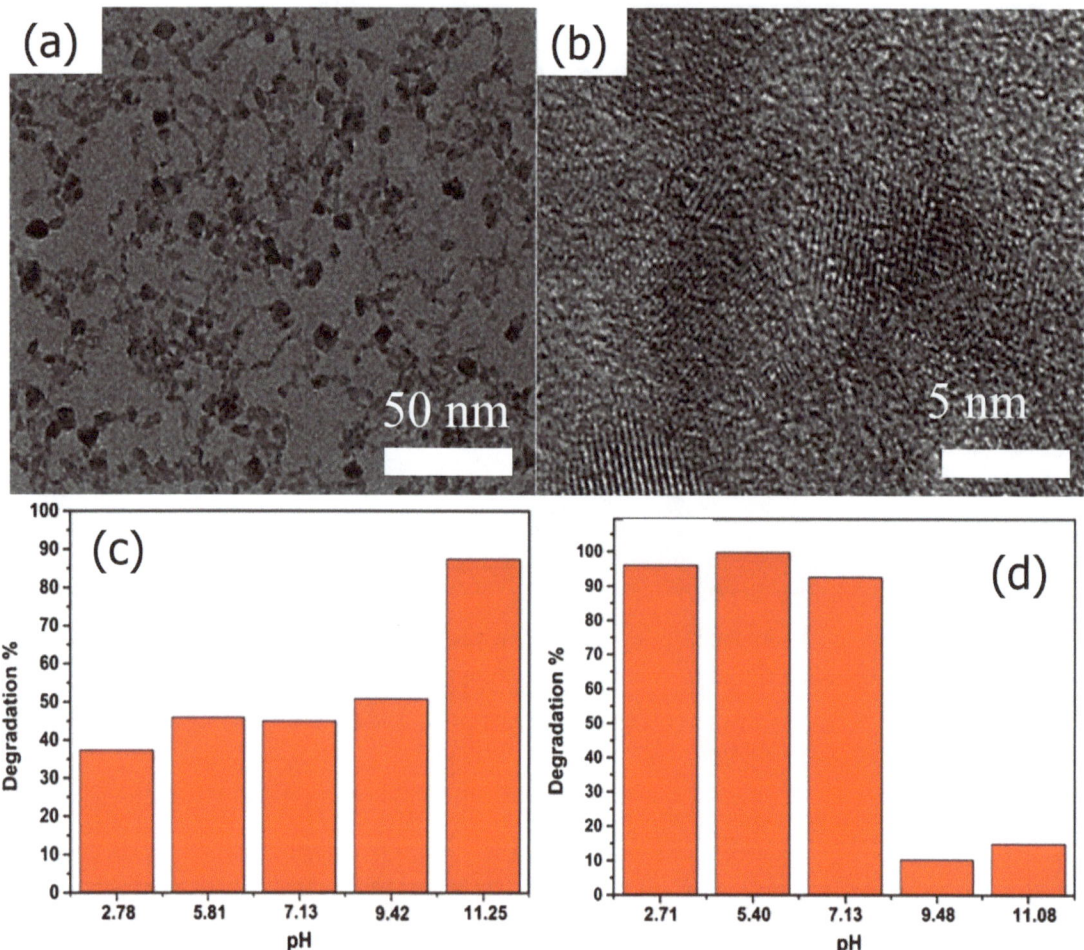

Figure 6. (**a**) TEM and (**b**) HR-TEM images of the prepared TiO$_2$ nanoparticles. Effect of pH on photodegradation of (**c**) MB and (**d**) CR dyes after under UV–Vis light illumination. Adopted with permission from Ref. [86] Copyright 2020, Elsevier.

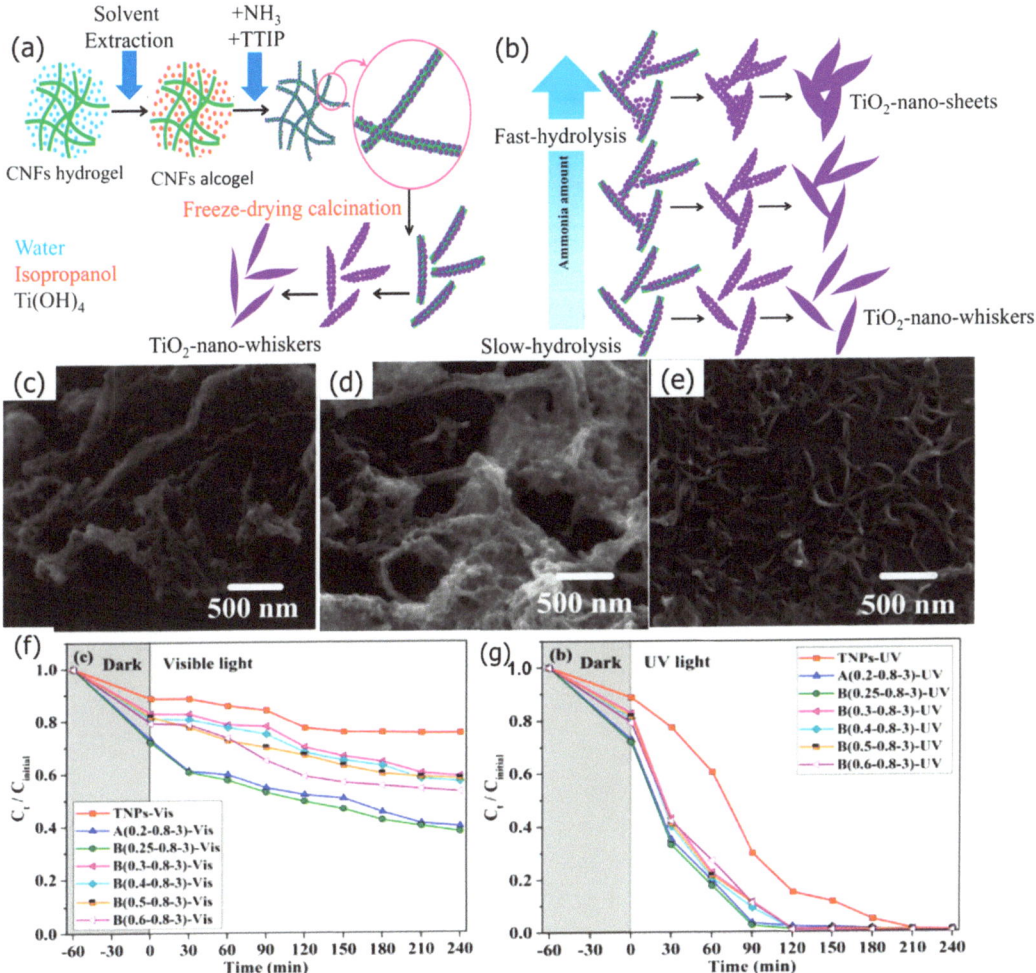

Figure 7. (**a**) Synthesis process of alcogel—TiO$_2$, (**b**) the shape evolution of alcogel—TiO$_2$ with an increasing ammonia amount. FESEM images of (**c**) hydrogel—TiO$_2$, (**d**) aerogel—TiO$_2$, (**e**) alcogel—TiO$_2$. Concentration changes in MB aqueous solution under (**f**) UV—and (**g**) visible light. Adopted with permission from Ref. [87] Copyright 2020, Elsevier.

Nozaki et al. [91] prepared TiO$_2$ nanosheet (TiO$_2$ NSs) photocatalysts for the degradation of MB when exposed to UV radiation (365 nm, 2.5 mW cm^{-2}). Ti(OBu)$_4$ and (NH$_4$)$_2$TiF$_6$ were both employed as starting materials, and a variety of samples with various side lengths were produced through hydrothermal synthesis by varying the F/Ti ratio between 0.3 and 2.0. Titania nanosheets produced with an F/Ti ratio of 0.3 led to the best degrading efficiency. The increase in surface area brought on by the reduction in size is what is responsible for the increased catalytic activity.

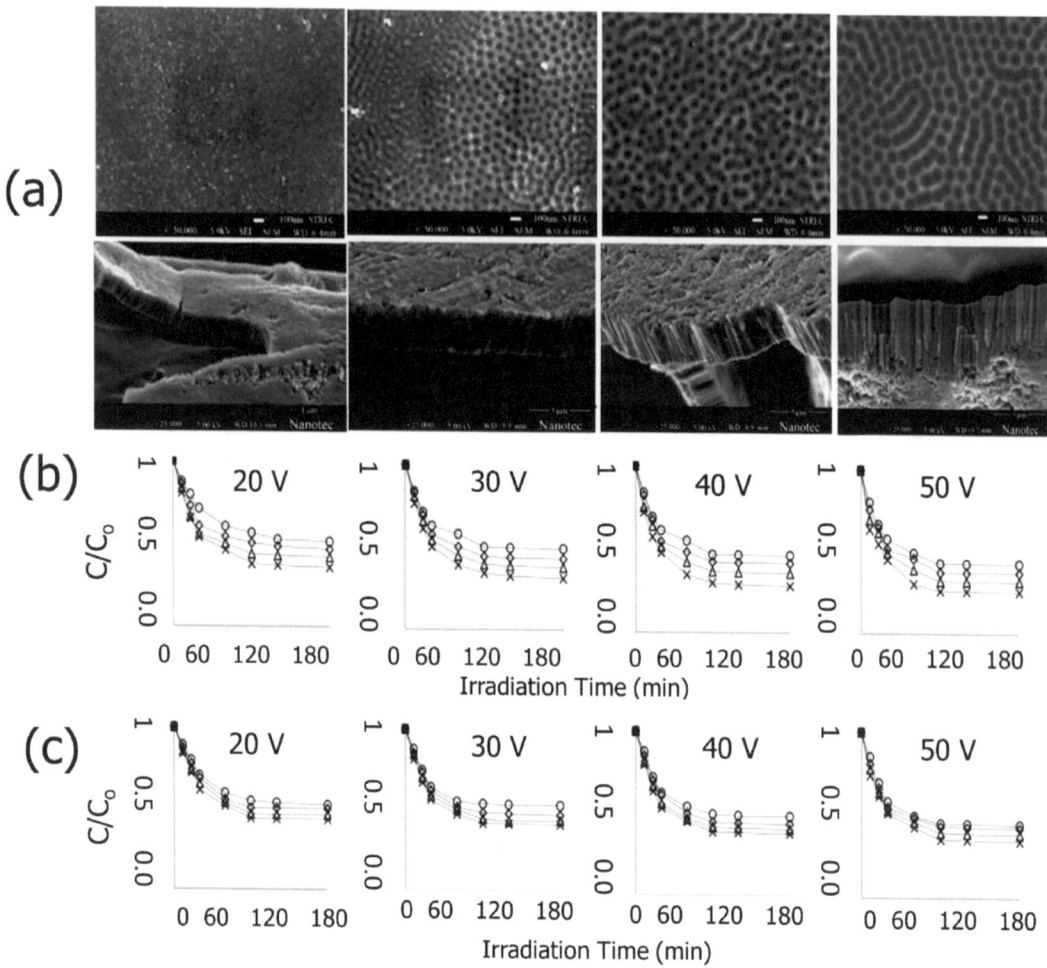

Figure 8. (a) FE-SEM images of TiO$_2$ NTs anodized at 20, 30, 40, and 50 V. Photocatalytic decolorization using TiO$_2$ NTs sheets anodized at 20–50 V for various initial concentrations of (b) IC and (c) RB5. Adopted with permission from Ref. [78] Copyright 2021, Elsevier.

Recently, Nair developed a floating photocatalyst to accelerate the photocatalytic degradation of a TiO$_2$ nanosheet in a cellulose acetate matrix and supplied support using ethylene vinyl acetate (Figure 9a–c) [92]. The degradation of CR dye by a floating photocatalyst made of TiO$_2$ nanosheets performed well in both UV and solar light (Figure 9d,e).

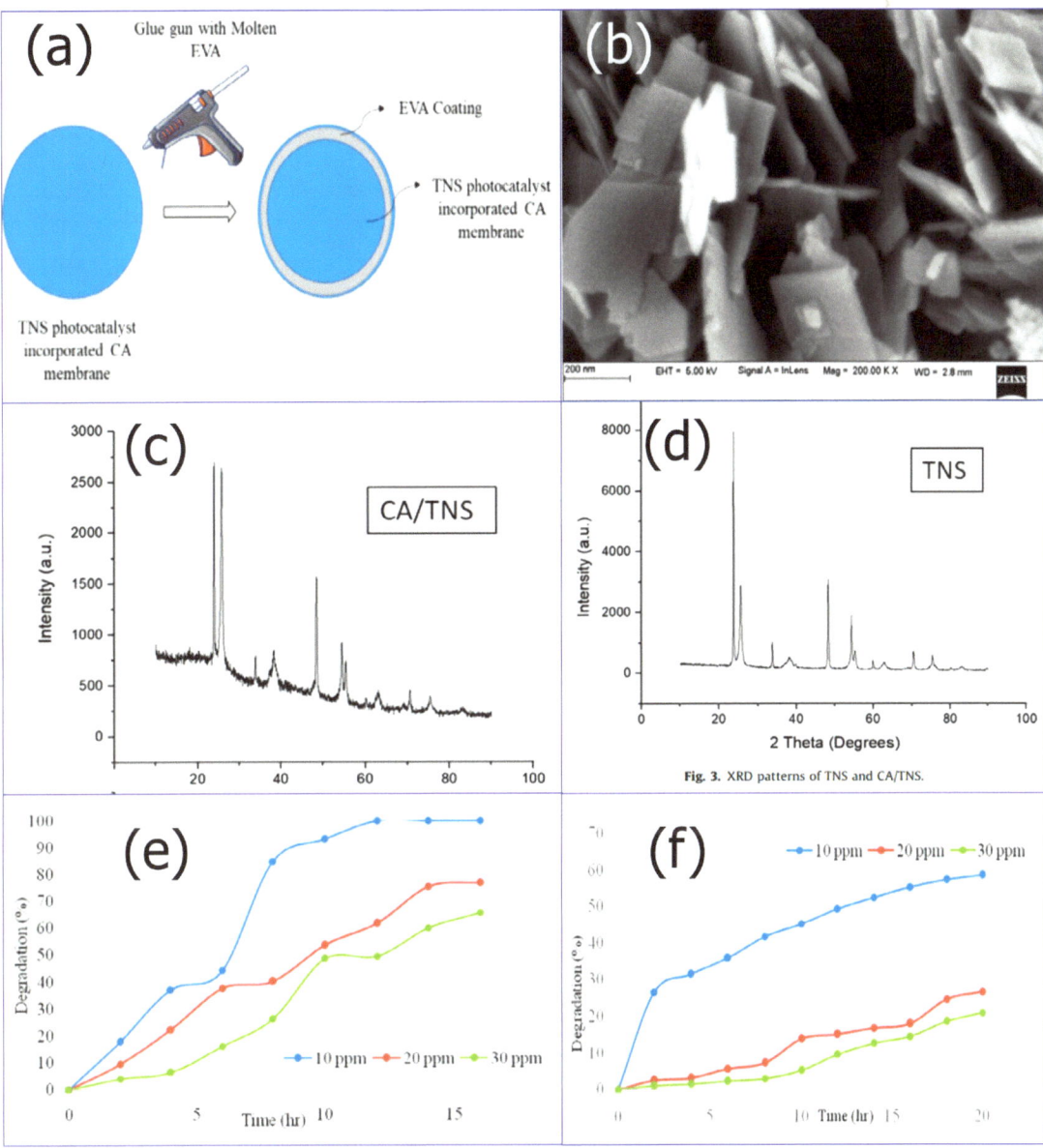

Figure 9. (**a**) Schematic representation of the synthesis of a floating photocatalyst, (**b**) FESEM image of TiO_2 NSs. XRD of TiO_2 NSs (**c**) and CA/TiO_2 NSs (**d**), (**e**) Solar degradation study of CR dye using a floating photocatalyst, and (**f**) UV degradation of CR dye using a floating photocatalyst. Adopted with permission from Ref. [92] Copyright 2021, Elsevier.

4.2. Zinc Oxide Photocatalysts

ZnO is commonly found in a hexagonal wurtzite crystal structure, which has a non-centrosymmetric lattice and exhibits piezoelectric and pyroelectric properties. The crystal structure also affects the photocatalytic activity of ZnO, with the (001) crystal facet showing the highest photocatalytic activity due to its high surface energy and abundance of surface defects [93,94]. The photocatalytic activity of ZnO is attributed to the formation of electron–hole pairs upon the absorption of light with energy equal to or greater than its bandgap energy. The photogenerated holes can oxidize water or hydroxide ions to produce hydroxyl radicals, while the photogenerated electrons can reduce oxygen or organic molecules [95]. However, the recombination of electron–hole pairs limits the efficiency of photocatalytic reactions, and various strategies have been employed to enhance the separation of photogenerated charge carriers and improve the photocatalytic activity of ZnO, including doping with transition metals, surface modification with noble metals or semiconductors, and use of a heterostructure with other semiconductors [95,96].

Until now, various morphologies of ZnO nanomaterials such as nanorods, nanowires, nanotubes, and hollow structures have been prepared [97]. ZnO nanomaterials are usually prepared via different techniques including evaporative decomposition of solution [98], solid state reaction [99], sol-gel [100], and so on. The preparation technique shows a noticeable effect on the particle size and the morphology, which directly affects the physical and chemical properties of the prepared ZnO nanostructured materials. As reported by Singh et al., the morphology of ZnO nanostructures can be controlled by adding additives or capping agents such as triethanolamine (TEA), oleic acid, and thioglycerol [101]. The creation of ZnO nanostructures with different morphologies requires the application of many surfactants, including sodium dodecyl sulfate (SDS), tetraethylammonium bromide (TEAB), and cetyltrimethylammonium bromide (CTAB). The photocatalytic activity of ZnO is significantly influenced by its structural morphology [102,103]. This influence stems from the crystal structure, which facilitates the separation of charge carriers. However, some studies have shown that non-spherical morphologies such as nanorods and nanowires can also exhibit enhanced photocatalytic performance due to their unique crystal facets and surface areas, which can promote charge transfer and improve catalytic efficiency. Ultimately, the selection of a particular morphology for a photocatalyst relies on the specific application and the desired properties. The flake-like structure, however, exhibits comparatively lower performance [102].

Although a great deal of attention has been devoted to the photocatalytic activity of ZnO, its photocatalytic activity suffers from the following drawbacks: (i) the large bandgap of ZnO (3.37 eV), which inhibits light absorption in the UV (380 nm) range, which means ZnO cannot absorb visible light, drastically lowering photocatalytic efficiency [104,105]; (ii) the degradation reactions at the semiconductor–liquid interface are slowed down by the quick recombination of the charge carriers; (iii) the difficulties of using a traditional filtration process to recover ZnO powder from a suspension; (iv) the propensity to clump together during catalytic processes and the vulnerability to UV-induced corrosion; and (v) photocorrosion is one of the main restrictions of ZnO as a photocatalyst for wastewater treatment [106]. The following four phases describe how photocorrosion occurs [107].

$$O^{2-}_{surface} + h^+_{VB} \rightarrow O^-_{surface} \tag{29}$$

$$O^-_{surface} + 3O^{2-} + 3h^+_{VB} \rightarrow 2\left(O-O^{2-}\right) \tag{30}$$

$$\left(O-O^{2-}\right) + 2h^+_{VB} \rightarrow O_2 \tag{31}$$

$$2Zn^{2+} \rightarrow 2Zn^{2+}aq \tag{32}$$

The net equation for the photocorrosion of ZnO is shown below.

$$ZnO + 2h^+_{VB} \rightarrow Zn^{2+} + \frac{1}{2}O_2 \tag{33}$$

Therefore, the majority of ZnO photocatalytic studies have been carried out under UV radiation. Recently, various studies have been conducted on utilizing ZnO with different morphologies for the elimination of organic pollutants. To use as a photodegradation for

MO dye, Bhatia and Verma [108] synthesized ZnO nanoparticles (ZnO NPs) with varying defect concentrations by burning at 700 °C and subsequent quenching in air.

From FESEM images (Figure 10a–d), ZnO NPs are less than 45 nm in size, and an uneven grain boundary distribution was discovered in the quenching-induced defects. The improvement of the photocatalytic effectiveness of synthesized nanoparticles is greatly influenced by their various morphologies. The photodegradation of MO dye has been proven to follow a first order model through kinetic analysis, with a rate constant of 0.0165 min^{-1}. (Figure 10f). Additionally,, it has been reported that approximately 100% MO degradation has been achieved in 150 min under UV irradiation (Figure 10e). Recently Dodoo-Arhin prepared ZnO NPs using the sol-gel method, with zinc acetate as a Zn source [109]. The effect of calcination temperature on the particle size of ZnO was investigated. As shown in SEM, the obtained wurtzite ZnO NPs showed crystallite sizes ranging from 16 nm to 30 nm. Increasing the calcination temperature increases the crystallite size and reduces the energy band gap of the obtained ZnO NPs. In addition, the prepared ZnO showed a rice-like microstructure morphology (Figure 10g–i). The prepared ZnO NPs were utilized as a photocatalyst toward the degradation of RhB dye under UV– light. The obtained analysis showed that ZnO NPs calcined at 400 °C obtained the highest degradation efficiency (95.41%), as shown in Figure 10h.

ZnO NPs also showed important catalytic activity under solar illumination. For example, ZnO NPs of various morphologies were developed by Saikia et al. [110] for the photodegradation of malachite green (MG) (Figure 11a–d). Under solar light, the flower-shaped homocentric bundles of hydrothermally synthesized ZnO nanorods (ZnO NRs) in the form of pencil-like structures demonstrate outstanding photocatalytic activity, as evidenced by the pseudo first-order kinetics of the Langmuir–Hinshelwood model in the photodegradation of MG dye (Figure 11e,f). Additionally, ZnO NPs and homocentric pencil-like ZNRs bundles in the shape of flowers were prepared.

Additionally, photocatalytic degradation of organic dyes by ZnO nanosheets (ZnO NSs) was also investigated. For example, Komarneni et al. [111] used a simple, ultra-rapid solution approach to create ZnO NSs with many oxygen-vacancy defects. The surface area was significantly increased from 6.7 to 34.5 m^2 g^{-1} by the addition of 1 mol L^{-1} Na$_2$SO$_4$. With a rate constant of 0.0179 min^{-1} under visible light (>420 nm), the as-prepared ZnO NSs showed abundant oxygen vacancies, which are crucial for improving visible light absorption and, consequently, high photocatalytic activities towards the degradation of RhB. These rates were about 13 and 11 times higher, respectively, than those of ZnO NPs with few oxygen defects.

Further enhancement with a rate constant k = 0.0421 min^{-1} was achieved by hybridized ZnO NSs with Ag$_3$PO$_4$ nanoparticles. A synergistic effect of surface oxygen vacancies and Ag$_3$PO$_4$ coupling was suggested by this augmentation, which was attributed to the improved visible light absorption as well as the well-matched energy level that is responsible for effective charge transfer between oxygen-vacancy-rich ZnO NSs and Ag$_3$PO$_4$. ZnO nanowires also display a significant superior photocatalytic activity toward organic dye degradation. For example, by using a low-cost, low-temperature hydrothermal approach, Leprince-Wang et al. [112] demonstrated an effective synthesis of nontoxic, biocompatible ZnO nanostructures only on the surface of commercially available concrete and tiling pavements (Figure 12a–e).

The obtained data showed an enhancement in photocatalytic activity for degrading organic dyes in aqueous media with high photocatalytic stability (Figure 12f–h).

The degradation of Acid Red 57 (AR57) under UV irradiation was studied further by El-Bindary et al. [113]. In this study, ZnO nanowires (ZnO NWs) were prepared by a low-temperature co-precipitation technique employing zinc sulfate as a precursor. After 190 min, the effectiveness of the photocatalytic degradation of ZnO NWs produced at 400, 500, and 600 °C was 90.03, 77.67, and 72.71%, respectively. Moreover, the degradation of AR57 fitted first-order kinetics. Recently, the effect of several kinds of organic dyes as a guiding agent for the formation of ZnO NWs was described by Yang et al. [114].

Figure 10. (a) Low and high magnification SEM images of ZnO, (b) air quenched ZnO, (c) photodegradation vs. irradiation time of MO dye for pristine ZnO and air quenched ZnO, (d) rate constant for after 150 min of UV exposure (Adopted with permission from Ref. [108] Copyright 2017, Elsevier), SEM images of ZnO calcined at (e) 400 °C, (f) 500 °C, and (g) 600 °C. (h) RhB degradation vs. irradiation time and (i) the effect of the catalyst load on RhB degradation (Adopted with permission from Ref. [109] Copyright 2019, Elsevier).

Figure 11. FE-SEM images of (**a**) ZnO NPs, (**b**) ZnO NPs (hydrothermal), (**c**) ZnO flowers, (**d**) one ZnO flower, (**e**) comparative photodegradation of MG by ZnO NPs, and (**f**) the estimated rate constant for photodegradation of MG by ZnO NPs under solar light. Adopted with permission from Ref. [103] Copyright 2014, Elsevier.

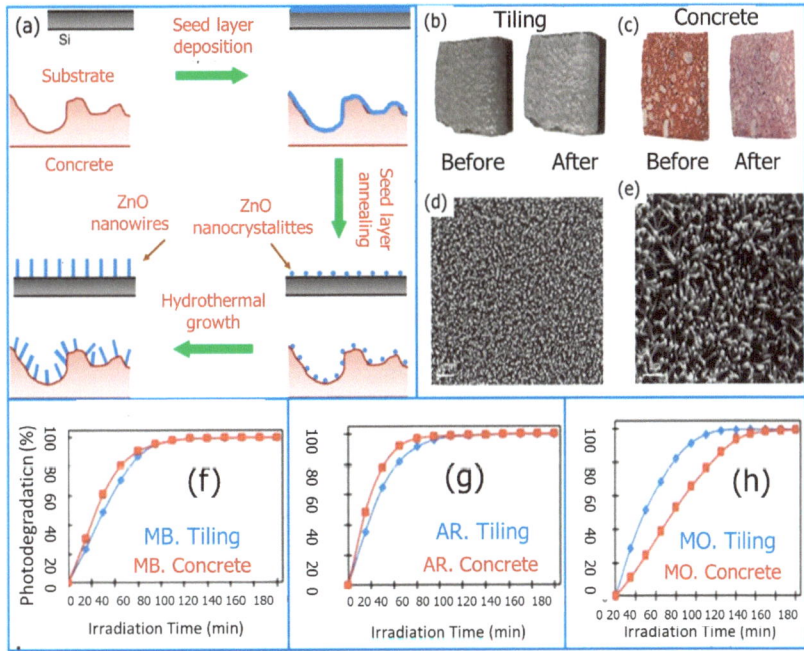

Figure 12. (**a**) Scheme of the ZnO NWs growth on a flat substrate (i.e., Si wafer) vs. a construction material (i.e., concrete), (**b**) top views of the gray tiling, and (**c**) red concrete pavement before and after the growth of ZnO nanostructures. (**d**) SEM top-view of ZnO NWs on the tiling surface, (**e**) ZnO NRs on the concrete surface, and (**f–h**) the photodegradation rate of MB, AR, and MO dyes vs. time. Adopted with permission from Ref. [112] Copyright 2019, Springer Nature.

EBT was found to effectively direct the growth of ZnO NWs, inhibiting their growth along the c-axis direction. Additionally, the introduction of UV light during growth significantly enhanced the photocatalytic degradation of EBT absorbed on the surface of the ZnO NWs. ZnO NWs arrays grown on glass fiber cloth with a dominant exposed polar (0001) facet exhibited superior photocatalytic performance compared to other arrays.

4.3. Copper Oxide Photocatalysts

Copper (II) oxide, also known as cupric oxide (CuO), is a naturally occurring p-type semiconducting metal oxide. This is due to the presence of oxygen vacancy defects in its crystal structure. CuO has an indirect small bandgap of 1.2 eV at room temperature [115]. Copper also exists in other polymorphs such as copper (I) oxide, commonly known as cuprous oxide (Cu_2O), and copper (III) oxide (Cu_4O_3). CuO displays higher thermal stability than Cu_2O, whereas Cu_4O_3 is a challenging metastable phase due to its mixed copper atom oxidation state, making its synthesis problematic [116,117]. CuO is a brownish-black powder that is used in various applications, such as catalysis, chemical and gas sensors, superconductors, and energy conversion and storage devices, as well as in biomedicine and textile production [118–120]. CuO has been extensively used as a heterogeneous catalyst in diverse chemical processes such as the oxidation of carbon monoxide (CO), hydrocarbons, and phenol in supercritical water, the selective catalytic reduction of nitric oxide with ammonia, and the breakdown of nitrous oxide [121].

Furthermore, CuO has been investigated as a photocatalytic material [122]. Additionally, it has been reported how the bandgap of CuO nanostructures is influenced by their morphology [123]. The subsequent section will demonstrate the effectiveness of CuO nanostructures as a photocatalyst for eliminating organic dyes from aqueous solutions, including an analysis of the removal mechanism and the impact of morphology on the removal efficiency.

CuO nanostructures are among the interesting photocatalytic materials usually utilized for the removal of various organic pollutants due to their high abundance, low cost, narrower bandgap, excellent chemical stability, and facile synthesis [124]. The narrow bandgap of CuO makes it active in the visible region of the electromagnetic spectrum. To enhance the catalytic activity of CuO, H_2O_2 is often added to the reaction mixture. H_2O_2 is a better electron acceptor than O_2, so it quickly captures the photogenerated electrons from the photocatalyst's surface, becoming reduced and forming hydroxyl radicals (•OH) [125]. This is important for offering more active radicals, as well as reducing the rate of electron–hole recombination, which boosts the utilization of holes during the photocatalytic process. Without H_2O_2, CuO displays an inability to generate a sufficient amount of •OH radicals, and thus CuO is considered as an ineffective photocatalyst for degrading organic pollutants [126]. This is because the redox potential required to produce •OH radicals is higher than the VBs of CuO. Thus, CuO has weaker oxidative capabilities for the breakdown of organic contaminants and cannot produce hydroxyl radicals when illuminated. There are many works that indicate that without H_2O_2, CuO in different morphologies displays no catalytic activity, as reported by He et al. [127]. Recently Latief et al. prepared CuO NPs with a range of sizes between 25 and 90 nm (Figure 13a,b) to deprecate CR dye from aqueous solutions. The analysis showed that the addition of H_2O_2 to CuO NPs significantly enhanced the degradation of CR dye under UV light, as presented in Figure 13c. In addition, increases in H_2O_2 result in improvements in the degradation rate (Figure 13d).

CuO nanostructures with various morphologies, including flower-like, boat-like, plate-like, and ellipsoid-like structures, were prepared and showed excellent catalytic activity towards the degradation of MB. However, without H_2O_2, the degradation stopped after 15 h, highlighting the significant dependence of CuO nanostructure photocatalytic activity on H_2O_2. Adsorption–oxidation–desorption is the postulated possible mechanism for the photocatalytic degradation of dyes by CuO in the presence of H_2O_2 under light illumination [127]. In this mechanism, various free radicals such as HO·, HOO·, and O_2^- are mainly responsible for the degradation of dyes. The first step is adsorption of dye and H_2O_2 molecules on the surface of the CuO nanostructure. The second step involves the decomposition of H_2O_2 into free radicals (HO·, HOO·, or O_2^-). These free radicals display a high oxidative ability to oxidize the organic dye. The third step involves desorption of the small molecules from the CuO surface, and finally the catalyst is recovered [128].

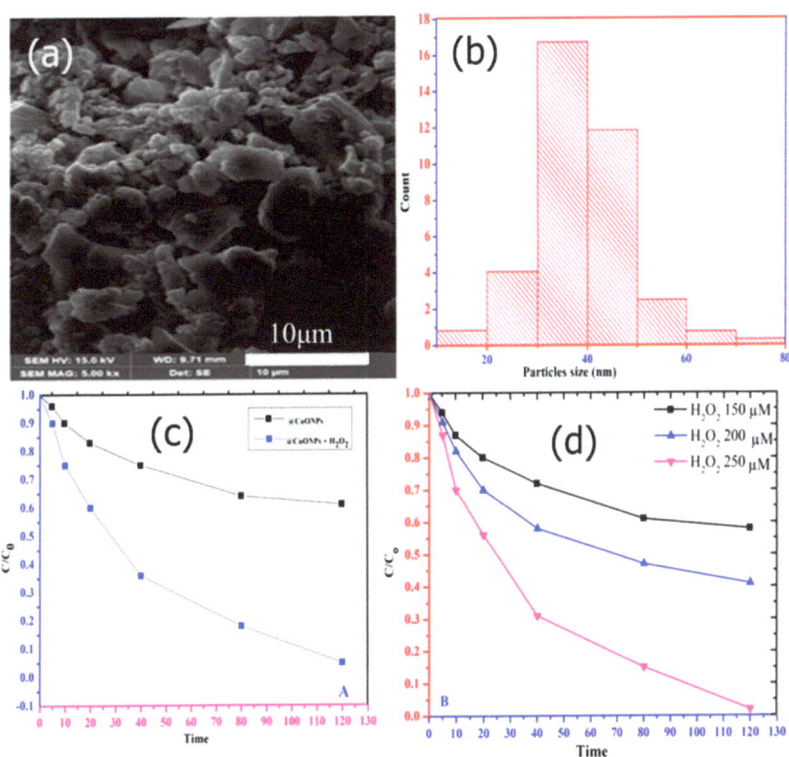

Figure 13. (a) SEM image of CuO NPs, (b) histogram spectrum of CuO NPs, (c) effect and removal of CR dye by UV/H_2O_2, and (d) effect of H_2O_2 concentration in the degradation of CR dyes (Adopted with permission from Ref. [125] Copyright 2023, MDPI).

Morphology is a critical factor that affects the photocatalytic activity of CuO nanostructures toward the degradation of organic dyes. For example, Wang et al. reported that CuO nanowires (Figure 14a,b) showed excellent catalytic activity toward the degradation of RhB [129]. The CuO nanowires demonstrated a total degradation efficiency of 95.5% after 9 h of UV light irradiation (as shown in Figure 14c), which was significantly higher than that of commercial CuO powders (which achieved only 39.6% degradation). In addition, Sadollahkhani et al. investigated the photocatalytic performance of CuO nanoparticles with various morphologies (as depicted in Figure 14d–f) for the degradation of CR dye under UV illumination [130]. The decomposition of CR dye takes place on the surface of CuO; thus, adsorption plays a critical role in the photocatalytic degradation. Samples doped with Zn appear to have improved degradation performance (63%) according to the photocatalytic examination of the degradation of MB dye (Figure 14g). The photocatalytic studies revealed that the degradation of CR for nanorods was the highest among the other prepared materials, with a total degradation efficiency of 67% after 210 min irradiation (Figure 14h). Moreover, the CR degradation reaction follows a first order kinetics model with the three CuO morphologies. On the other hand, Anandan et al. [131] prepared dandelion-like CuO microspheres with sizes of ~1–2 μm through the ultrasound process without any surfactants. The photocatalytic performance of the CuO nanoparticles that were synthesized was investigated for the degradation of Reactive Black-5. The study found that the CuO microspheres with a dandelion-like morphology exhibited excellent photocatalytic activity, with approximately 76% of the dye degraded in just 5 h under visible light exposure. The degradation reaction followed a pseudo-first-order kinetic model, with a rate constant of 0.312 h^{-1}. Recently, George et al. prepared flowers similar to CuO 3D nanostructures doped with nickel, zinc, and iron (Figure 14h) [132].

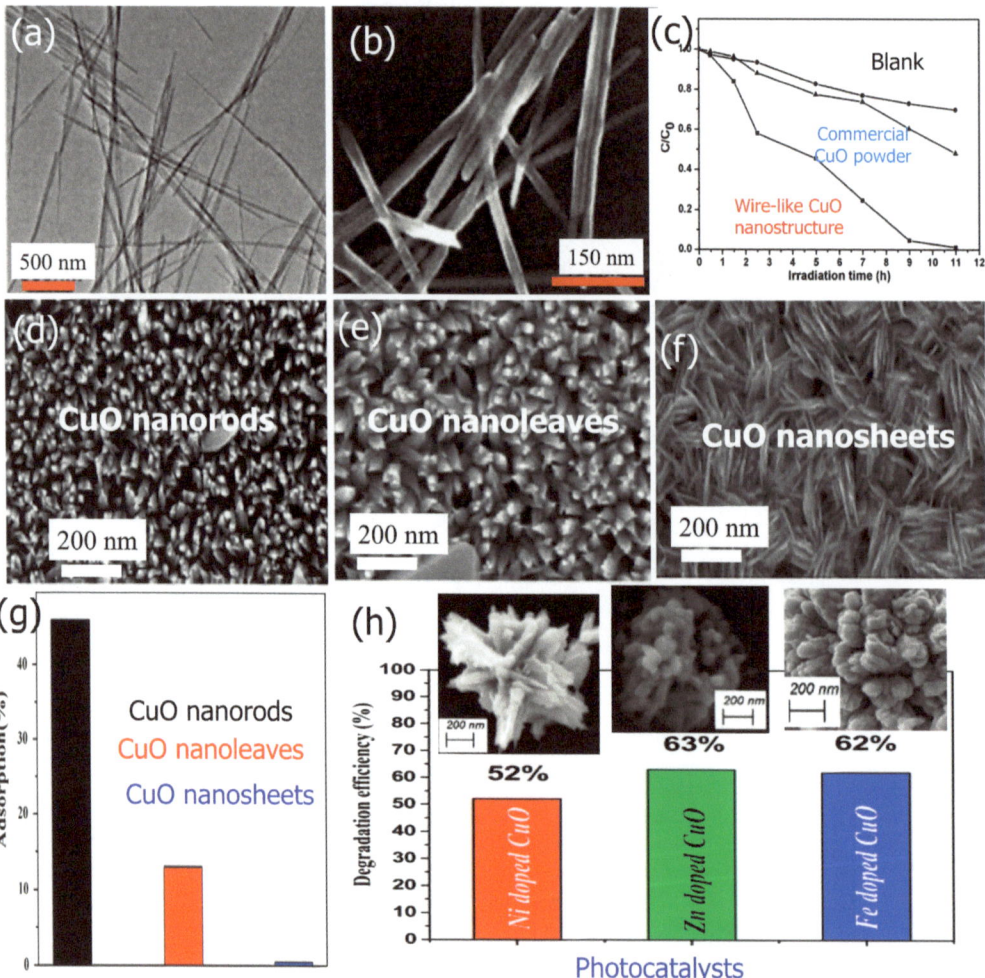

Figure 14. (**a**) TEM and (**b**) SEM images of CuO nanowires, (**c**) the photocatalytic efficiency of RhB degradation on different catalysts in the presence of CuO nanowire catalyst (Adopted with permission from Ref. [129] Copyright 2012, Elsevier), (**d**–**f**) SEM images of different CuO morphologies, (**g**) adsorption of CR on the surface of different morphologies of CuO [Initial CR concentration is 20 mg L^{-1}, dose of CuO is 0.05 g, and stirring time is 30 min] (Adopted with permission from Ref. [130] Copyright 2014, Elsevier), and (**h**) photocatalytic degradation efficiency of MB dye using Ni, Zn, and Fe-doped CuO (together with their SEM images) (Adopted with permission from Ref. [130] Copyright 2021, Elsevier).

Using NaBH$_4$ as q reducing agent in an aqueous medium, Nazim et al. [133] recently synthesized porous CuO nanosheets (Figure 15a) for use as a photocatalyst towards the degradation of food dye. The optical energy band gap of the prepared CuO was found to be approximately 1.92 eV. The CuO nanosheets were tested as photocatalysts for the degradation of Allura Red AC (AR) dye and showed excellent photocatalytic degradation efficiency of around 96.99% in just 6 min under visible light irradiation at room temperature (Figure 15b,c). The photodegradation kinetics of AR followed a pseudo-first-order reaction model, with a rate constant of 0.524 min^{-1}. Additionally, the CuO nanosheets exhibited remarkable recycling ability for AR degradation (Figure 15d).

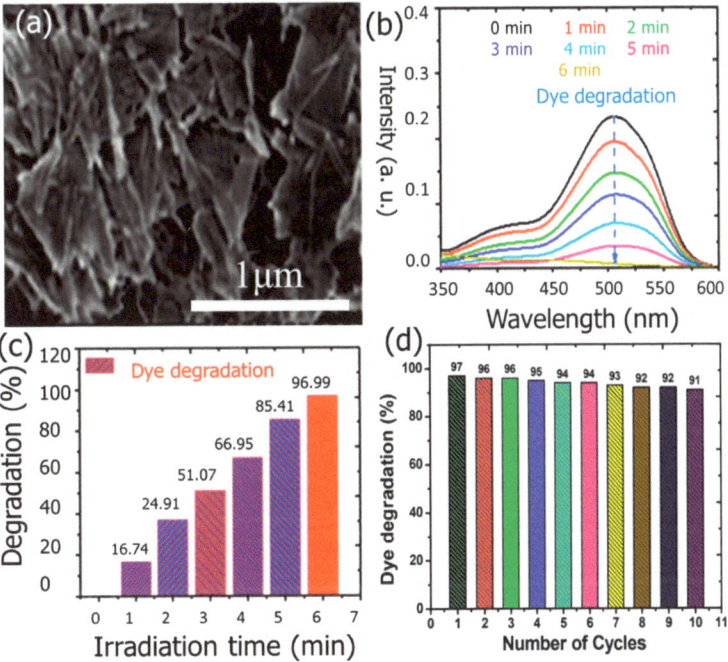

Figure 15. (**a**) FESEM images of CuO nanosheets, (**b**) UV–Vis spectrum of dye degradation at different times, (**c**) gradual degradation histogram against time of the AR dye in aqueous medium, and (**d**) recyclability of the CuO nanosheets after ten degradation cycles for AR dye. Adopted with permission from Ref. [133] Copyright 2021, American Chemical Society.

4.4. Nickle Oxide Photocatalysts

NiO is a broad bandgap (3.6–4.0 eV) p-type semiconducting oxide. NiO has attracted a lot of attention due to its application in numerous areas, including fuel cell electrodes, battery cathodes, dye-sensitized solar cells, etc. [134–137].

In addition, NiO displays a promising ability to produce OH radicals and thus is a potential candidate toward the degradation of organic pollutants. For example, Jayakumar et al. [138] prepared NiO nanoparticles through a chemical precipitation method for use as photocatalysts for the degradation of MB dye in aqueous media. The photocatalytic degradation results showed that the NiO nanoparticles are potential photocatalysts for the degradation of MB dye.

Saeed et al. utilized a chemical reduction process to prepare NiO NPs and NiO/nanoclay nanocomposite (NiO/Nc) and evaluated their photocatalytic efficiency for the degradation of orange II dyes in aqueous solution [139]. According to the SEM examination, the

NiO NPs were spherical with variable forms and diameters in the range of 100–400 nm (Figure 16a,b). The photodegradation investigation showed that, within 20 min, orange II in aqueous medium was degraded by the NiO NPs and NiO/Nc by 93% and 96%, respectively (Figure 16c). Green methods are successfully applied for preparing NiO as a photocatalyst. For example, using the antioxidant property of *Punica granatum* L. (pomegranate) juice extract and its bio-reducing ability for MO breakdown in water, Barzinjy et al. [140] prepared NiO NPs. The biosynthesized NiO nanoparticles displayed an active catalytic ability toward the degradation of MO from media with a total degradation performance of 96%. Furthermore, Sarani et al. [141] prepared NiO nanoparticles via a green method by using extract as a stabilizing agent. The average crystal size of prepared NiO nanoparticles was approximately 54–58 nm, and the estimated energy band gap was 3–3.7. The prepared NiO nanoparticles were investigated as photocatalysts for degradation of acid orange 7 (AO7) dye in aqueous solution under visible light. The NiO nanoparticles exhibited excellent photocatalytic performance (90.2%) toward the AO7 dye and displayed excellent re-usability several times. Table 1 summarizes different metal oxide morphologies and their photodegradation performances.

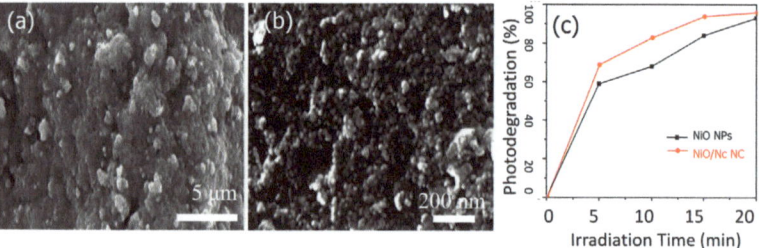

Figure 16. SEM image of (**a**) NiO NPs and (**b**) NiO/Nc composite, and (**c**) percent of degradation for orange II dye. Adopted with permission from Ref. [139] Copyright 2022, Springer Nature.

Table 1. Summary of different metal oxide morphologies and their photodegradation performances.

Photocatalyst	Morphology	Pollutant	Degradation Conditions	Degradation Rate (%)	Ref.
TiO$_2$	TiO$_2$ Nanowires	MB	UV light Time 80 min Pseudo-first order (k = 0.0098 min^{-1})	75.00	[83]
	TiO$_2$ Nanotubes		UV light Time 80 min Pseudo-first order (k = 0.0426 min^{-1})	99.00	[83]
	TiO$_2$ Nanowires	RhB	UV light Time 10 min	50.00	[84]

Table 1. *Cont.*

Photocatalyst	Morphology	Pollutant	Degradation Conditions	Degradation Rate (%)	Ref.
ZnO	ZnO Nanoparticles	MO	UV light Time 150 min Pseudo-first order (k = 0.01659 cm^{-1})	99.00	[108]
	ZnO Nanoparticles	RhB	UV light Time 160 min	95.40	[109]
	ZnO Nanoparticles	RhB	UV light Pseudo-first order (k = 0.014 min^{-1})	80.44	[110]
	ZnO Nanoparticles	MG	UV light Pseudo-first order (k = 0.017 min^{-1})	83.68	
	ZnO Nanoflower	MG	UV light Pseudo-first order (k = 0.023 min^{-1})	90.87	[110]
CuO	CuO nanowires	RhB	UV light Time 9 h	95.00	[129]
	CuO nanorods	CR	UV light Time 210 min	67.00	[130]

Table 1. Cont.

Photocatalyst	Morphology	Pollutant	Degradation Conditions	Degradation Rate (%)	Ref.
CuO	NiO/CuO nanoflower	CR	UV light Time 210 min	48.00	[130]
	CuO nanosheets		UV light Time 210 min	12.00	
	Porous CuO Nanosheets	AR	UV-vis light Pseudo-first order (k = 0.524 min^{-1})	96.99	[133]
NiO	NiO/Clay Nanocomposite	O II	UV–Vis light Time 20 min	96.00	[139]
WO$_3$	WO$_3$@400 °C	MB	UV light Time 160 min	11.00	[142]
	WO$_3$@450 °C		UV light Time 16 min	14.30	[142]
	WO$_3$@550 °C		UV light Time 160 min	20.00	[142]

Table 1. Cont.

Photocatalyst	Morphology	Pollutant	Degradation Conditions	Degradation Rate (%)	Ref.
WO$_3$	WO$_3$@300 °C	RhB	Visible light Time 240 min	96.10	[143]

4.5. Tungsten Oxide Photocatalysts

In recent decades, tungsten-based oxides (WO$_3$) have been the subject of substantial research, with a variety of morphologies being exhibited. A twisted WO$_6$ octahedron connects the crystal in the stoichiometric ratio of the WO$_3$ structure to generate a perovskite crystal structure. Its crystal forms are hexagonal, orthorhombic, and monoclinic. At the same time, oxygen vacancies and unsaturated, coordinated W atoms might arise from the easy loss of the oxygen lattice. Consequently, there are numerous non-stoichiometric compounds in tungsten oxide, including WO$_{2.72}$, WO$_{2.8}$, WO$_{2.83}$, and WO$_{2.9}$.

Among the low-cost semiconductors with potential photocatalytic activity is tungsten oxide (WO$_3$) [56]. Its highly adjustable stoichiometries and structures, together with its Earth-abundance and strong sensitivity to the solar spectrum due to its low band gap of 2.5–3.0 eV [144], make it a popular choice for photocatalysis under visible light [145]. Furthermore, WO$_3$ has stable physical and chemical characteristics, little toxicity, and a strong capacity to oxidize valence band holes [39]. Nanoparticles, nanowires, nanosheets, and nanospheres are among the frequently occurring morphologies. There are many ways to produce different WO$_3$ dimensions: 0 dimensional (0D), 1D, 2D, and 3D WO$_3$, in that order. Characteristics vary among dimensions. It is possible to create 0D WO$_3$ monodisperse monoclinic WO$_3$ quantum dots by breaking down the ammonium tungstate oxide complex, which is created hydrothermally using hydrazine hydrate and WCl$_6$ [146]. It is possible to carefully regulate the particle size distribution of WO$_{3-x}$ QDS within the range of 1.3–4.5 nm by varying the reaction temperature [147]. 1D WO$_3$ is widely available and simple to synthesize. Today, various structures of 1D WO$_3$ have been discovered, such as those of nanofibers, nanotubes, nanorods, and nanowires. As for 2D WO$_3$, thin films, nanosheets, nanoplates, and so on, these have garnered a great deal of interest because of their high surface volume ratio, surface polarization, modulated surface activity, and oxygen-rich vacancies. The majority of 2D WO$_3$ structures are thin layers. As previously reported by Yin et al., 3D WO$_3$ is a layered structure made of nanoparticles, nanoplates, nanorods, and nanosheets. It can take the form of irregular structures such as mesoporous structures, microspheres, micro flowers, and sea urchin-like formations. Typically, 3D WO$_3$ displays high porosity, a large specific surface area, and a distinctive shape [148].

For photocatalytic applications, WO$_3$ shows some advantages, such as its high physiochemical catalyst stability that is limited to photo-corrosion and has strong solar spectrum absorption. Under UV–Vis light, WO$_3$ is a highly reactive catalyst that effectively oxidizes various organic and inorganic pollutants in wastewater. WO$_3$ is a photocatalyst with a wide range of hues that can change from yellow to green to bluish to grayish depending on its oxidation state. There are many techniques that have been used to enhance the catalytic activity of WO$_3$. For example, doping with other materials such as dyes to form dye-doped WO$_3$ as reported by Tahir et al. [149], or doping with Pt to form Pt–WO$_3$ composite [150].

One of the promising strategies is to design WO$_3$ with a different morphology, and we will present the recent progress in utilizing WO$_3$ with different morphologies in degradation some organic dyes. Ojha et al. [142] used the sol-gel method to prepare WO$_3$ nanorods and sheets in various crystal phases. As presented in Figure 17a, they showed that the calcination temperature plays a critical role in the shapes and crystal phases of the WO$_3$ nanostructures, and thus in the photocatalytic activity toward degradation of MB dye due to the modified electronic structure, which causes a variation in the value of the band gap (Figure 17b). The synthesized WO$_3$ nanosheets showed improved photocatalytic activity for the photodegradation of MB dye compared to WO$_3$ nanorods (Figure 17c). The sheet-type structure provides more active surface for the interaction of dye molecules compared to the rods.

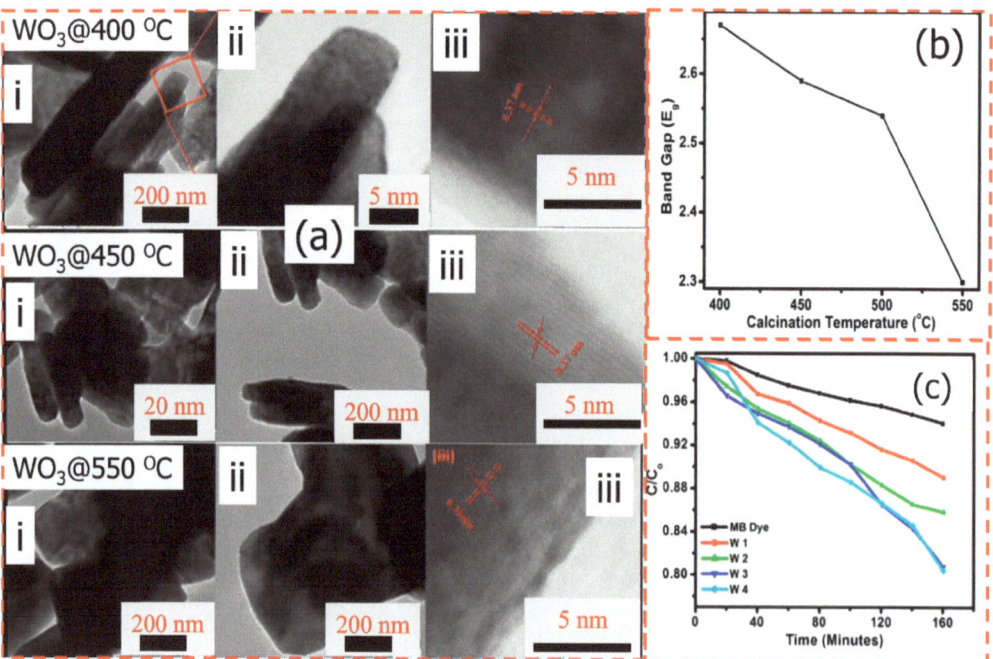

Figure 17. (**a**) TEM images with various resolutions of WO$_3$ calcined at different temperatures, (**b**) variation in band gap with calcination temperature, and (**c**) C/Co vs. UV irradiation time plot for degradation of MB dye with W1 (at 400 °C), W2 (at 450 °C), W3 (at 550 °C), and W4 (at 550 °C). Adopted with permission from Ref. [142] Copyright 2022, Springer Nature.

Yin et al. [148] prepared WO$_3$ photocatalyst through a hydrothermal process. The prepared WO$_3$ nanoparticles showed promising absorption of UV, visible, and near-infrared (NIR) bandwidths. WO$_3$ photocatalyst exhibited excellent catalytic activity toward the degradation of MB. Furthermore, they confirmed that the temperature showed different morphology (Figure 18a–d) and that the higher temperature displayed better catalytic activity (Figure 18e). Recently, Mzimela et al. [143] prepared highly agglomerated WO$_3$ nanoparticles through the facile acid precipitation method at various temperature (Figure 18f–h) for degradation of RhB dye. The results showed that WO$_3$ calcined at 300 °C, 5 g L^{-1} catalyst dose, 5 ppm RhB concentration, and pH of 9.5, while the catalyst showed an excellent degradation efficiency of 96.1% after 4 h under visible light irradiation (Figure 18i). Furthermore, the degradation kinetics obeys the L–H model, which describes heterogenous photocatalytic surface reactions.

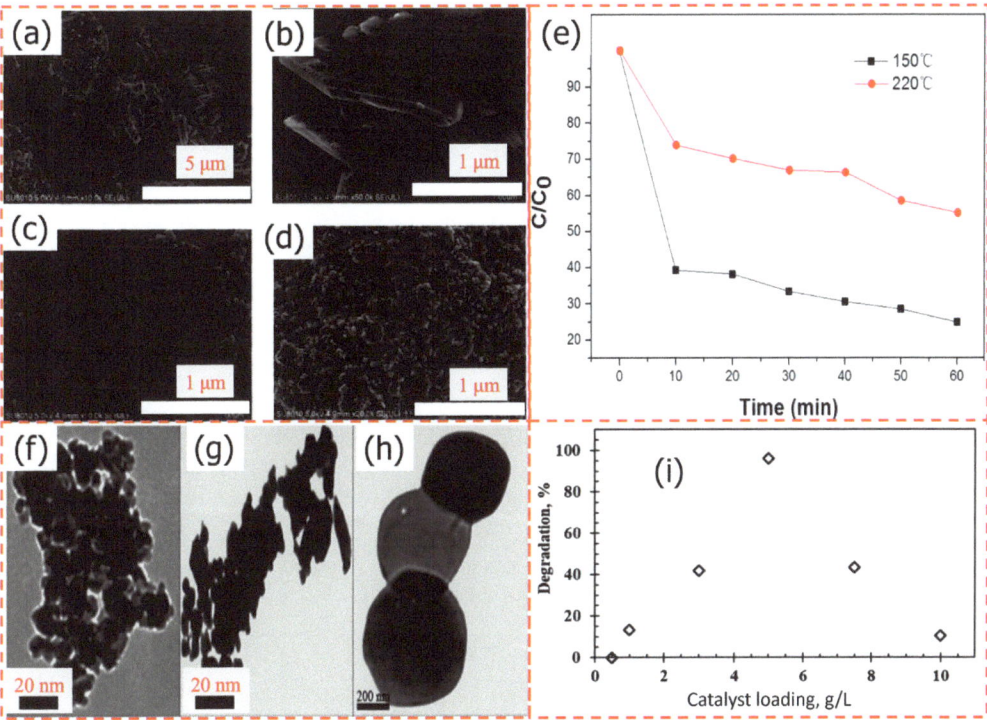

Figure 18. (**a**,**b**) SEM images of WO$_3$ calcined at 150 °C, and (**c**,**d**) at 220 °C. (**e**) The photodegradation profile of MB in the presence of the WO$_3$ photocatalyst under NIR laser irradiation. Adopted with permission from Ref. [148] Copyright 2022, Springer Nature. TEM images of WO$_3$ particles calcined at (**f**) 300 °C, (**g**) 500 °C, and (**h**) 700 °C, and (**i**) effect of catalyst loading on the photocatalytic degradation of RhB at a pH 9.5 and 5 ppm RhB at a calcination temperature of 300 °C. Adopted with permission from Ref. [143] Copyright 2022, Royal Society of Chemistry.

5. Conclusions and Future Perspectives

Due to the industrial revolution in the modern world, various pollutants are emitted into the environment, both directly and indirectly. Thus, great efforts have been devoted to develop advanced technology for the elimination of such environmental pollutants. One of the most promising methods for meeting global needs in an environmentally responsible manner is photocatalysis. Metal oxides are found to be potential photocatalysts because of their affordability, effectiveness, and environmental friendliness. Furthermore, metal oxides may be widely used in a wide range of applications because of their large surface area, simplicity of fabrication, and sufficient supply. Moreover, metal oxides are abundant, highly active, and could be prepared either through eco-friendly or conventional methods in large quantities. Unfortunately, a few drawbacks prevent them from being used practically, including a large bandgap, a high rate of photogenerated electron–hole pair recombination, and catalyst deactivation. Therefore, various strategies have been developed to overcome these limitations; one of these techniques is the design of metal oxides with unique morphologies. The morphology of metal oxides displayed a significant influence on the overall degradation performance, as indicated by shifting the photocatalytic degradation toward various kinds of dyes in aqueous solutions. Therefore, in the current study, the recent progress in the photocatalytic degradation of organic dyes by prominent metal oxides with different morphologies is discussed. The metal oxide nanoparticles, namely, TiO$_2$, ZnO,

CuO, NiO, and WO_3, all are widely utilized for the photodegradation of various organic dyes. The non-spherical morphologies such as nanorods and nanowires exhibit enhanced photocatalytic performance due to their unique crystal facets and surface areas, which can promote charge transfer and improve catalytic efficiency. Ultimately, the selection of a particular morphology for a photocatalyst relies on the specific application and the desired properties. The porous design and the substantial specific surface area are responsible for the increased photocatalytic activity. On the other hand, the flake-like structure exhibits comparatively lower performance.

In the future, its highly recommended to do more research to more deeply understand the degradation mechanisms of metal oxides with different morphologies. Until now, most of the prepared metal oxides with different morphologies were obtained by using conventional methods on their appropriate precursors. Some of them are toxic or require the use of a high temperature or organic solvents during the procedure, which is not preferred for sustainability. Thus, it is important to find alternative, sustainable methods for preparing these oxides with different morphologies. Presently, there are some studies reporting the preparation of NiO using green methods; however, they are limited in number and need to be more accurate, as well as to prepare other metal oxides such as TiO_2, ZnO, CuO, and WO_3 for organic dye degradation in aqueous solutions.

Author Contributions: G.A.M.A. and A.S.A.A. designed the work and collected the data. A.H.N., A.S.A.A., and T.A.S.E.-N. wrote the main manuscript text. A.S.A.A. and G.A.M.A. prepared the figures. G.A.M.A., N.F.A., and K.F.C. revised and edited the manuscript. All authors have read and agreed to the published version of the manuscript.

Funding: This work was funded by the Deanship of Scientific Research at Jouf University under Grant No. (DSR2022-RG-0139).

Data Availability Statement: Data is contained within the article.

Acknowledgments: The authors would like to thank the Deanship of Scientific Research at Jouf University under Grant No. (DSR2022-RG-0139).

Conflicts of Interest: The authors declare no conflict of interest.

References

1. Maheshwari, K.; Solanki, Y.S.; Ridoy, M.S.H.; Agarwal, M.; Dohare, R.; Gupta, R. Ultrasonic treatment of textile dye effluent utilizing microwave-assisted activated carbon. *Environ. Prog. Sustain. Energy* **2020**, *39*, e13410. [CrossRef]
2. Ali, H. Biodegradation of Synthetic Dyes—A Review. *Water Air Soil Pollut.* **2010**, *213*, 251–273. [CrossRef]
3. Muhammad; Norzahir, S. A review on the water problem associate with organic pollutants derived from phenol, methyl orange, and remazol brilliant blue dyes. *Mater. Today Proc.* **2020**, *31*, A141–A150. [CrossRef]
4. Sharifi, A.; Montazerghaem, L.; Naeimi, A.; Abhari, A.R.; Vafaee, M.; Ali, G.A.M.; Sadegh, H. Investigation of photocatalytic behavior of modified ZnS:Mn/MWCNTs nanocomposite for organic pollutants effective photodegradation. *J. Environ. Manag.* **2019**, *247*, 624–632. [CrossRef] [PubMed]
5. Couto, S.R. Dye removal by immobilised fungi. *Biotechnol. Adv.* **2009**, *27*, 227–235. [CrossRef] [PubMed]
6. Pavithra, K.G.; Senthil Kumar, P.; Jaikumar, V.; Sundar Rajan, P. Removal of colorants from wastewater: A review on sources and treatment strategies. *J. Ind. Eng. Chem.* **2019**, *75*, 1–19. [CrossRef]
7. Solehudin, M.; Sirimahachai, U.; Ali, G.A.M.; Chong, K.F.; Wongnawa, S. One-pot synthesis of isotype heterojunction g-C_3N_4-MU photocatalyst for effective tetracycline hydrochloride antibiotic and reactive orange 16 dye removal. *Adv. Powder Technol.* **2020**, *31*, 1891–1902. [CrossRef]
8. Pathan, A.; Bhatt, S.H.; Vajapara, S.; Bhasin, C.P. Solar Light Induced Photo Catalytic Properties of α-Fe_2O_3 Nanoparticles for Degradation of Methylene Blue Dye. *Int. J. Thin Film Sci. Technol.* **2022**, *11*, 213–224.
9. Pointing, S. Feasibility of bioremediation by white-rot fungi. *Appl. Microbiol. Biotechnol.* **2001**, *57*, 20–33.
10. Chandralata, R. Fungi from marine habitats: An application in bioremediation1 1Contribution No. 3538 of the National Institute of Oceanography. *Mycol. Res.* **2000**, *104*, 1222–1226.
11. Narayan, R.B.; Goutham, R.; Srikanth, B.; Gopinath, K.P. A novel nano-sized calcium hydroxide catalyst prepared from clam shells for the photodegradation of methyl red dye. *J. Environ. Chem. Eng.* **2018**, *6*, 3640–3647. [CrossRef]
12. Pathan, A.; Prajapati, C.G.; Dave, R.P.; Bhasin, C.P. Effective and Feasible Photocatalytic Degradation of Janus Green B dye in Aqueous Media using PbS/CTAB Nanocomposites. *Int. J. Thin Film Sci. Technol.* **2022**, *11*, 245–255.

13. O'Neill, C.; Hawkes, F.R.; Hawkes, D.L.; Lourenço, N.D.; Pinheiro, H.M.; Delée, W. Colour in textile effluents—Sources, measurement, discharge consents and simulation: A review. *J. Chem. Technol. & Biotechnol.* **1999**, *74*, 1009–1018.
14. Jin, X.-C.; Liu, G.-Q.; Xu, Z.-H.; Tao, W.-Y. Decolorization of a dye industry effluent by Aspergillus fumigatus XC6. *Appl. Microbiol. Biotechnol.* **2007**, *74*, 239–243. [CrossRef] [PubMed]
15. Maafa, I.M.; Ali, M.A. Enhanced Organic Pollutant Removal Efficiency of Electrospun $NiTiO_3/TiO_2$-Decorated Carbon Nanofibers. *Polymers* **2023**, *15*, 109. [CrossRef] [PubMed]
16. Mohamad Amran Mohd, S.; Dalia Khalid, M.; Wan Azlina Wan Abdul, K.; Azni, I. Cationic and anionic dye adsorption by agricultural solid wastes: A comprehensive review. *Desalination* **2011**, *280*, 1–13.
17. Yuxing, W.; Jian, Y. Laccase-catalyzed decolorization of synthetic dyes. *Water Res.* **1999**, *33*, 3512–3520.
18. Esther, F.; Tibor, C.; Gyula, O. Removal of synthetic dyes from wastewaters: A review. *Environ. Int.* **2004**, *30*, 953–971.
19. Pourbabaee, A.A.; Malekzadeh, F.; Sarbolouki, M.N.; Najafi, F. Aerobic Decolorization and Detoxification of a Disperse Dye in Textile Effluent by a New Isolate of *Bacillus* sp. *Biotechnol. Bioeng.* **2006**, *93*, 631–635. [CrossRef]
20. Singh, H. Fungal decolorization and degradation of dyes. In *Mycoremediation: Fungal Bioremediation*; Wiley: Hoboken, NJ, USA, 2006; pp. 420–483.
21. Ahmed, A.S.A.; Sanad, M.M.S.; Kotb, A.; Negm, A.N.R.M.; Abdallah, M.H. Removal of methyl red from wastewater using a NiO@Hyphaene thebaica seed-derived porous carbon adsorbent: Kinetics and isotherm studies. *Mater. Adv.* **2023**, *4*, 2981–2990. [CrossRef]
22. Yuzhu, F.; Viraraghavan, T. Fungal decolorization of dye wastewaters: A review. *Bioresour. Technol.* **2001**, *79*, 251–262.
23. Haoran, D.; Guangming, Z.; Lin, T.; Changzheng, F.; Chang, Z.; Xiaoxiao, H.; Yan, H. An overview on limitations of TiO_2-based particles for photocatalytic degradation of organic pollutants and the corresponding countermeasures. *Water Res.* **2015**, *79*, 128–146.
24. Xi, Z.; Jing, W.; Xing-Xing, D.; Yun-Kai, L. Functionalized metal-organic frameworks for photocatalytic degradation of organic pollutants in environment. *Chemosphere* **2020**, *242*, 125144.
25. Pelizzetti, E.; Minero, C. Metal Oxides as Photocatalysts for Environmental Detoxification. *Comments Inorg. Chem.* **1993**, *15*, 297–337. [CrossRef]
26. Li, L.; Zhong, Y.; Hu, Y.; Bai, J.; Qiao, F.; Ahmed, A.S.A.; Ali, G.; Zhao, X.; Xie, Y. Room-temperature synthesis of Ag- and Mn-doped $Cs_2NaBiCl_6$ octahedrons for dye photodegradation. *CrystEngComm* **2023**, *25*, 4355–4363. [CrossRef]
27. Nigora, T.; Irma, K. Effects of electronic structure of catalytic nanoparticles on carbon nanotube growth. *Carbon Trends* **2021**, *5*, 100092.
28. Muhammad Sohail, B.; Naveed, R.; Tayyaba, N.; Ghulam, A.; Xiangling, G.; Muhammad, A.; Muhammad, Q.; Humaira, B.; Syed Shoaib Ahmad, S.; Mika, S. Metallic nanoparticles for catalytic reduction of toxic hexavalent chromium from aqueous medium: A state-of-the-art review. *Sci. Total Environ.* **2022**, *829*, 154475.
29. Chen, H.; Nanayakkara, C.E.; Grassian, V.H. Titanium Dioxide Photocatalysis in Atmospheric Chemistry. *Chem. Rev.* **2012**, *112*, 5919–5948. [CrossRef]
30. Ansari, S.A.; Khan, M.M.; Kalathil, S.; Nisar, A.; Lee, J.; Cho, M.H. Oxygen vacancy induced band gap narrowing of ZnO nanostructures by an electrochemically active biofilm. *Nanoscale* **2013**, *5*, 9238–9246. [CrossRef]
31. Wang, H.; Rogach, A.L. Hierarchical SnO_2 Nanostructures: Recent Advances in Design, Synthesis, and Applications. *Chem. Mater.* **2014**, *26*, 123–133. [CrossRef]
32. Sun, C.; Li, H.; Chen, L. Nanostructured ceria-based materials: Synthesis, properties, and applications. *Energy Environ. Sci.* **2012**, *5*, 8475–8505. [CrossRef]
33. Oturan, M.A.; Aaron, J.J. Advanced Oxidation Processes in Water/Wastewater Treatment: Principles and Applications. A Review. *Crit. Rev. Environ. Sci. Technol.* **2014**, *44*, 2577–2641. [CrossRef]
34. Rauf, M.A.; Meetani, M.A.; Hisaindee, S. An overview on the photocatalytic degradation of azo dyes in the presence of TiO_2 doped with selective transition metals. *Desalination* **2011**, *276*, 13–27. [CrossRef]
35. Laouini, S.E.; Bouafia, A.; Soldatov, A.V.; Algarni, H.; Tedjani, M.L.; Ali, G.A.M.; Barhoum, A. Green Synthesized of Ag/Ag_2O Nanoparticles Using Aqueous Leaves Extracts of Phoenix dactylifera L. and Their Azo Dye Photodegradation. *Membranes* **2021**, *11*, 468. [CrossRef] [PubMed]
36. Mohammad Mansoob, K.; Syed Farooq, A.; Abdullah, A.-M. Metal oxides as photocatalysts. *J. Saudi Chem. Soc.* **2015**, *19*, 462–464.
37. Moniz, S.J.A.; Zhu, J.; Tang, J. 1D Co-Pi Modified $BiVO_4/ZnO$ Junction Cascade for Efficient Photoelectrochemical Water Cleavage. *Adv. Energy Mater.* **2014**, *4*, 1301590. [CrossRef]
38. Rauf, M.A.; Ashraf, S.; Alhadrami, S.N. Photolytic oxidation of Coomassie Brilliant Blue with H_2O_2. *Dye. Pigment.* **2005**, *66*, 197–200. [CrossRef]
39. Theerthagiri, J.; Chandrasekaran, S.; Salla, S.; Elakkiya, V.; Senthil, R.; Nithyadharseni, P.; Maiyalagan, T.; Micheal, K.; Ayeshamariam, A.; Arasu, M.V. Recent developments of metal oxide based heterostructures for photocatalytic applications towards environmental remediation. *J. Solid State Chem.* **2018**, *267*, 35–52. [CrossRef]
40. Danish, M.S.; Estrella, L.L.; Alemaida, I.M.A.; Lisin, A.; Moiseev, N.; Ahmadi, M.; Nazari, M.; Wali, M.; Zaheb, H.; Senjyu, T. Photocatalytic Applications of Metal Oxides for Sustainable Environmental Remediation. *Metals* **2021**, *11*, 80. [CrossRef]
41. Asma, R.; Muhammad, I.; Ali, S.; Faiza, N.; Maaz, K.; Qasim, K.; Muhammad, M. Photocatalytic degradation of dyes using semiconductor photocatalysts to clean industrial water pollution. *J. Ind. Eng. Chem.* **2021**, *97*, 111–128.

42. Ajmal, A.; Majeed, I.; Malik, R.N.; Idriss, H.; Nadeem, M.A. Principles and mechanisms of photocatalytic dye degradation on TiO_2 based photocatalysts: A comparative overview. *RSC Adv.* **2014**, *4*, 37003–37026. [CrossRef]
43. Ioannis, K.K.; Triantafyllos, A.A. TiO_2-assisted photocatalytic degradation of azo dyes in aqueous solution: Kinetic and mechanistic investigations: A review. *Appl. Catal. B Environ.* **2004**, *49*, 1–14.
44. Meng, A.; Zhou, S.; Wen, D.; Han, P.; Su, Y. $g-C_3N_4/CoTiO_3$ S-scheme heterojunction for enhanced visible light hydrogen production through photocatalytic pure water splitting. *Chin. J. Catal.* **2022**, *43*, 2548–2557. [CrossRef]
45. Alvarez-Ramirez, J.; Femat, R.; Meraz, M.; Ibarra-Valdez, C. Some remarks on the Langmuir–Hinshelwood kinetics. *J. Math. Chem.* **2016**, *54*, 375–392. [CrossRef]
46. Tran, H.D.; Nguyen, D.Q.; Do, P.T.; Tran, U.N.P. Kinetics of photocatalytic degradation of organic compounds: A mini-review and new approach. *RSC Adv.* **2023**, *13*, 16915–16925. [CrossRef]
47. Barkha, R.; Arpan Kumar, N.; Niroj Kumar, S. Fundamentals principle of photocatalysis. In *Micro and Nano Technologies*; Elsevier: Amsterdam, The Netherlands, 2022; pp. 1–22. [CrossRef]
48. Kumar, K.V.; Porkodi, K.; Rocha, F. Langmuir–Hinshelwood kinetics—A theoretical study. *Catal. Commun.* **2008**, *9*, 82–84. [CrossRef]
49. Swenson, H.; Stadie, N.P. Langmuir's Theory of Adsorption: A Centennial Review. *Langmuir* **2019**, *35*, 5409–5426. [CrossRef]
50. Sun, P.; Zhang, J.; Liu, W.; Wang, Q.; Cao, W. Modification to L-H Kinetics Model and Its Application in the Investigation on Photodegradation of Gaseous Benzene by Nitrogen-Doped TiO_2. *Catalysts* **2018**, *8*, 326. [CrossRef]
51. Zhan, X.; Yan, C.; Zhang, Y.; Rinke, G.; Rabsch, G.; Klumpp, M.; Schäfer, A.I.; Dittmeyer, R. Investigation of the reaction kinetics of photocatalytic pollutant degradation under defined conditions with inkjet-printed TiO_2 films—From batch to a novel continuous-flow microreactor. *React. Chem. Eng.* **2020**, *5*, 1658–1670. [CrossRef]
52. Huang, C.; Chen, L.; Li, H.; Mu, Y.; Yang, Z. Synthesis and application of Bi_2WO_6 for the photocatalytic degradation of two typical fluoroquinolones under visible light irradiation. *RSC Adv.* **2019**, *9*, 27768–27779. [CrossRef]
53. Amir, Z.; Muhammad, K.; Muhammad Asim, K.; Qasim, K.; Aziz, H.-Y.; Alei, D.; Muhammad, M. Review on the hazardous applications and photodegradation mechanisms of chlorophenols over different photocatalysts. *Environ. Res.* **2021**, *195*, 110742.
54. Yue, B.; Zhou, Y.; Xu, J.; Wu, Z.; Zhang, X.; Zou, Y.; Jin, S. Photocatalytic Degradation of Aqueous 4-Chlorophenol by Silica-Immobilized Polyoxometalates. *Environ. Sci. Technol.* **2002**, *36*, 1325–1329. [CrossRef] [PubMed]
55. Kumar, K.V.A.; Chandana, L.; Ghosal, P.; Ch, S. Simultaneous photocatalytic degradation of p-cresol and Cr (VI) by metal oxides supported reduced graphene oxide. *Mol. Catal.* **2018**, *451*, 87–95. [CrossRef]
56. Naeimi, A.; Sharifi, A.; Montazerghaem, L.; Abhari, A.R.; Mahmoodi, Z.; Bakr, Z.H.; Soldatov, A.V.; Ali, G.A.M. Transition metals doped WO_3 photocatalyst towards high efficiency decolourization of azo dye. *J. Mol. Struct.* **2022**, *1250*, 131800. [CrossRef]
57. Hisatomi, T.; Kubota, J.; Domen, K. Recent advances in semiconductors for photocatalytic and photoelectrochemical water splitting. *Chem. Soc. Rev.* **2014**, *43*, 7520–7535. [CrossRef] [PubMed]
58. Hoffmann, M.R.; Martin, S.T.; Choi, W.; Bahnemann, D.W. Environmental Applications of Semiconductor Photocatalysis. *Chem. Rev.* **1995**, *95*, 69–96. [CrossRef]
59. Fujishima, A.; Honda, K. Electrochemical Photolysis of Water at a Semiconductor Electrode. *Nature* **1972**, *238*, 37–38. [CrossRef]
60. Ryu, A. Recent progress on photocatalytic and photoelectrochemical water splitting under visible light irradiation. *J. Photochem. Photobiol. C Photochem. Rev.* **2010**, *11*, 179–209.
61. Inoue, T.; Fujishima, A.; Konishi, S.; Honda, K. Photoelectrocatalytic reduction of carbon dioxide in aqueous suspensions of semiconductor powders. *Nature* **1979**, *277*, 637–638. [CrossRef]
62. Kazuya, N.; Akira, F. TiO_2 photocatalysis: Design and applications. *J. Photochem. Photobiol. C Photochem. Rev.* **2012**, *13*, 169–189.
63. Dahl, M.; Liu, Y.; Yin, Y. Composite Titanium Dioxide Nanomaterials. *Chem. Rev.* **2014**, *114*, 9853–9889. [CrossRef] [PubMed]
64. Ethiraj, A.S.; Rhen, D.S.; Soldatov, A.V.; Ali, G.A.M.; Bakr, Z.H. Efficient and recyclable Cu incorporated TiO_2 nanoparticle catalyst for organic dye photodegradation. *Int. J. Thin Film Sci. Technol.* **2021**, *10*, 169–182.
65. Bajpai, S.; Tiwary, S.K.; Sonker, M.; Joshi, A.; Gupta, V.; Kumar, Y.; Shreyash, N.; Biswas, S. Recent Advances in Nanoparticle-Based Cancer Treatment: A Review. *ACS Appl. Nano Mater.* **2021**, *4*, 6441–6470. [CrossRef]
66. Giahi, M.; Pathania, D.; Agarwal, S.; Ali, G.A.M.; Chong, K.F.; Gupta, V.K. Preparation of Mg-doped TiO_2 nanoparticles for photocatalytic degradation of some organic pollutants. *Stud. Univ. Babes-Bolyai Chem.* **2019**, *64*, 7–18. [CrossRef]
67. Nair, A.; PonnanEttiyappan, J. Ag–TiO_2 Nanofiber Membranes for Photocatalytic Degradation of Dyes. *Adv. Sci. Lett.* **2018**, *24*, 5764–5767. [CrossRef]
68. Nair, A.K.; JagadeeshBabu, P.E. TiO_2 nanosheet-graphene oxide based photocatalytic hierarchical membrane for water purification. *Surf. Coat. Technol.* **2017**, *320*, 259–262. [CrossRef]
69. Tomoaki, T.; Haruka, N.; Motoki, M.; Akihide, I.; Akihiko, K. Photocatalytic CO_2 reduction using water as an electron donor over Ag-loaded metal oxide photocatalysts consisting of several polyhedra of Ti^{4+}, Zr^{4+}, and Ta^{5+}. *J. Photochem. Photobiol. A Chem.* **2018**, *358*, 416–421.
70. Wang, R.; Hashimoto, K.; Fujishima, A.; Chikuni, M.; Kojima, E.; Kitamura, A.; Shimohigoshi, M.; Watanabe, T. Light-induced amphiphilic surfaces. *Nature* **1997**, *388*, 431–432. [CrossRef]
71. Ye, S.; Sun, H.; Wu, J.; Wan, L.; Ni, Y.; Wang, R.; Xiang, Z.; Deng, X. Supercritical CO_2 Assisted TiO_2 Preparation to Improve the UV Resistance Properties of Cotton Fiber. *Polymers* **2022**, *14*, 5513. [CrossRef]

72. Bahareh Ghorbani, A.; Neda, G.; Javad Vahabzade, P.; Azadeh Ebrahimian, P. Boosting the photoconversion efficiency of TiO_2 nanotubes using UV radiation-assisted anodization as a prospective method: An efficient photocatalyst for eliminating resistant organic pollutants. *Ceram. Int.* **2020**, *46*, 19942–19951.
73. Reza, A.; Ali Reza, M.; Lotf Ali, S.; Soheila, S. Characterization and optical properties of spherical WO_3 nanoparticles synthesized via the reverse microemulsion process and their photocatalytic behavior. *Mater. Lett.* **2014**, *133*, 208–211.
74. Bai, H.; Liu, Z.; Sun, D.D. Hierarchically multifunctional TiO_2 nano-thorn membrane for water purification. *Chem. Commun.* **2010**, *46*, 6542–6544. [CrossRef] [PubMed]
75. Liu, B.; Nakata, K.; Sakai, M.; Saito, H.; Ochiai, T.; Murakami, T.; Takagi, K.; Fujishima, A. Mesoporous TiO_2 Core–Shell Spheres Composed of Nanocrystals with Exposed High-Energy Facets: Facile Synthesis and Formation Mechanism. *Langmuir* **2011**, *27*, 8500–8508. [CrossRef] [PubMed]
76. Ata, U.; Lutfur, R.; Syed Zajif, H.; Wasim, A.; Abdul, T.; Asim, J.; Sadia Zafar, B.; Waheed, S.K.; Rabia, R.; Irshad, H.; et al. Mechanistic insight of dye degradation using TiO_2 anchored α-MnO_2 nanorods as promising sunlight driven photocatalyst. *Mater. Sci. Eng. B* **2021**, *271*, 115341.
77. Sarah Mozzaquatro, P.; Alexsandra, V.; Guilin, Y.; Jingfeng, W.; Selene, M.A.G.U.; Dachamir, H.; Antônio Augusto, U. An overview on nanostructured TiO_2–containing fibers for photocatalytic degradation of organic pollutants in wastewater treatment. *J. Water Process Eng.* **2021**, *40*, 101827.
78. Thammasak, R.; Orawan, R.; Sanya, S.; Sivakorn, A. Application of TiO_2 nanotubes as photocatalysts for decolorization of synthetic dye wastewater. *Water Resour. Ind.* **2021**, *26*, 100163.
79. Lee, S.Y.; Kang, D.; Jeong, S.; Do, H.T.; Kim, J.H. Photocatalytic Degradation of Rhodamine B Dye by TiO_2 and Gold Nanoparticles Supported on a Floating Porous Polydimethylsiloxane Sponge under Ultraviolet and Visible Light Irradiation. *ACS Omega* **2020**, *5*, 4233–4241. [CrossRef] [PubMed]
80. Yao, L.; Haas, T.W.; Guiseppi-Elie, A.; Bowlin, G.L.; Simpson, D.G.; Wnek, G.E. Electrospinning and stabilization of fully hydrolyzed poly (vinyl alcohol) fibers. *Chem. Mater.* **2003**, *15*, 1860–1864. [CrossRef]
81. Katsumata, K.-i.; Okazaki, S.; Cordonier, C.E.J.; Shichi, T.; Sasaki, T.; Fujishima, A. Preparation and Characterization of Self-Cleaning Glass for Vehicle with Niobia Nanosheets. *ACS Appl. Mater. Interfaces* **2010**, *2*, 1236–1241. [CrossRef]
82. Zhang, M.; Wang, C.; Wang, Y.; Li, S.; Zhang, X.; Liu, Y. Tunable bismuth doping/loading endows $NaTaO_3$ nanosheet highly selective photothermal reduction of CO_2. *Nano Res.* **2023**, *16*, 2142–2151. [CrossRef]
83. Zhen, J.; Fan-Li, M.; Yong, J.; Tao, L.; Jin-Yun, L.; Bai, S.; Jin, W.; Jin-Huai, L.; Xing-Jiu, H. Porous TiO_2 nanowires derived from nanotubes: Synthesis, characterzation and their enhanced photocatalytic properties. *Microporous Mesoporous Mater.* **2013**, *181*, 146–153.
84. Wu, H.B.; Hng, H.H.; Lou, X.W. Direct Synthesis of Anatase TiO_2 Nanowires with Enhanced Photocatalytic Activity. *Adv. Mater.* **2012**, *24*, 2567–2571. [CrossRef] [PubMed]
85. Sadaf Bashir, K.; Mengjing, H.; Shuang, S.; Zhengjun, Z. Morphological influence of TiO_2 nanostructures (nanozigzag, nanohelics and nanorod) on photocatalytic degradation of organic dyes. *Appl. Surf. Sci.* **2017**, *400*, 184–193.
86. Krishnamoorthy, S.; Ronen, B.-Z.; Orit, M.; Tomer, Z. Controllable synthesis of TiO_2 nanoparticles and their photocatalytic activity in dye degradation. *Mater. Res. Bull.* **2020**, *126*, 110842.
87. Saba, H.; Masoud, J. Synthesis of TiO_2 nanoparticles coated on cellulose nanofibers with different morphologies: Effect of the template and sol-gel parameters. *Mater. Sci. Semicond. Process.* **2020**, *109*, 104927.
88. Gihoon, C.; Patrik, S.; Marco, A. Anodic TiO_2 nanotube membranes: Site-selective Pt-activation and photocatalytic H2 evolution. *Electrochim. Acta* **2017**, *258*, 302–310.
89. Macak, J.M.; Tsuchiya, H.; Taveira, L.; Aldabergerova, S.; Schmuki, P. Smooth Anodic TiO_2 Nanotubes. *Angew. Chem. Int. Ed.* **2005**, *44*, 7463–7465. [CrossRef]
90. Kar, A.; Smith, Y.R.; Subramanian, V. Improved Photocatalytic Degradation of Textile Dye Using Titanium Dioxide Nanotubes Formed Over Titanium Wires. *Environ. Sci. Technol.* **2009**, *43*, 3260–3265. [CrossRef]
91. Kowaka, Y.; Nozaki, K.; Mihara, T.; Yamashita, K.; Miura, H.; Tan, Z.; Ohara, S. Development of TiO_2 Nanosheets with High Dye Degradation Performance by Regulating Crystal Growth. *Materials* **2023**, *16*, 1229. [CrossRef]
92. Nair, A.K.; Roy George, D.; Jos Baby, N.; Reji, M.; Joseph, S. Solar dye degradation using TiO_2 nanosheet based nanocomposite floating photocatalyst. *Mater. Today Proc.* **2021**, *46*, 2747–2751. [CrossRef]
93. Faisal, S.; Jan, H.; Shah, S.A.; Shah, S.; Khan, A.; Akbar, M.T.; Rizwan, M.; Jan, F.; Wajidullah; Akhtar, N.; et al. Green Synthesis of Zinc Oxide (ZnO) Nanoparticles Using Aqueous Fruit Extracts of Myristica fragrans: Their Characterizations and Biological and Environmental Applications. *ACS Omega* **2021**, *6*, 9709–9722. [CrossRef] [PubMed]
94. Manikanika; Lalita, C. Photocatalytic activity of zinc oxide for dye and drug degradation: A review. *Mater. Today Proc.* **2022**, *52*, 1653–1656. [CrossRef]
95. Uribe-López, M.C.; Hidalgo-López, M.C.; López-González, R.; Frías-Márquez, D.M.; Núñez-Nogueira, G.; Hernández-Castillo, D.; Alvarez-Lemus, M.A. Photocatalytic activity of ZnO nanoparticles and the role of the synthesis method on their physical and chemical properties. *J. Photochem. Photobiol. A Chem.* **2021**, *404*, 112866. [CrossRef]
96. Borysiewicz, M.A. ZnO as a Functional Material, a Review. *Crystals* **2019**, *9*, 505. [CrossRef]
97. Ewelina, G.; Martyna, M.; Marta, P.-G.; Adriana, Z.-M. Metal oxide photocatalysts. In *Metal Oxides*; Elsevier: Amsterdam, The Netherlands, 2018; pp. 51–209. [CrossRef]

98. Sotirios, B.; Panayotis, G.; Spyros, N.Y.; Vassilios, D.; Lajos, T.; Athanassios, C.; Nikolaos, B. Preparation of ZnO nanoparticles by thermal decomposition of zinc alginate. *Thin Solid Film.* **2007**, *515*, 8461–8464.
99. Yin, X.; Wang, B.; He, M.; He, T. Facile synthesis of ZnO nanocrystals via a solid state reaction for high performance plastic dye-sensitized solar cells. *Nano Res.* **2012**, *5*, 1–10. [CrossRef]
100. Hasnidawani, J.N.; Azlina, H.N.; Norita, H.; Bonnia, N.N.; Ratim, S.; Ali, E.S. Synthesis of ZnO Nanostructures Using Sol-Gel Method. *Procedia Chem.* **2016**, *19*, 211–216. [CrossRef]
101. Singh, A.K.; Viswanath, V.; Janu, V.C. Synthesis, effect of capping agents, structural, optical and photoluminescence properties of ZnO nanoparticles. *J. Lumin.* **2009**, *129*, 874–878. [CrossRef]
102. Clament Sagaya Selvam, N.; Vijaya, J.J.; Kennedy, L.J. Effects of Morphology and Zr Doping on Structural, Optical, and Photocatalytic Properties of ZnO Nanostructures. *Ind. Eng. Chem. Res.* **2012**, *51*, 16333–16345. [CrossRef]
103. Xu, L.; Hu, Y.-L.; Pelligra, C.; Chen, C.-H.; Jin, L.; Huang, H.; Sithambaram, S.; Aindow, M.; Joesten, R.; Suib, S.L. ZnO with Different Morphologies Synthesized by Solvothermal Methods for Enhanced Photocatalytic Activity. *Chem. Mater.* **2009**, *21*, 2875–2885. [CrossRef]
104. Sakthivel, S.; Neppolian, B.; Shankar, M.V.; Arabindoo, B.; Palanichamy, M.; Murugesan, V. Solar photocatalytic degradation of azo dye: Comparison of photocatalytic efficiency of ZnO and TiO_2. *Sol. Energy Mater. Sol. Cells* **2003**, *77*, 65–82. [CrossRef]
105. Ramesh, V.; Raja, S.; Pugazhendhi, A.; Thivaharan, V. Synthesis, characterization and photocatalytic dye degradation capability of Calliandra haematocephala-mediated zinc oxide nanoflowers. *J. Photochem. Photobiol. B Biol.* **2020**, *203*, 111760.
106. Kian Mun, L.; Chin Wei, L.; Koh Sing, N.; Joon Ching, J. Recent developments of zinc oxide based photocatalyst in water treatment technology: A review. *Water Res.* **2016**, *88*, 428–448.
107. Gerischer, H. Electrochemical behavior of semiconductors under illumination. *J. Electrochem. Soc.* **1966**, *113*, 1174. [CrossRef]
108. Sonik, B.; Neha, V. Photocatalytic activity of ZnO nanoparticles with optimization of defects. *Mater. Res. Bull.* **2017**, *95*, 468–476.
109. Dodoo-Arhin, D.; Asiedu, T.; Agyei-Tuffour, B.; Nyankson, E.; Obada, D.; Mwabora, J.M. Photocatalytic degradation of Rhodamine dyes using zinc oxide nanoparticles. *Mater. Today Proc.* **2021**, *38*, 809–815. [CrossRef]
110. Lakshi, S.; Diganta, B.; Mrinal, S.; Banajit, M.; Dipak Kumar, D.; Pinaki, S. Photocatalytic performance of ZnO nanomaterials for self sensitized degradation of malachite green dye under solar light. *Appl. Catal. A Gen.* **2015**, *490*, 42–49.
111. Jing, W.; Yi, X.; Yan, D.; Ruosong, C.; Lan, X.; Sridhar, K. Defect-rich ZnO nanosheets of high surface area as an efficient visible-light photocatalyst. *Appl. Catal. B Environ.* **2016**, *192*, 8–16.
112. Le Pivert, M.; Poupart, R.; Capochichi-Gnambodoe, M.; Martin, N.; Leprince-Wang, Y. Direct growth of ZnO nanowires on civil engineering materials: Smart materials for supported photodegradation. *Microsyst. Nanoeng.* **2019**, *5*, 57. [CrossRef]
113. Kiwaan, H.A.; Atwee, T.M.; Azab, E.A.; El-Bindary, A.A. Efficient photocatalytic degradation of Acid Red 57 using synthesized ZnO nanowires. *J. Chin. Chem. Soc.* **2019**, *66*, 89–98. [CrossRef]
114. Wang, H.; Cai, Y.; Wang, C.; Xu, H.; Fang, J.; Yang, Y. Seeded growth of ZnO nanowires in dye-containing solution: The submerged plant analogy and its application in photodegradation of dye pollutants. *CrystEngComm* **2020**, *22*, 4154–4161. [CrossRef]
115. Dhineshbabu, N.R.; Rajendran, V.; Nithyavathy, N.; Vetumperumal, R. Study of structural and optical properties of cupric oxide nanoparticles. *Appl. Nanosci.* **2016**, *6*, 933–939. [CrossRef]
116. Murali, D.S.; Aryasomayajula, S. Thermal conversion of Cu_4O_3 into CuO and Cu_2O and the electrical properties of magnetron sputtered Cu_4O_3 thin films. *Appl. Phys. A* **2018**, *124*, 279. [CrossRef]
117. Wang, Y.; Lany, S.; Ghanbaja, J.; Fagot-Revurat, Y.; Chen, Y.P.; Soldera, F.; Horwat, D.; Mücklich, F.; Pierson, J.F. Electronic structures of Cu_2O, Cu_4O_3, and CuO: A joint experimental and theoretical study. *Phys. Rev. B* **2016**, *94*, 245418. [CrossRef]
118. Weldegebrieal, G.K. Photocatalytic and antibacterial activity of CuO nanoparticles biosynthesized using Verbascum thapsus leaves extract. *Optik* **2020**, *204*, 164230. [CrossRef]
119. Phang, Y.-K.; Aminuzzaman, M.; Akhtaruzzaman, M.; Muhammad, G.; Ogawa, S.; Watanabe, A.; Tey, L.-H. Green Synthesis and Characterization of CuO Nanoparticles Derived from Papaya Peel Extract for the Photocatalytic Degradation of Palm Oil Mill Effluent (POME). *Sustainability* **2021**, *13*, 796. [CrossRef]
120. Akintelu, S.A.; Folorunso, A.S.; Folorunso, F.A.; Oyebamiji, A.K. Green synthesis of copper oxide nanoparticles for biomedical application and environmental remediation. *Heliyon* **2020**, *6*, e04508. [CrossRef]
121. Liu, J.; Jin, J.; Deng, Z.; Huang, S.Z.; Hu, Z.Y.; Wang, L.; Wang, C.; Chen, L.-H.; Li, Y.; Tendeloo, G.V.; et al. Tailoring CuO nanostructures for enhanced photocatalytic property. *J. Colloid Interface Sci.* **2012**, *384*, 1–9. [CrossRef]
122. Lufeng, Y.; Deqing, C.; Limin, W. CuO core–shell nanostructures: Precursor-mediated fabrication and visible-light induced photocatalytic degradation of organic pollutants. *Powder Technol.* **2016**, *287*, 346–354.
123. Chiang, C.-Y.; Kosi, A.; Nicholas, F.; Vibha Rani, S.; Sahab, D.; Sheryl, E. Copper oxide nanoparticle made by flame spray pyrolysis for photoelectrochemical water splitting—Part II. Photoelectrochemical study. *Int. J. Hydrog. Energy* **2011**, *36*, 15519–15526. [CrossRef]
124. Sapkota, K.P.; Lee, I.; Hanif, M.A.; Islam, M.A.; Akter, J.; Hahn, J.R. Enhanced Visible-Light Photocatalysis of Nanocomposites of Copper Oxide and Single-Walled Carbon Nanotubes for the Degradation of Methylene Blue. *Catalysts* **2020**, *10*, 297. [CrossRef]
125. Latif, S.; Abdulaziz, F.; Alanazi, A.M.; Alsehli, A.H.; Alsowayigh, M.M.; Alanazi, A.A. Effect of H_2O_2 @CuONPs in the UV Light-Induced Removal of Organic Pollutant Congo Red Dye: Investigation into Mechanism with Additional Biomedical Study. *Molecules* **2023**, *28*, 410. [CrossRef]

126. Miyauchi, M.; Nakajima, A.; Watanabe, T.; Hashimoto, K. Photocatalysis and Photoinduced Hydrophilicity of Various Metal Oxide Thin Films. *Chem. Mater.* **2002**, *14*, 2812–2816. [CrossRef]
127. Yang, M.; He, J. Fine tuning of the morphology of copper oxide nanostructures and their application in ambient degradation of methylene blue. *J. Colloid Interface Sci.* **2011**, *355*, 15–22. [CrossRef]
128. Zhang, W.; Yang, Z.; Wang, X.; Zhang, Y.; Wen, X.; Yang, S. Large-scale synthesis of β-MnO_2 nanorods and their rapid and efficient catalytic oxidation of methylene blue dye. *Catal. Commun.* **2006**, *7*, 408–412. [CrossRef]
129. Wang, L.; Zhou, Q.; Zhang, G.; Liang, Y.; Wang, B.; Zhang, W.; Lei, B.; Wang, W. A facile room temperature solution-phase route to synthesize CuO nanowires with enhanced photocatalytic performance. *Mater. Lett.* **2012**, *74*, 217–219. [CrossRef]
130. Sadollahkhani, A.; Hussain Ibupoto, Z.; Elhag, S.; Nur, O.; Willander, M. Photocatalytic properties of different morphologies of CuO for the degradation of Congo red organic dye. *Ceram. Int.* **2014**, *40*, 11311–11317. [CrossRef]
131. Rao, M.P.; Wu, J.J.; Syed, A.; Ameen, F.; Anandan, S. Synthesis of Dandelion—Like CuO microspheres for photocatalytic degradation of reactive black-5. *Mater. Res. Express* **2018**, *5*, 015053. [CrossRef]
132. George, A.; Raj DM, A.; Venci, X.; Raj, A.D.; Irudayaraj, A.A.; Josephine, R.L.; Kaviyarasu, K. Photocatalytic effect of CuO nanoparticles flower-like 3D nanostructures under visible light irradiation with the degradation of methylene blue (MB) dye for environmental application. *Environ. Res.* **2022**, *203*, 111880. [CrossRef]
133. Nazim, M.; Khan, A.A.P.; Asiri, A.M.; Kim, J.H. Exploring Rapid Photocatalytic Degradation of Organic Pollutants with Porous CuO Nanosheets: Synthesis, Dye Removal, and Kinetic Studies at Room Temperature. *ACS Omega* **2021**, *6*, 2601–2612. [CrossRef]
134. Adawiya, J.H.; Riyad; Hiba, M.S.; Mohammed, J.H. Photocatalytic Activity of Nickel Oxide. *J. Mater. Res. Technol.* **2019**, *8*, 2802–2808.
135. Ahmed, A.S.A.; Xiang, W.; Abdelmotalleib, M.; Zhao, X. Efficient NiO Impregnated Walnut Shell-Derived Carbon for Dye-Sensitized Solar Cells. *ACS Appl. Electron. Mater.* **2022**, *4*, 1063–1071. [CrossRef]
136. Wei, S.; Di Lecce, D.; Brescia, R.; Pugliese, G.; Shearing, P.R.; Hassoun, J. Electrochemical behavior of nanostructured NiO@C anode in a lithium-ion battery using $LiNi_{1/3}Co_{1/3}Mn_{1/3}O_2$ cathode. *J. Alloys Compd.* **2020**, *844*, 155365. [CrossRef]
137. Ethiraj, A.S.; Uttam, P.; Varunkumar, K.; Chong, K.F.; Ali, G.A. Photocatalytic performance of a novel semiconductor nanocatalyst: Copper doped nickel oxide for phenol degradation. *Mater. Chem. Phys.* **2020**, *242*, 122520. [CrossRef]
138. Jayakumar, G.; Albert Irudayaraj, A.; Dhayal Raj, A. Photocatalytic Degradation of Methylene Blue by Nickel Oxide Nanoparticles. *Mater. Today Proc.* **2017**, *4*, 11690–11695. [CrossRef]
139. Khan, N.A.; Saeed, K.; Khan, I.; Gul, T.; Sadiq, M.; Uddin, A.; Zekker, I. Efficient photodegradation of orange II dye by nickel oxide nanoparticles and nanoclay supported nickel oxide nanocomposite. *Appl. Water Sci.* **2022**, *12*, 131. [CrossRef]
140. Barzinjy, A.A.; Hamad, S.M.; Aydın, S.; Ahmed, M.H.; Hussain, F.H.S. Green and eco-friendly synthesis of Nickel oxide nanoparticles and its photocatalytic activity for methyl orange degradation. *J. Mater. Sci. Mater. Electron.* **2020**, *31*, 11303–11316. [CrossRef]
141. Hamidian, K.; Rigi, A.H.; Najafidoust, A.; Sarani, M.; Miri, A. Study of photocatalytic activity of green synthesized nickel oxide nanoparticles in the degradation of acid orange 7 dye under visible light. *Bioprocess Biosyst. Eng.* **2021**, *44*, 2667–2678. [CrossRef]
142. Ahmed, B.; Kumar, S.; Ojha, A.K.; Donfack, P.; Materny, A. Facile and controlled synthesis of aligned WO_3 nanorods and nanosheets as an efficient photocatalyst material. *Spectrochim. Acta Part A Mol. Biomol. Spectrosc.* **2017**, *175*, 250–261. [CrossRef]
143. Mzimela, N.; Tichapondwa, S.; Chirwa, E. Visible-light-activated photocatalytic degradation of rhodamine B using WO_3 nanoparticles. *RSC Adv.* **2022**, *12*, 34652–34659. [CrossRef]
144. Acedo-Mendoza, A.; Infantes-Molina, A.; Vargas-Hernández, D.; Chávez-Sánchez, C.; Rodríguez-Castellón, E.; Tánori-Córdova, J. Photodegradation of methylene blue and methyl orange with CuO supported on ZnO photocatalysts: The effect of copper loading and reaction temperature. *Mater. Sci. Semicond. Process.* **2020**, *119*, 105257. [CrossRef]
145. Dong, P.; Hou, G.; Xi, X.; Shao, R.; Dong, F. WO_3-based photocatalysts: Morphology control, activity enhancement and multifunctional applications. *Environ. Sci. Nano* **2017**, *4*, 539–557. [CrossRef]
146. Ma, H.; Tsai, S.-B. Design of research on performance of a new iridium coordination compound for the detection of Hg^{2+}. *Int. J. Environ. Res. Public Health* **2017**, *14*, 1232. [CrossRef] [PubMed]
147. Cao, Z.; Qin, M.; Jia, B.; Gu, Y.; Chen, P.; Volinsky, A.A.; Qu, X. One pot solution combustion synthesis of highly mesoporous hematite for photocatalysis. *Ceram. Int.* **2015**, *41*, 2806–2812. [CrossRef]
148. Yin, X.; Liu, L.; Ai, F. Enhanced Photocatalytic Degradation of Methylene Blue by WO_3 Nanoparticles Under NIR Light Irradiation. *Front. Chem.* **2021**, *9*, 683765. [CrossRef]
149. Tahir, M.; Nabi, G.; Hassan, M.; Iqbal, T.; Kiran, H.; Majid, A. Morphology tailored synthesis of C-WO_3 nanostructures and its photocatalytic application. *J. Inorg. Organomet. Polym. Mater.* **2018**, *28*, 738–745. [CrossRef]
150. Kumar, S.G.; Devi, L.G. Review on modified TiO_2 photocatalysis under UV/visible light: Selected results and related mechanisms on interfacial charge carrier transfer dynamics. *J. Phys. Chem. A* **2011**, *115*, 13211–13241. [CrossRef]

Disclaimer/Publisher's Note: The statements, opinions and data contained in all publications are solely those of the individual author(s) and contributor(s) and not of MDPI and/or the editor(s). MDPI and/or the editor(s) disclaim responsibility for any injury to people or property resulting from any ideas, methods, instructions or products referred to in the content.

Article

The Influence of Annealing Temperature on the Microstructure and Electrical Properties of Sputtered ZnO Thin Films

Adil Alshoaibi

Department of Physics, College of Science, King Faisal University, P.O. Box 400, Al Ahsa 31982, Saudi Arabia; adshoaibi@kfu.edu.sa

Abstract: Thin films are the backbone of the electronics industry, and their widespread application in heat sensors, solar cells, and thin-film transistors has attracted the attention of researchers. The current study involves the deposition of a hetero-structured (ZnO/Zn/ZnO) thin film on a well-cleaned glass substrate using the DC magnetron sputtering technique. The samples were then annealed at 100, 200, 300, 400, and 500 °C. The structural, morphological, and electrical characteristics of the annealed samples as well as one as-deposited sample were then examined using atomic force microscopy (AFM), scanning electron microscopy (SEM), X-ray diffraction (XRD), and a Hall effect measuring apparatus. XRD analysis showed a hexagonal ZnO crystal structure for the samples annealed at 300 and 400 °C, whereas the samples annealed at 100 and 200 °C showed metallic zinc and hexagonal ZnO, and the crystallinity decreased for the sample annealed at 500 °C with pure hexagonal crystal symmetry. According to the AFM study, as the annealing temperature increases, the average roughness (R_a) decreases. Temperature has an inverse relationship with particle size. The optimal annealing temperature was determined to be 400 °C. Over this temperature range, the average roughness and particle size increased. Similarly, when R_a decreased, the conductivity increased and the resistance decreased. A fundamental difficulty is that the heating of the heterostructure to 400 °C melts the Zn-based intermediate layer, which alters the Zn phase and disrupts the sample homogeneity.

Keywords: ZnO; Zn; heterostructure; annealing; sputtering

Citation: Alshoaibi, A. The Influence of Annealing Temperature on the Microstructure and Electrical Properties of Sputtered ZnO Thin Films. *Inorganics* **2024**, *12*, 236. https://doi.org/10.3390/inorganics12090236

Academic Editors: Roberto Nisticò and Silvia Mostoni

Received: 21 July 2024
Revised: 18 August 2024
Accepted: 26 August 2024
Published: 29 August 2024

Copyright: © 2024 by the author. Licensee MDPI, Basel, Switzerland. This article is an open access article distributed under the terms and conditions of the Creative Commons Attribution (CC BY) license (https:// creativecommons.org/licenses/by/ 4.0/).

1. Introduction

Semiconductor nanomaterials play a significant role in renewable energy, notably in thin-film solar cells, and have attracted the attention of scientists and researchers worldwide [1]. Many elemental and compound semiconductor nanomaterials have been investigated, with ZnO being one of the most prominent and promising [2]. ZnO is a binary II–VI compound semiconductor material having Wurtzite crystal structure, a wide and direct bandgap of 3.3 eV at ambient temperature, and an exciton binding energy of 60 meV [3–5]. It can be grown in several different types of nanoscales, thus making it possible to obtain various novel products [6]. Moreover, the properties of ZnO can be altered by fabricating thin films [7]. ZnO thin films have been extensively researched in various areas due to their high bond strength [8], good optical performance [9], severe exciton stabilisation [10], and outstanding piezoelectric properties [11], and they have many prospective applications in multiple technological areas, such as clear film/electrodes [12] in screen systems and solar energy [13]. Another benefit of ZnO compared to other metals is its substantial cost, which makes it an extremely prospective candidate for industrial applications [14]. ZnO is currently one of the leading materials used as a window layer [9], transparent conducting oxide (TCO) [15,16], and buffer layer in the solar cell industry [17].

The transparent conductive oxides (TCOs) thin films have several options in optoelectronic devices. Such thin films can be used especially for organic light-emitting diodes (OLEDs) [16], solar cells, heat sensors [18,19], and thin-film transistors (TFTs) [20]. One of the most studied and industrially used TCO is tin-doped indium oxide (ITO) [17,21]. The latest rise in consumption by the optoelectronic devices sector ITO has become rare and more expensive [6,22]. Because of its advantages in several fields and its unique characteristics like low price, non-toxicity, and good chemical stability in maintaining plasma, an extremely successful substitute for indium tin oxide (ITO) is ZnO [21,23]. Native and extrinsic defects in ZnO nanostructures, however, are considered to reach profound concentrations that limit their application performance [24]. A detailed understanding of the type, composition, and electronic parameters of deep-level facilities is the key to understanding and controlling electronic characteristics [7,25]. Understanding surface defect behaviour is essential for ZnO to be successfully applied [26]. Inherent defects and vacancies in ZnO are mainly classified into four kinds: surface defects including O vacancies (V_O) and Zn vacancies (V_{Zn}) and interstitials (Zn and O), which are part of the majority of the fabric [27,28]. ZnO's large-scale development process regulates the development of its inherent flaws [29,30]. It is recognized that the number of failures depends on a post-growth sample therapy, which can change its characteristics significantly [31,32]. However, in many practical systems, no control over defects is one of the major problems in using ZnO [33,34]. Thus, one way to control point and surface defects is to post-anneal the ZnO nanoparticles to obtain high UV photodetection in a short interval of time [26,32]. Higher temperature treatment in air and N_2 results in excellent monitoring of surface-related abnormalities such as V_O and V_{Zn} and decreases radiative recombination of the surface defects [22,35]. To achieve a defect-free ZnO thin film, we used the magnetron sputtering method to deposit thin film at room temperature with a ZnO/Zn/ZnO heterostructure accompanied by thermal annealing at various temperatures. The objective of the ZnO/Zn/ZnO heterostructure was to obtain a defect-free ZnO thin film by harnessing the characteristics of the intermediate Zn layer. The metallic Zn layer exhibits a 'surfactant effect', promoting recrystallization during the deposition process. In conjunction with moderate-temperature annealing, the Zn layer facilitates the 'therapy' of surrounding ZnO crystallites, enhancing the overall crystalline quality by addressing lattice mismatches and thermal expansion differences, thereby reducing defects in the ZnO layers [36,37]. Furthermore, the Zn layer facilitates grain coalescence during thermal treatment, resulting in larger grain sizes and fewer grain boundaries, which are crucial factors in achieving a high-quality, defect-free ZnO thin film.

2. Results and Discussion

2.1. Surface Morphology

2.1.1. Atomic Force Microscopy (AFM) Analysis

Figure 1 shows atomic force microscopy (AFM) images of annealed thin films in comparison to as-deposited films, whereas Figure 2 describes the influence of annealing temperature on the average roughness (R_a) of the films. R_a measures the average length between the peaks and valleys and the deviation from the mean line on the entire surface within the sampling length. From Figure 1, one can observe that Ra first increases with an increase in the annealing temperature and then decreases to a minimum value. A minimum roughness was observed for the sample annealed at 400 °C, which is about 4.95 nm. A further increase in the annealing temperature results in an increase in the Ra value. A maximum R_a value was obtained for the sample annealed at 500 °C, which is about 10.60 nm. The initial decrease in roughness with increasing annealing temperature was attributed to the coalescence of the grains. A further increase in roughness with an increase in the annealing temperature may be due to the further increase in grain size. Moreover, the heterostructure is composed of three layers; that is, Zn is sandwiched between the ZnO layers. The melting point of Zn is 420 °C, whereas that of ZnO is 1975 °C. Increasing the annealing temperature from 400 °C to 500 °C changed the Zn phase from solid to liquid. The liquid Zn layer might penetrate or percolate into the interstitial pores in the underlying ZnO layers. This penetration can lead to a disruption in the uniformity of the ZnO structure, as the liquid Zn may fill voids and create localized regions of different densities or compositions. As can be seen, an increase in the peaks and valleys in the film results in an increase in the average roughness above the optimum value of the annealing temperature. The infiltration of liquid Zn into the ZnO layers can also promote the formation of microcracks or other structural imperfections as the film cools down and as Zn solidifies, leading to an increase in the R_a of the film. Furthermore, the increase in R_a at higher annealing temperatures can be linked to the thermodynamic instability introduced by the liquid phase of Zn. As Zn re-solidifies upon cooling, it may not return to its original crystalline orientation, leading to further imperfections and non-uniformities in the ZnO/Zn/ZnO heterostructure.

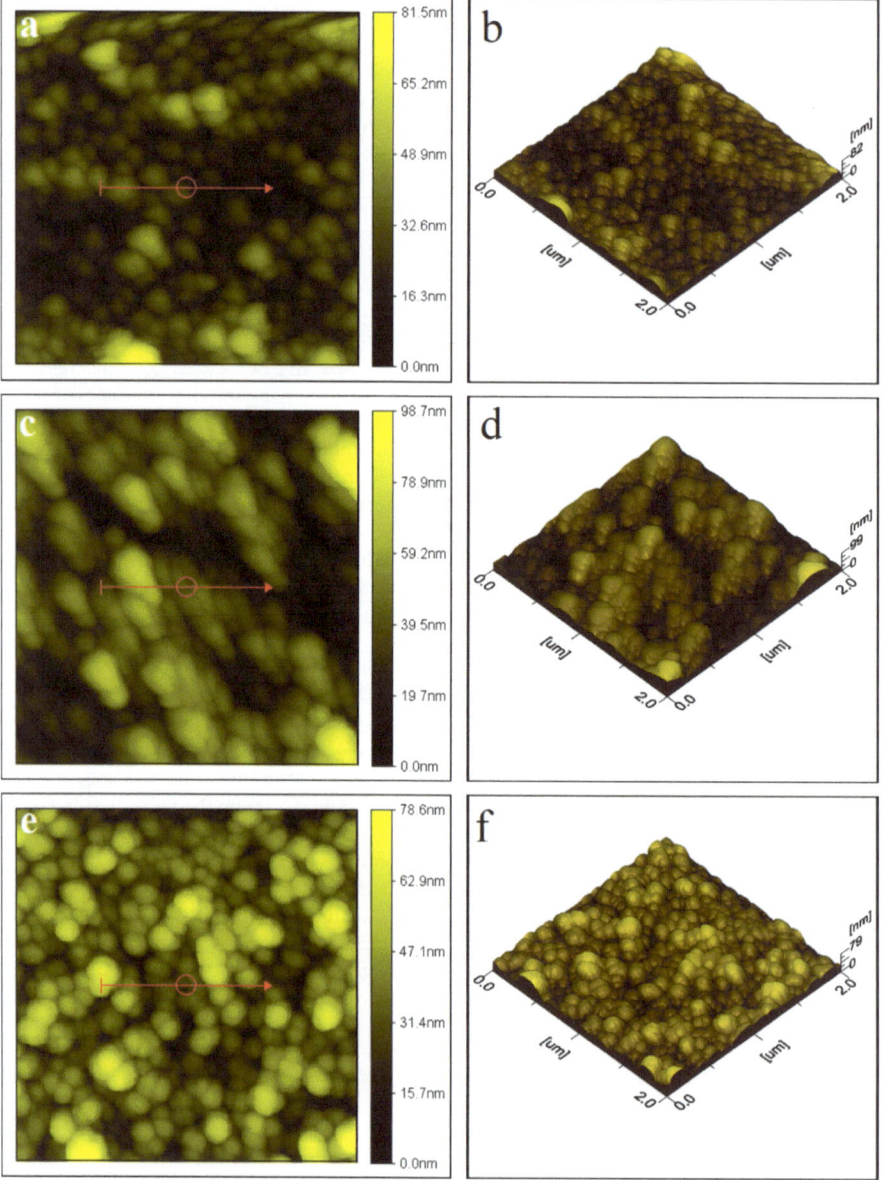

Figure 1. *Cont.*

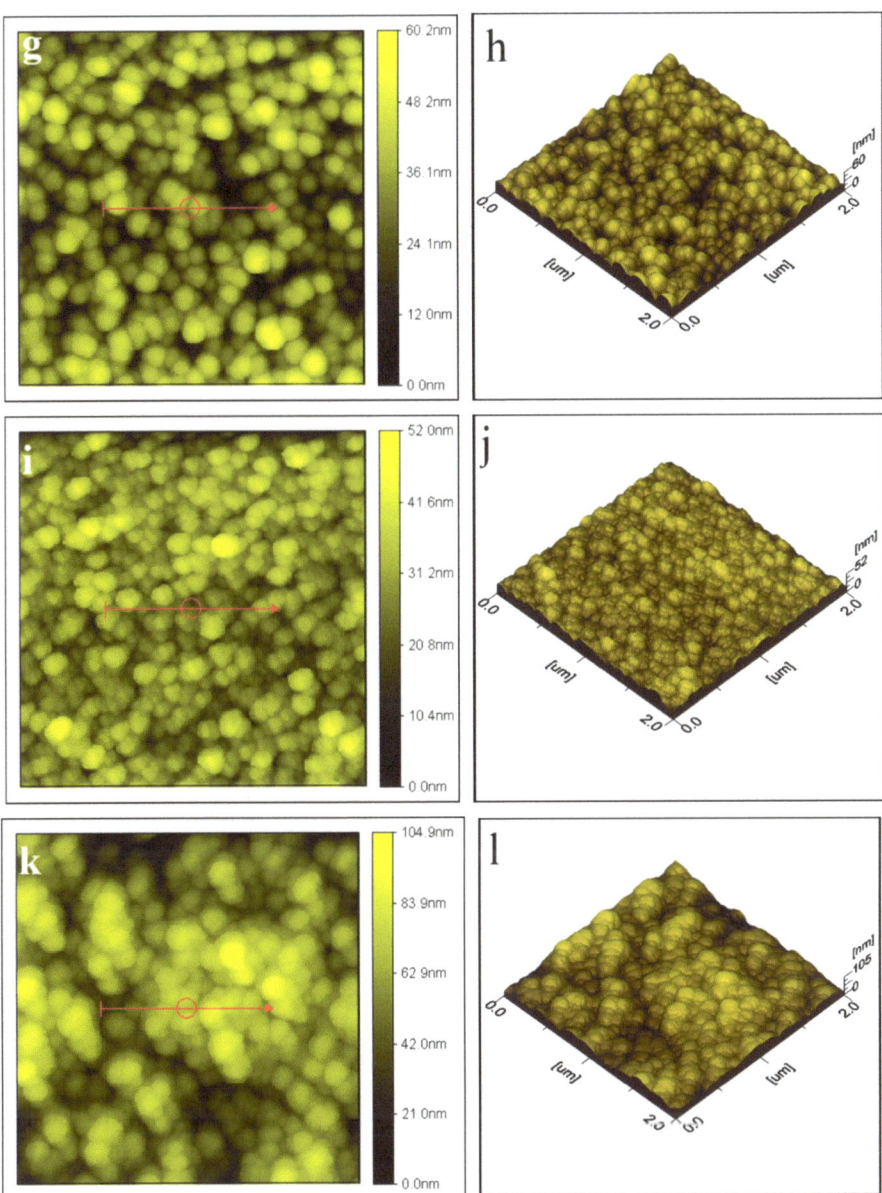

Figure 1. AFM 2D and 3D images of thin films: (**a**,**b**) as deposited, (**c**,**d**) annealed at 100 °C, (**e**,**f**) annealed at 200 °C, (**g**,**h**) annealed at 300 °C, (**i**,**j**) annealed at 400 °C, and (**k**,**l**) annealed at 500 °C.

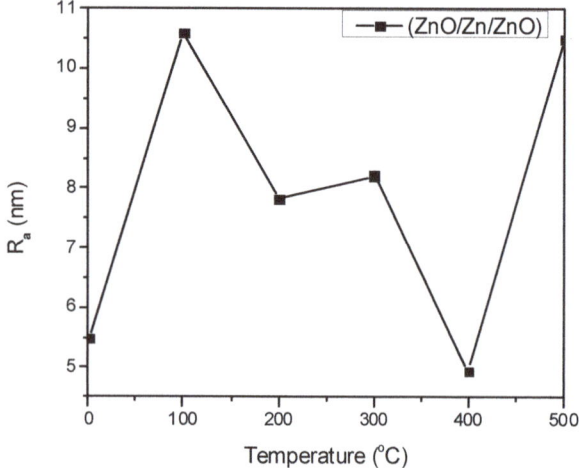

Figure 2. Average roughness (Ra) vs. annealing temperature of thin films.

2.1.2. Scanning Electron Microscopy (SEM) Analysis

Figure 3 shows the scanning electron microscopy (SEM) images of the ZnO/Zn/ZnO heterostructure of the as-deposited thin film in comparison with annealed thin films. The annealed samples have a smooth surface morphology compared to the as-deposited thin films. However, the surface smoothness is disturbed by the higher annealing temperature. Moreover, the average particle size also has a great impact on the annealing temperature. Table 1 lists the average particle size calculated for all the thin films along with the annealing temperature.

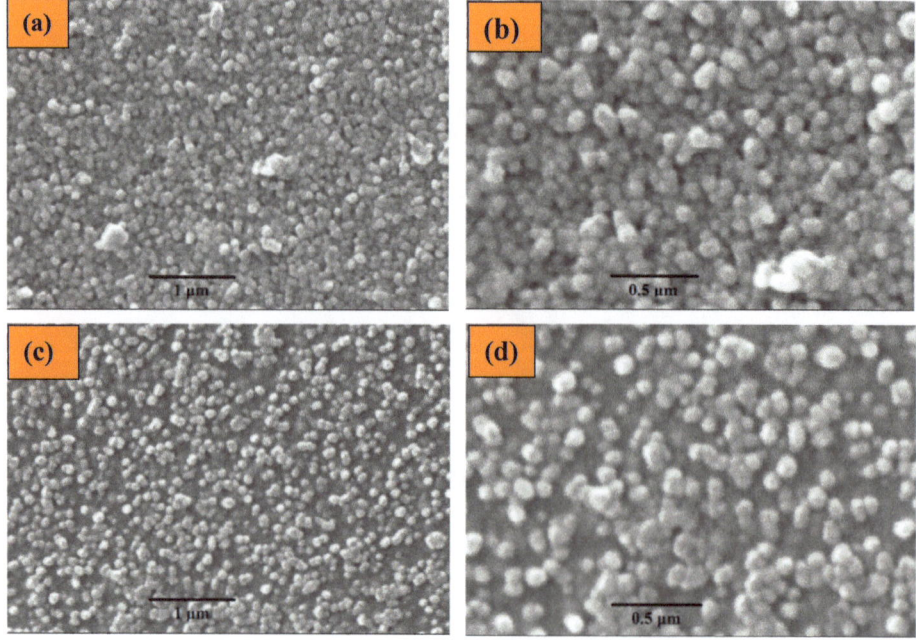

Figure 3. *Cont.*

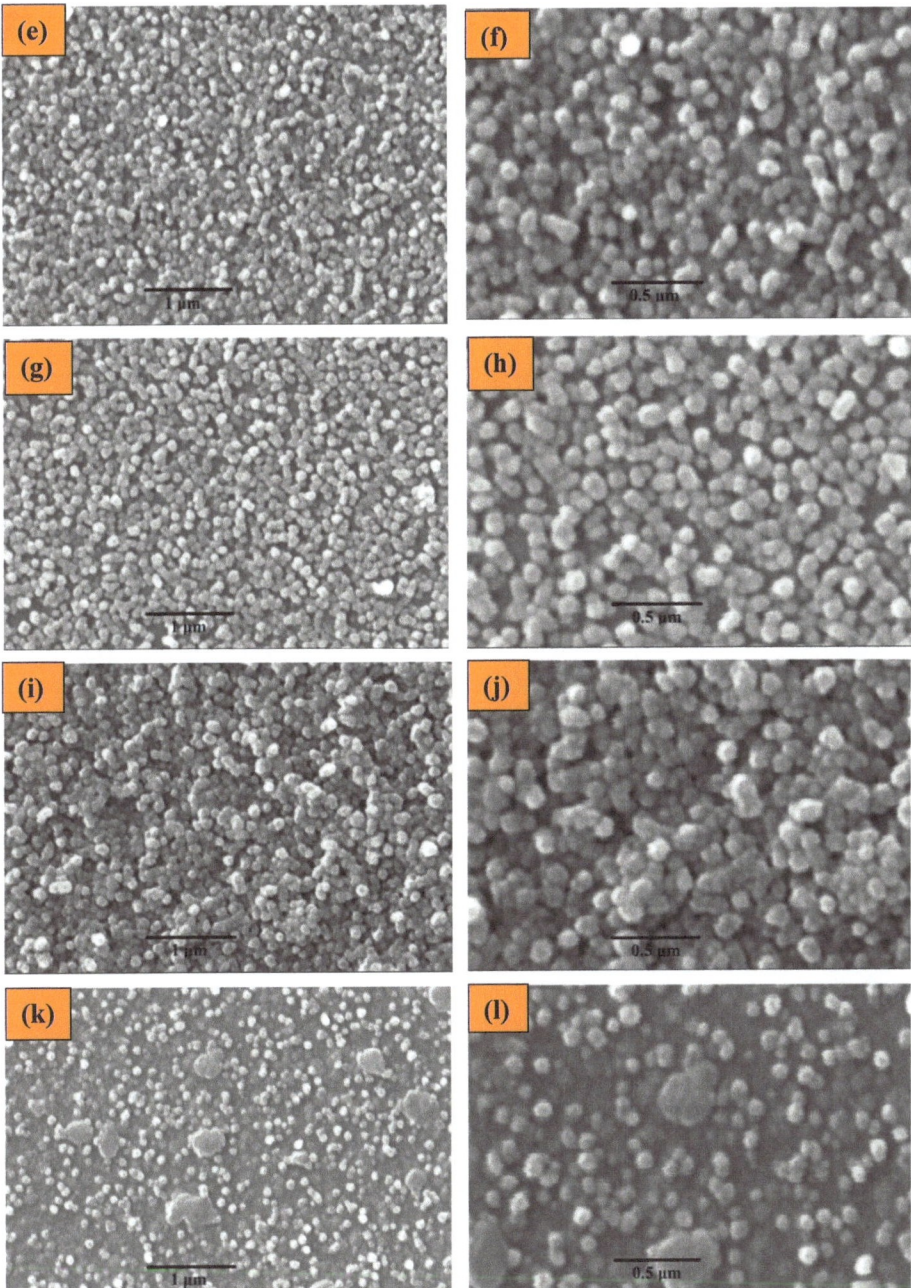

Figure 3. SEM images of thin films with low and high magnifications: (**a,b**) as deposited, (**c,d**) annealed at 100 °C, (**e,f**) annealed at 200 °C, (**g,h**) annealed at 300 °C, (**i,j**) annealed at 400 °C, and (**k,l**) annealed at 500 °C.

Table 1. Average particle size of thin films annealed at different temperatures in comparison to the as-deposited sample.

S. No.	Thin Film	Average Particle Size
1	As deposited	74.41 ± 1.75 nm
2	100 °C	78.08 ± 1.83 nm
3	200 °C	66.85 ± 1.62 nm
4	300 °C	54.46 ± 1.23 nm
5	400 °C	50.67 ± 1.02 nm
6	500 °C	163.00 ± 2.51 nm

From the table, we can see that the average particle size decreases with an increase in the post-heat treatment (annealing temperature). An optimum value of the particle size was calculated for the sample annealed at 400 °C. A further increase in temperature in the middle layer, which is comprised of Zn, affects the surface morphology and smoothness of the films. Figure 4 shows the energy dispersive X-ray (EDX) spectra of the film annealed at 400 °C, confirming that ZnO film is deposited.

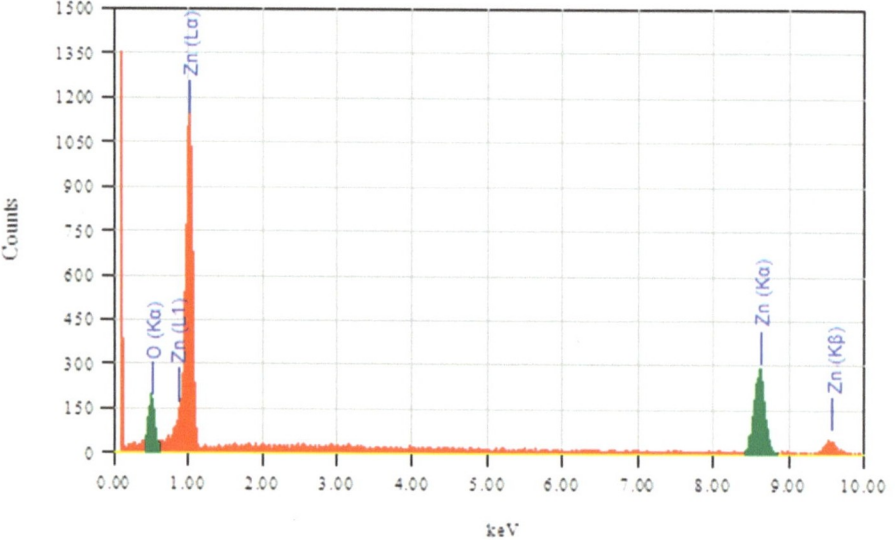

Figure 4. EDX spectra of hetero-structured ZnO/Zn/ZnO thin film annealed at 400 °C.

2.2. Structural Analysis

Figure 5 displays the XRD patterns of the as-prepared ZnO/Zn/ZnO thin films that were prepared and subjected to annealing temperatures of 100, 200, 300, 400, and 500 °C. XRD offers valuable information regarding the crystalline structure and phase composition of the samples at various annealing temperatures. The XRD patterns of the sample that underwent annealing at 100 °C indicated the existence of a zincite hexagonal crystal structure. The diffraction peaks observed in the figure match those of the hexagonal zincite ZnO. However, there was also an extra peak corresponding to metallic Zn. This suggests that a portion of the Zn still existed in its metallic form. The XRD spectrum of the sample annealed at 200 °C exhibited peaks corresponding to both Zn and ZnO. The identification of the ZnO peaks provides evidence for the creation of the hexagonal structure of zincite, whereas the Zn peak indicates that a portion of the zinc did no undergone conversion at this particular temperature. The samples annealed at 300 and 400 °C displayed XRD patterns indicative of pure ZnO with a hexagonal crystal structure. There were no discernible peaks in the spectra corresponding to metallic Zn. This indicates the complete transformation of Zn into ZnO, leading to the formation of a ZnO crystal structure with only one phase. Significantly, the strength of the diffraction peaks increased as the annealing temperature increased, culminating in the highest intensity at 400 °C. These findings indicate that the quality of the ZnO crystal structure is enhanced as the temperature increases, reaching its peak at 400 °C. However, the sample annealed at 500 °C exhibited a reduction in the intensity of the ZnO diffraction peaks in comparison to the sample annealed at 400 °C while still maintaining the pristine hexagonal structure of ZnO. The decrease in the maximum intensity at 500 °C can be ascribed to various factors. At elevated temperatures, excessive grain growth occurs, as discussed in the SEM and AFM results, and the particle size and roughness increase, resulting in the formation of imperfections within the crystal structure. The presence of these imperfections causes X-rays to disperse more widely, resulting in a decrease in the overall intensity of the diffraction patterns. Extended exposure to elevated temperatures can generate thermal stress in the thin film, which may lead to the formation of microcracks or other structural flaws that diminish the crystalline quality of the material.

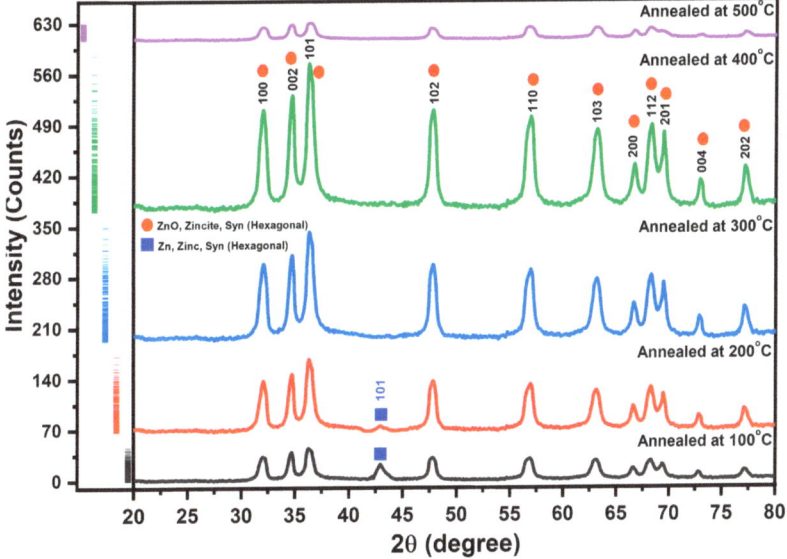

Figure 5. XRD spectra of ZnO/Zn/ZnO thin film annealed at 100, 200, 300, 400, and 500 °C.

2.3. Electrical Properties

Figure 6 shows the electrical resistivity and conductivity of the thin films annealed at different temperatures. In addition, Table 2 lists the electrical resistivity and conductivity values of the annealed and deposited thin films. From the figure and table, we can see that the electrical resistivity/conductivity of the film changes with an increase in the annealing temperature. The electrical behaviour demonstrates a complex relationship among particle size, surface roughness, and electrical conductivity, which is closely connected to the microstructural evolution during thermal annealing. The electrical conductivity of the thin films exhibited a substantial increase from the as-deposited state to a thin film annealed at 400 °C, as displayed in Figure 6. This improvement in conductivity is linked to the reduction in particle size and enhancement in crystalline quality, as demonstrated by the XRD analysis, as well as a corresponding decrease in surface roughness, which facilitates crystalline quality. The decrease in surface roughness from the as-deposited thin film to the 400 °C annealed film indicates a smoother film surface, which minimizes scattering sites for charge carriers, thus enhancing conductivity. Smaller particle sizes and improved crystal structures decrease the number of grain boundaries, which are known to act as scattering centres for charge carriers (electrons and holes) [38–40]. The reduction in grain boundary scattering allows for more efficient carrier transport, thereby increasing the electrical conductivity of the films. However, a significant change occurred at 500 °C, where the particle size increased dramatically and the conductivity dropped sharply. This decline in conductivity can be ascribed to the formation of larger grains, which may introduce structural imperfections such as microcracks. These imperfections likely result from thermal stress or excessive grain growth, which disrupts the crystal structure and increases the scattering of carriers, ultimately leading to a higher electrical resistance [41]. This phenomenon aligns with the established understanding that while smaller grains typically increase resistance due to grain boundary scattering, excessively large grains can introduce new defects that also adversely affect conductivity.

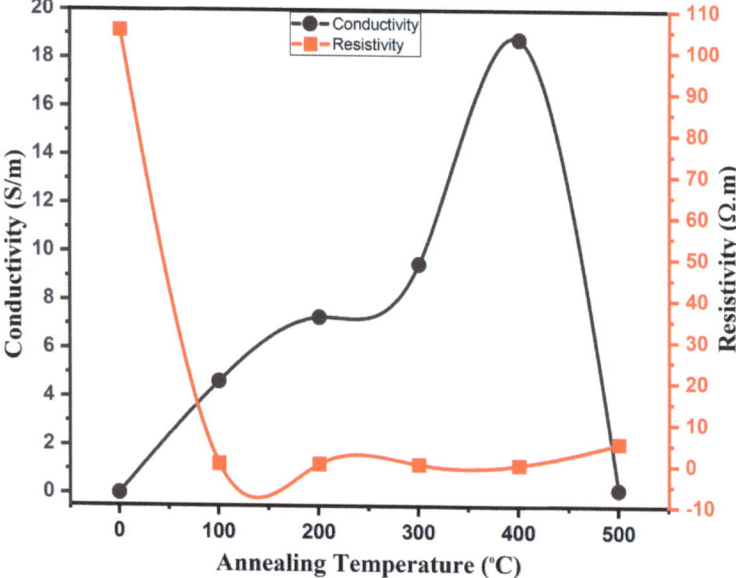

Figure 6. Resistivity/conductivity vs. annealing temperature of thin films.

Table 2. Electrical resistivity and conductivity values of thin films annealed at different temperatures in comparison to a deposited thin film.

S. No.	Thin Film	Electrical Conductivity (S/m)	Electrical Resistivity (Ω.m)
1	As deposited	0.0095	105
2	100 °C	4.637	0.2157
3	200 °C	7.292	0.1371
4	300 °C	9.49	0.105
5	400 °C	18.79	0.05321
6	500 °C	0.187	5.348

3. Conclusions

In summary, magnetron sputtering was used to deposit hetero-structured (ZnO/Zn/ZnO) thin films on clean glass substrates. The resulting samples were annealed in an inert environment at 100, 200, 300, 400, and 500 °C. XRD analysis confirmed the presence and highly crystalline hexagonal structure of ZnO at 300 and 400 °C, whereas at other annealing temperatures, the sample either possessed metallic zinc or low crystallinity. Atomic force microscopy (AFM) and scanning electron microscopy (SEM) were used to effectively study the surface morphology and particle size. A Hall effect measurement device was used to analyse the electrical characteristics. The annealed thin-film data were plotted in comparison with the deposited thin film. AFM analysis indicated that the average roughness (Ra) of the film decreased with an increase in the annealing temperature, reaching an optimum value of 4.95 nm for the sample annealed at 400 °C, and then increased with a further increase in the annealing temperature. SEM analysis showed that the particle sizes initially decreased with increasing annealing temperature but increased at higher temperatures. The Hall effect measurement system showed that the electrical resistivity decreased for the annealed samples. In addition, the resistivity showed a good relationship with the surface roughness of the thin films; the lower the surface roughness with lower scattering centres of charge carriers, the lower the resistivity. The sample annealed at 400 °C exhibited the lowest resistivity.

4. Experimental

This section discusses the fabrication process of a hetero-structured (ZnO/Zn/ZnO) thin film and its thermal annealing at different temperatures and gives the description used to analyse the morphological, topographical, structural, and electrical properties of the samples. Magnetron sputtering was used to deposit a hetero-structured (ZnO/Zn/ZnO) thin film on a glass substrate, which was subsequently annealed at 100, 200, 300, 400, and 500 °C.

4.1. Films Deposition

4.1.1. Substrate Preparation

The cleanliness of the substrate is crucial for achieving high-quality thin films. Soda-lime glass slides (SLG) (cat. No. 7105) were cut into small squares (10 cm × 10 cm × 1 cm) using a diamond saw cutter. The cleaning process involved several steps. First, the substrates were immersed in methanol and cleaned ultrasonically for 30 min to remove organic residues. Washing with the soap solution was then carried out, followed by rinsing in deionised water. Subsequently, the substrates were immersed in a chromic acid solution for 20 min. Finally, the substrates were ultrasonically cleaned with water for 30 min. Subsequently, the substrate surface was dried by blowing pressurised inert gas and immediately transferred to the deposition chamber.

4.1.2. Sputtering System and Film Deposition Process

Deposition of ZnO and Zn thin films over SLG substrates was performed using a disk-shaped ZnO and Zn target (100 mm diameter, 8.25 mm thickness) in a magnetron sputtering system (Alliance Concept, DP650, Annecy, France). The system consisted of mechanical and turbomolecular pumps, sputtering guns, RF and DC power supplies, and heating and bias capabilities for substrate support. Each film deposition cycle included the following steps: After loading the substrates, the chamber was evacuated to a base pressure of a few microtorr, followed by the introduction of argon gas. The argon flow rate was maintained at 20 sccm to achieve a pressure of 10 mTorr. The DC power was then turned on, and its value was adjusted to the desired level. Initially, the target was sputtered with its shutter closed for approximately 15 min to remove any native oxide layer present on its surface. Subsequently, ZnO (bottom layer), Zn (middle layer), and ZnO (top layer) film deposition was performed to fabricate films with a thickness of 1.5 μm, 30 nm, and 75 nm, respectively. The base pressure was maintained at 7.49×10^{-6} bar, and a 300 W DC power source was employed to keep the potential difference between the target and glass substrate constant. During growth, the film thickness was monitored using a quartz crystal microbalance. The sputtered samples were annealed in an inert argon atmosphere for 1 h at 100, 200, 300, 400, and 500 °C. The deposition parameters are listed in Table 3.

Table 3. Deposition parameters of ZnO/Zn/ZnO thin films heterostructure.

Deposition Parameters		ZnO	Zn	ZnO
Pressure (mTorr)		25	50	50
Power (Watt)		300	300	300
Thickness		1.5 µm	30 nm	75 nm
Annealing Temperature (°C)				
100	200	300	400	500

4.2. Characterization Tools

The surface morphologies of the as-grown and annealed multilayer structures were examined using scanning electron microscopy (SEM) (JSM-840, JEOL, Tokyo, Japan) with an operating voltage of 15–20 kV, spot size of 30–50 nm, and working distance of 10 mm. For SEM analysis, samples were coated by sputtering 100 nm gold, and small silver strips were placed on the sample to make it conductive. Atomic force microscopy (AFM) (JEOL SPM-5200) was used to examine the topography and measure the surface roughness of the multilayer structure while operating in the non-contact mode with scan sizes of (10 × 10) and (3 × 3) µm^2. X-ray diffraction (XRD) was used to determine the structural and phase purity of the prepared samples using a D8 ADVANCE, Bruker, Germany instrument over the range of 20–80° with Cu Kα radiation (λ = 1.5418Å) and a step size and dwell time of 0.041 and 3 s, respectively. The electrical conductivity was measured using either a four-point electrical resistance probe or a Hall automated measuring system utilizing the Van-Der-Pauw method (Ecopia HMS-5000, Bridge Technology, USA). Four electrodes of gold paste were placed at four different equal points, and four needle tips were placed to measure the voltage and current.

Funding: This work was supported by the Deanship of Scientific Research, Vice Presidency for Graduate Studies and Scientific Research, King Faisal University, Saudi Arabia (Grant No. 3,061).

Data Availability Statement: The original contributions presented in the study are included in the article, further inquiries can be directed to the author.

Conflicts of Interest: The author declares that he has no known competing financial interests or personal relationships that could have appeared to influence the work reported in this paper.

References

1. Khan, M.; Islam, M. Deposition and Characterization of Molybdenum Thin Films Using DC Plasma Magnetron Sputtering 1. *Semiconductors* **2013**, *47*, 1610–1615. [CrossRef]
2. Hussein, A.K. Applications of nanotechnology in renewable energies—A comprehensive overview and understanding. *Renew. Sustain. Energy Rev.* **2015**, *42*, 460–476. [CrossRef]
3. Ali, S.; Saleem, S.; Salman, M.; Khan, M. Synthesis, structural and optical properties of ZnS–ZnO nanocomposites. *Mater. Chem. Phys.* **2020**, *248*, 122900. [CrossRef]
4. Sharma, D.K.; Shukla, S.; Sharma, K.K.; Kumar, V. A review on ZnO: Fundamental properties and applications. *Mater. Today Proc.* **2020**, *49*, 3028–3035. [CrossRef]
5. Kumar, V.P.S.; Manikandan, N.; Nagaprasad, N.; Letatesfaye, J.; Krishnaraj, R. Analysis of the Performance Characteristics of ZnO Nanoparticles' Dispersed Polyester Oil. *Adv. Mater. Sci. Eng.* **2022**, *2022*, 4844979. [CrossRef]
6. Liu, C.; Burghaus, U.; Besenbacher, F.; Wang, Z.L. Preparation and Characterization of Nanomaterials for Sustainable Energy Production. *ACS Nano* **2010**, *4*, 5517–5526. [CrossRef]
7. Sanchez, C.; Lebeau, B.; Chaput, F.; Boilot, J. Optical Properties of Functional Hybrid Organic-Inorganic Nanocomposites. *Adv. Mater.* **2004**, *15*, 1969–1994. [CrossRef]
8. Du, Y.; Zhao, F.; Liu, L.; Gao, Y.; Xing, L.; Li, Q.; Fu, C.; Zhong, Z.; Zhang, X. Improvement of bond strength between ZnO nanorods and carbon fibers using magnetron sputtered ZnO films as the interphase. *CrystEngComm* **2017**, *19*, 868–875. [CrossRef]
9. Ferhati, H.; Djeffal, F.; Kacha, K. Optimizing the optical performance of ZnO/Si-based solar cell using metallic nanoparticles and interface texturization. *Optik* **2018**, *153*, 43–49. [CrossRef]
10. Klingshirn, C.; Priller, H.; Decker, M.; Bruckner, J.; Kalt, H.; Hauschild, R.; Zeller, J.; Waag, A.; Bakin, A.; Wehmann, H.; et al. Excitonic properties of ZnO. *Adv. Solid State Phys.* **2005**, *45*, 275–287.

11. Kou, L.Z.; Guo, W.L.; Li, C. Piezoelectricity of ZNO and its nanostructures. In *2008 Symposium on Piezoelectricity, Acoustic Waves, and Device Applications*; IEEE: New York, NY, USA, 2008; pp. 354–359. [CrossRef]
12. Chen, Z.; Wang, J.; Wu, H.; Yang, J.; Wang, Y.; Zhang, J.; Bao, Q.; Wang, M.; Ma, Z.; Tress, W.; et al. A Transparent Electrode Based on Solution-Processed ZnO for Organic Optoelectronic Devices. *Nat. Commun.* **2022**, *13*, 4387. [CrossRef]
13. Blom, F.R.; Bauhuis, G. 365 RF Planar Magnetron Sputtered ZnO Films II: Electrical Properties. *Thin Solid Film.* **1991**, *204*, 365–376. [CrossRef]
14. Mahmood, F.S.; Gould, R.D.; Hassan, A.K.; Salih, H.M. DC properties of ZnO thin films prepared by rf magnetron sputtering. *Thin Solid Film.* **1995**, *770*, 95–98.
15. Ponja, S.D.; Sathasivam, S.; Parkin, I.P.; Carmalt, C.J. Highly conductive and transparent gallium doped zinc oxide thin films via chemical vapor deposition. *Sci. Rep.* **2020**, *10*, 1–7.
16. Sharma, R.; Yoo, S. ZnO in organic electronics. In *Nanostructured Zinc Oxide: Synthesis, Properties and Applications*; Elsevier: Amsterdam, The Netherlands, 2021. [CrossRef]
17. Saitou, M.; Makabe, A.; Tomoyose, T. AFM study of surface roughening in sputter-deposited nickel films on ITO glasses. *Europhys. Lett.* **2000**, *52*, 185–188. [CrossRef]
18. Hasan, S.A.; Gibson, D.; Song, S.; Wu, Q.; Ng, W.P.; McHale, G.; Dean, J.; Fu, Y.Q. ZnO thin film based flexible temperature sensor. *Proc. IEEE Sens.* **2017**, *2017*, 1–3.
19. Xuan, J.; Zhao, G.; Sun, M.; Jia, F.; Wang, X.; Zhou, T.; Yin, G.; Liu, B. Low-temperature operating ZnO-based NO_2 sensors: A review. *RSC Adv.* **2020**, *10*, 39786–39807. [CrossRef]
20. Nomura, K.; Ohta, H.; Ueda, K.; Kamiya, T.; Hirano, M.; Hosono, H. Thin-Film Transistor Fabricated in Single-Crystalline Transparent Oxide Semiconductor. *Nanotechnology* **2003**, *300*, 1269–1272. [CrossRef]
21. Bel Hadj Tahar, R.; Ban, T.; Ohya, Y.; Takahashi, Y. Tin doped indium oxide thin films: Electrical properties. *J. Appl. Phys.* **1998**, *83*, 2631–2645. [CrossRef]
22. Kind, B.H.; Yan, H.; Messer, B.; Law, M.; Yang, P. Nanowire Ultraviolet Photodetectors and Optical. *Adv. Mater.* **2002**, *14*, 200–202. [CrossRef]
23. Zaier, A. Annealing effects on the structural, electrical and optical properties of ZnO thin films prepared by thermal evaporation technique. *J. King Saud Univ. Sci.* **2015**, *27*, 356–360. [CrossRef]
24. Husna, J.; Aliyu, M.M.; Islam, M.A.; Chelvanathan, P. Influence of Annealing Temperature on the Properties of ZnO Thin Films Grown by Sputtering. *Energy Procedia* **2012**, *25*, 55–61. [CrossRef]
25. Shivaraj, B.W.; Murthy, H.N.N.; Krishna, M.; Satyanarayana, B.S. Effect of Annealing Temperature on Structural and Optical properties of Dip and Spin coated ZnO Thin Films. *Procedia Mater. Sci.* **2015**, *10*, 292–300. [CrossRef]
26. Films, T.S.; Engineering, E.; Enschede, A.E. 1.1. Sputtered thin film ZnO. *Thin Solid Film.* **1991**, *204*, 349–364.
27. Soci, C.; Zhang, A.; Xiang, B.; Dayeh, S.A.; Aplin, D.P.R.; Park, J.; Bao, X.Y.; Lo, Y.H.; Wang, D. ZnO Nanowire UV Photodetectors with High Internal Gain. *Nano Lett.* **2007**, *7*, 1003–1009. [CrossRef]
28. Gurylev, V.; Perng, T.P. Defect engineering of ZnO: Review on oxygen and zinc vacancies. *J. Eur. Ceram. Soc.* **2021**, *41*, 4977–4996. [CrossRef]
29. Heremans, J.P. The ugly duckling. *Nature* **2014**, *508*, 327–328. [CrossRef]
30. McCluskey, M.D. Defects in ZnO. In *Defects in Advanced Electronic Materials and Novel Low Dimensional Structures*; Elsevier: Amsterdam, The Netherlands, 2018.
31. Gorla, C.R.; Emanetoglu, N.W.; Liang, S.; Mayo, W.E.; Lu, Y.; Wraback, M.; Shen, H.; Gorla, C.R.; Emanetoglu, N.W.; Liang, S.; et al. Structural, optical, and surface acoustic wave properties of epitaxial ZnO films grown on (0112) sapphire by metalorganic chemical vapor deposition. *J. Appl. Phys.* **2009**, *85*, 2595–2602. [CrossRef]
32. Huang, P.; Huang, C.; Lin, M.; Chou, C.; Hsu, C.; Kuo, C. The Effect of Sputtering Parameters on the Film Properties of Molybdenum Back Contact for CIGS Solar Cells. *Int. J. Photoenergy* **2013**, *2013*, 390824. [CrossRef]
33. Kumar, S.; Gupta, V.; Sreenivas, K. Synthesis of photoconducting ZnO nano-needles using an unbalanced magnetron sputtered ZnO/Zn/ZnO multilayer structure. *Nanotechnology* **2005**, *16*, 1167. [CrossRef]
34. McCluskey, M.D.; Jokela, S.J. Defects in ZnO. *J. Appl. Phys.* **2009**, *106*, 071101. [CrossRef]
35. Mohamed, D. Influence of Low Ag Doping on Structural, Morphological and Optical Properties of Sol-Gel Dip-Coated Nanostructured ZnO Thin Films. *Int. J. Nat. Sci.* **2016**, *4*, 15–28.
36. Chen, W.-T.; Fang, P.-C.; Chen, Y.-W.; Chiu, S.-J.; Ku, C.-S.; Brahma, S.; Lo, K.-Y. Zn dots coherently grown as the seed and buffer layers on Si(111) for ZnO thin film: Mechanism, in situ analysis, and simulation. *J. Vac. Sci. Technol. A* **2022**, *40*, 063403. [CrossRef]
37. Fu, Z.; Lin, B.; Liao, G.; Wu, Z. The effect of Zn buffer layer on growth and luminescence of ZnO films deposited on Si substrates. *J. Cryst. Growth* **1998**, *193*, 316–321. [CrossRef]
38. Seto, J.Y.W. The electrical properties of polycrystalline silicon films. *J. Appl. Phys.* **1975**, *46*, 5247–5254. [CrossRef]
39. Bhuvana, K.P.; Elanchezhiyan, J.; Gopalakrishnan, N.; Balasubramanian, T. Influence of grain size on the properties of AlN doped ZnO thin film. *Mater. Sci. Semicond. Processing.* **2011**, *14*, 84–88. [CrossRef]

40. Zhu, M.W.; Wang, Z.J.; Chen, Y.N.; Wang, H.L.; Zhang, Z.D. Effect of grain boundary on electrical properties of polycrystalline lanthanum nickel oxide thin films. *Appl. Phys. A* **2013**, *112*, 1011–1018. [CrossRef]
41. Imajo, T.; Suemasu, T.; Toko, K. Strain effects on polycrystalline germanium thin films. *Sci. Rep.* **2021**, *11*, 8333. [CrossRef]

Disclaimer/Publisher's Note: The statements, opinions and data contained in all publications are solely those of the individual author(s) and contributor(s) and not of MDPI and/or the editor(s). MDPI and/or the editor(s) disclaim responsibility for any injury to people or property resulting from any ideas, methods, instructions or products referred to in the content.

Article

Synthesis and Redox Properties of Iron and Iron Oxide Nanoparticles Obtained by Exsolution from Perovskite Ferrites Promoted by Auxiliary Reactions

Dmitrii Filimonov *, Marina Rozova, Sergey Maksimov and Denis Pankratov

Department of Chemistry, Lomonosov Moscow State University, 119991 Moscow, Russia; mgrozova@yahoo.com (M.R.); pankratov@radio.chem.msu.ru (D.P.)
* Correspondence: dfilin@gmail.com

Abstract: Nanoparticles of iron and iron oxides, as well as their composites, are of great scientific and technological interest. However, their properties and sustainability strongly depend on the preparation methods. Here, we present an original approach to synthesizing Fe and FeNi$_x$ metal nanoparticles by exsolution, in a reducing environment at elevated temperatures from perovskite ferrites (La$_{1-x}$Ca$_x$FeO$_{3-\gamma}$, CaFeO$_{2.5}$, etc.). This approach is made possible by the auxiliary reactions of non-reducible A-site cations (in ABO$_3$ notation) with the constituents of reducing compounds (h-BN etc.). The nanoparticles exsolved by our process are embedded in oxide matrices in individual voids formed in situ. They readily undergo redox cycling at moderate temperatures, while maintaining their localization. Fe nanoparticles can be obtained initially and after redox cycling in the high-temperature γ-form at temperatures below equilibrium. Using their redox properties, a new route to producing hollow and layered oxide magnetic nanoparticles (Fe$_3$O$_4$, Fe$_3$O$_4$/La$_{1-x}$Ca$_x$FeO$_{3-\gamma}$), by separating the oxidized exsolved particles, was developed. Our approach provides greater flexibility in controlling exsolution reactions and matrix compositions, with a variety of possible starting compounds and exsolution degrees, from minimal up to ~100% (in some cases). The described strategy is highly important for the development of a wide range of new functional materials.

Keywords: transition metal-embedded nanoparticles; metal exsolution; ^{57}Fe Mössbauer spectroscopy; redox behavior; hollow and layered oxide nanoparticles; cup-shaped nanoparticles; nano zero valent iron (nZVI) particles; γFe nanoparticles

1. Introduction

Nanomaterials based on nanoparticles of transition metals and/or their oxides are of great interest from both scientific and practical viewpoints. Their applications are very diverse and include catalytic, magnetic, electronic, sorption, and pharmaceutical functional materials, utilizing various types of nanoparticles (supported, embedded, individual nanoparticles of various shapes) [1–3]. Many transition metal nanoparticles are chemically active and, therefore, suffer from a lack of stability, which is further exacerbated by their tendency for recrystallization, agglomeration, and growth, etc., especially at elevated temperatures. Supported/embedded nanocomposite materials have the potential to combine useful properties of both nanoparticles and matrices and maintain long-term stability and durability [1,3,4]. Moreover, those materials can develop new features due to the synergy between nanoparticles and the support, which is attributed to the intrinsic properties of the components and the strength of their interaction [4–7].

There are a wide variety of ways to produce these diverse nanoparticle-based nanomaterials [5,8,9]. One of the actively developing strategies for the synthesis of nanocomposite materials with stronger metal nanoparticle–substrate interactions (and, therefore, enhanced functionality and stability) is based on the thermal decomposition of complex oxides under reducing conditions. This kind of decomposition can be complete or partial and, especially the latter, is termed metal exsolution. In this approach, the complex oxides must contain both reducible and non-reducible metals under conditions of synthesis [10–13]. The main progress in this field is associated with the use of perovskite-related oxides of transition metals. It is a vast family of oxides with various stoichiometry (ABO_3, $A_{n+1}B_nO_{3n+1}$ Ruddlesden–Popper (RP) phases, etc.), whose crystal structures are based on [AO_3] close-packed layers, where the main constituents of A-sites are non-reducible rare-earth or alkaline-earth cations, while B-sites, in the interstices, can contain reducible cations of transition metals (Fe, Co, Ni, Cu, etc.) [14,15]. A series of Ln_2O_3/M nanocomposites, obtained by complete redox decomposition of the perovskite precursors of easily reducible transition metals (M = Ni, Co, etc.), were reported as metal nanoparticle-supported catalytic materials with enhanced productivity and stability [16,17]. However, the synthesis of nanocomposites by partial redox decomposition of perovskites requires methods to control the processes of metal nanoparticle exsolution and matrix transformations. Several approaches are proposed in the literature to facilitate metal exsolution in this manner [10–12,18]. One approach includes modification of the initial oxides, such as introducing an A-site deficiency [19,20], doping of B-sites with easily reduced cations [19,21,22], and promoting the formation of RP phases by adjusting the composition of A-sites [13], as well as others. Another approach is to use special reduction conditions, such as lattice strains, voltage biasing, and plasma assistance, as well as varying the temperature and pO_2 [13,18].

Metal reduction and exsolution is accompanied by changes in the parent perovskite matrices, which can occur in different ways. One such change is that the starting A-site-deficient perovskites are converted into stoichiometric ones, with minimal topotactical changes [21,23]. Another possible change is the formation of the corresponding oxides of non-reducible metals (La_2O_3, etc.) [17]. Yet another change is that the initial ABO_3-based perovskites can be transformed into RP phases, with different A to B ratios for some A-site compositions (containing Sr^{2+}, etc.) [13].

Although the metal exsolution process is widely employed to enhance surface properties, it can also be used to modify the bulk properties of functional nanocomposites [24–26]. The exsolution at the interior grain and phase boundaries, together with the simultaneous transformation of host oxide matrices, provide a new pathway for the modification of a wide range of electrical, optical, and magnetic properties [25,26].

Overall, metal exsolution is a smart and effective strategy, with great potential for the preparation of new functional materials, with enhanced and unique properties. Its implementation, however, largely depends on the specific parent perovskite oxides and the appropriate reaction conditions [10,13,25]. Consequently, further development of methods for tuning the properties of the resulting nanocomposites, by controlling the exsolution process, is needed.

Herein, we present a novel approach to metal exsolution using perovskite-related ferrites, where B-metal exsolution is promoted by the additional auxiliary reactions involving non-reducible A-site cations. We have explored the behavior of a number of La and Ca-based ferrites in this new type of reaction, resulting in the formation of nanocomposites with nanoparticles embedded in individual voids, which are significantly different from the exsolved materials reported previously. The unique properties of the nanoparticles obtained in this manner, as well as the transformation of oxide matrices, will be discussed. It was shown that exsolved metal nanoparticles can be reversibly transformed into oxides in redox cycling, while maintaining their location inside the matrices. Moreover, the use of the chemical properties of nanocomposite matrices derived from alkaline earth-based ferrites makes it possible to separate and investigate the oxide nanoparticles obtained by the oxidation of exsolved metal nanoparticles.

2. Results and Discussion

2.1. The Main Concept of Metal Exsolution Promoted by Auxiliary Reactions

In this study, the synthesis of Fe-containing nanoparticles and nanocomposites is based on a specially developed process for metal exsolution from complex perovskite-related oxides of reducible transition metals (Fe, Ni, etc.). Our new strategy is to facilitate B-site reducible metal exsolution through additional auxiliary reactions (ARs) of non-reducible A-site (in the ABO_3 notation) cations with the constituents of reducing compounds, resulting in new complex oxides. This strategy, in a general simplified form, is described by schematic reaction (1):

$$ABO_{3-\gamma} + [R^n] + \{H_2\} \rightarrow B + AR^1_\eta O_y + \{H_2O\} + \{R^2\} \quad (1)$$

Here, $[R^n]$ is a reducing compound that reduces B cations, and at the same time, some (or all) of its constituents (R^1) form new complex oxides with non-reducible A-site cations that do not contain reducible B cations. In general, $[R^n]$ (R^1R^2, etc.) compounds can have a different chemical nature and aggregate states. For clarity of the explanation, the overall reaction can be nominally split into two consecutive pseudo-reactions, (2) and (3):

$$ABO_{3-\gamma} + [R^n] + \{H_2\} \rightarrow B + AO_{y1} + \eta R^1 O_{y2} + \{H_2O\} + \{R^2\} \quad (2)$$

$$AO_{y1} + \eta R^1 O_{y2} \rightarrow AR^1_\eta O_y \quad (3)$$

Chemical reactions for the formation of complex oxides containing R^1 and non-reducible A-site cations (1b type) are designated here as auxiliary reactions. Curly brackets {} denote possible compounds. The reaction proceeds at elevated temperatures and under low pO_2 conditions, which are required for the formation and preservation of the B components in metal form. An important feature of reaction (1) is that the main source of oxygen for the formation of the resulting oxides is the initial oxides. This imposes limitations on the possible ratios of R^1 and A cations in the resulting complex oxides. If the initial oxides include two or more different B cations, then reaction (1) will be more complex depending on the B cation properties, as shown schematically below:

$$AB'_x B''_\psi O_{3-\gamma} + [R^n] + \{H_2\} \rightarrow B' + \{B''\} + AR^1_\eta O_y + AB'_{x-\varepsilon} B''_{\psi+\phi} O_{3-\beta} + \{H_2O\} + \{R^2\} \quad (4)$$

Reaction (4) illustrates that all the different B cations can be exsolved to various degrees, or that the irreducible or less reducible ones can remain in perovskite oxides. The exsolved metals can also form alloys.

2.2. $LaFeO_3$ and $Ca_2Fe_2O_5$-Based Nanocomposites

For $LaFeO_3$ as a starting material and hexagonal BN as a reducing compound, the A-site cation AR (ACAR)-promoted exsolution can be written as reaction (5) (for clarity purposes, along with the nominal pseudo-reactions (6) and (7)):

$$LaFeO_3 + BN \rightarrow Fe + LaBO_3 + 1/2 N_2 \quad (5)$$

$$\{2 LaFeO_3 + 2 BN \rightarrow 2 Fe + La_2O_3 + B_2O_3 + N_2\} \quad (6)$$

$$\{La_2O_3 + B_2O_3 \rightarrow 2 LaBO_3\} \quad (7)$$

Herein, the processes involving reaction (5) were carried out at elevated temperatures of 700–750 °C, in a reducing atmosphere, provided by a 10% H_2/Ar flow. Although hydrogen is not formally involved in the reaction, its role here is to maintain low pO_2 conditions, essential for the existence of metal Fe nanoparticles. The formation of lanthanum borate $LaBO_3$ provides an additional driving force for the reduction of $LaFeO_3$ and decreases the reaction temperature compared to a reduction with hydrogen. The specified temperatures of reaction (5) are below the decomposition temperature of LaFeO$_3$ in a hydrogen-containing atmosphere, which is reported to be >850 °C [27,28]. Thus, its decomposition, described by reaction (3), does not occur under these conditions.

$$LaFeO_3 + 3/2H_2 \rightarrow Fe + 1/2La_2O_3 + 3/2H_2O \tag{8}$$

It is important to maintain reaction (5) as a main process. It should be noted that the higher temperature stability under reducing conditions makes ferrites more preferable starting materials compared to other perovskite-like compounds of reducible transition metals (cobaltites, nickelates, cuprates, etc.) [27].

Moreover, h-BN has a number of properties making it suitable for reactions of type (1, 5). First of all, it has decent reactivity in reactions of type (1, 4) with rare-earth (La, Y, etc.) and alkali-earth (Ca, Sr, etc.) perovskite-related ferrites. At the same time, it is exceptionally stable in inert and reducing atmospheres, and is also quite resistant to oxidation at moderate temperatures in air.

Figure 1 shows the powder XRD patterns of the LaFeO$_3$ ferrite-based samples after interaction with h-BN at ~750 °C with different amounts of exsolved Fe, depending on the LaFeO$_3$/h-BN ratio. As the reaction proceeds, the peaks corresponding to LaBO$_3$ and metal Fe appear in patterns, which are consistent with reaction (5) (Figure 1(a1)). The metal Fe peaks are distinguishable, but they strongly overlap with the LaBO$_3$ ones. The corresponding Mössbauer spectra are shown in Figure 1(b1,b2). Each of the spectra consist of two magnetically split sextet components and a paramagnetic singlet component. All the components are well resolved and do not broaden. According to their hyperfine parameters (Table 1), the first of two sextets with an isomer shift (δ) of 0.37 mm s^{-1} and a hyperfine magnetic field (H$_{hf}$) of 52.4 T correspond to LaFeO$_3$ and the second, with δ ~0 mm s^{-1} and H$_{hf}$ = 33 T, correspond to αFe. The paramagnetic singlet, according to its hyperfine parameters (δ~−0.1 mm s^{-1} at 298 K), corresponds to metal γFe. Moreover, γFe is a high-temperature paramagnetic form, with a close-packed fcc crystal structure. It is important to note that γFe is metastable at temperatures below 910 °C [29], while the synthesis temperature was ~750 °C. The metal γFe nanoparticles, however, may be undetectable in the XRD patterns (Figure 1(a2,b3)). The SEM images of LaFeO$_3$ after the ACAR exsolution display smooth surfaces, without distinguishable metal Fe nanoparticles distributed on them (Figure 2a). Figure 2b,c shows typical TEM images of such samples. For the samples at the low or moderate extent of reaction (Figure 2b,c), the images reveal agglomerated grains of different contrast, probably due to variations in thickness, but exsolved metal Fe particles are not clearly distinguishable. For the samples at the high extent of reaction >40%, obtained after prolonged heating for more than 15 h, the metal Fe particles are also largely undistinguishable, but some whisker-like metal Fe agglomerates are visible (Figure 2d) in small amounts with respect to the reaction extent. Additionally, the EDX analysis revealed numerous La-rich oxide grains, which, according to the XRD analysis, are actually LaBO$_3$, since boron is undetectable. Therefore, the metal Fe particles are embedded in the oxide matrix after the ACAR-promoted exsolution, which makes them hard to distinguish, except for the formation of whiskers, if any.

Table 1. Hyperfine parameters of Mössbauer spectra of the LaFeO$_3$/h-BN-derived nanocomposites measured at RT (except Figure 1(b2) at 78 K): δ—isomer shift, ΔEQ—quadrupole splitting, H—hyperfine magnetic field, A—relative area, —linewidth.

Sample	Component	δ (mm s^{-1}) ±0.01	ΔEQ (mm s^{-1}) ±0.01	H (T) ±0.1	A (%) ±0.5	(mm s^{-1}) ±0.01	Comments
In Figure 1(b1)	s11	0.37	−0.07	52.4	71	0.26	Fe^{3+} oct. in LaFeO$_3$
	s21	−0.01	−0.01	33.0	18	0.24	αFe
	d11	−0.10	0.00	-	11	0.25	γFe

Table 1. Cont.

Sample	Component	δ (mm s⁻¹) ±0.01	ΔEQ (mm s⁻¹) ±0.01	H (T) ±0.1	A (%) ±0.5	(mm s⁻¹) ±0.01	Comments
In Figure 1(b2) 78 K	s12	0.48	−0.07	56.2	71	0.25	Fe^{3+} oct. in $LaFeO_3$
	s22	0.11	0.01	33.8	18	0.25	αFe
	d12	0.01	0.00	-	11	0.25	γFe
In Figure 1(b3)	s13	0.37	−0.07	52.4	46	0.27	Fe^{3+} oct. in $LaFeO_3$
	s23	0.00	0.00	33.0	45	0.24	αFe
	d13	−0.10	0.00	-	8	0.25	γFe
In Figure 1(b4)	s15	0.37	−0.07	52.3	72	0.25	Fe^{3+} oct. in $LaFeO_3$
	s25	0.00	0.00	33.0	28	0.24	αFe
In Figure 3(1)	s123	0.37	−0.06	52.4	68	0.27	Fe^{3+} oct. in $LaFeO_3$
	s223	0.00	0.01	33.1	23	0.23	αFe
	d123	−0.10	0.00	-	9	0.26	γFe
In Figure 3(2)	s124	0.37	−0.07	52.2	70	0.26	Fe^{3+} oct. in $LaFeO_3$
	s224	−0.01	0.01	33.0	17	0.21	αFe
	d124	−0.10	0.00	-	13	0.23	γFe
In Figure 3(3)	s125	0.37	−0.07	52.2	71	0.25	Fe^{3+} oct. in $LaFeO_3$
	s225	0.35	−0.17	51.0	5	0.33	Fe^{3+} oct. in αFe_2O_3
	s325	0.00	0.01	33.0	23	0.23	αFe
	d125	−0.10	0.00	-	1	0.26	γFe
In Figure 3(4)	s126	0.37	−0.07	52.2	69	0.26	Fe^{3+} oct. in $LaFeO_3$
	s226	0.00	0.00	33.0	30	0.22	αFe
	d126	−0.12	0.00	-	1	0.25	γFe
In Figure 3(5)	s127	0.37	−0.08	52.3	70	0.25	Fe^{3+} oct. in $LaFeO_3$
	s227	0.38	−0.18	51.4	30	0.25	Fe^{3+} oct. in αFe_2O_3
In Figure 3(6)	s128	0.37	−0.06	52.2	70	0.26	Fe^{3+} oct. in $LaFeO_3$
	s228	0.00	0.00	33.0	14	0.21	αFe
	d128	−0.10	0.00	-	16	0.24	γFe

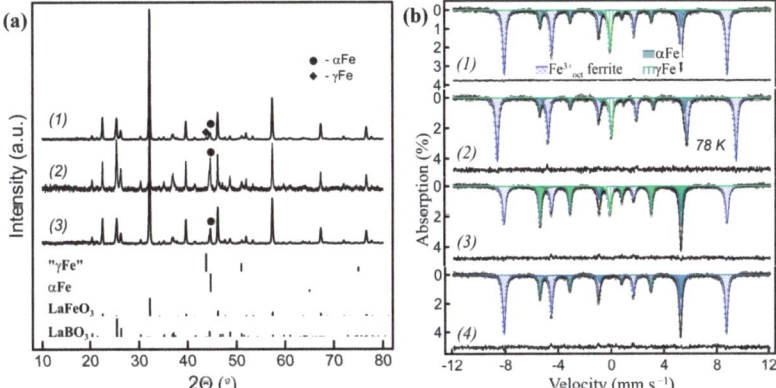

Figure 1. (**a**) Powder XRD patterns of the $LaFeO_3$/h-BN-derived nanocomposites: (a1) 29% of the total Fe amount exsolved, where 11% is in γ form; (a2) 54% of Fe exsolved (8% γFe); (a3) sample (a1) oxidized at 300 °C and subsequently reduced at 700 °C, all 29% of Fe exsolved is in α form (all the Fe contributions were evaluated by Mössbauer spectroscopy). (**b**) Corresponding Mössbauer spectra: (b1,b2) of sample (a1) at RT and 78 K, respectively; (b3) of sample (a2); (b4) of sample (a3) at RT.

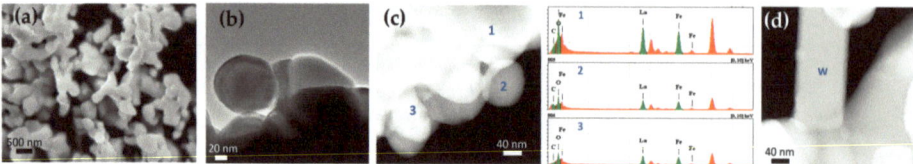

Figure 2. SEM and TEM images of the LaFeO$_3$/h-BN-derived nanocomposites: (**a**,**b**) SEM and bright field (BF) TEM images of the nanocomposites with ~45% of Fe total exsolved; (**c**,**d**) high-angle annular dark-field scanning TEM (HAADF-STEM) image and energy-dispersive X-ray (EDX) elemental analysis in selected locations of the nanocomposites with ~12% of Fe total (~10% γFe); (**d**) formation of the Fe whisker (w) in sample (**a**,**b**).

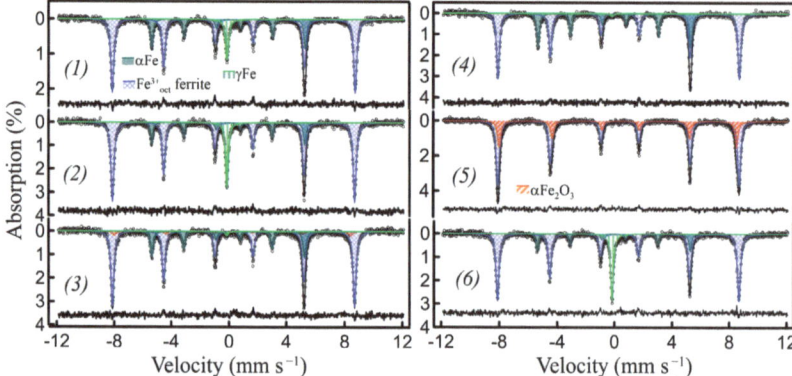

Figure 3. RT Mössbauer spectra of the LaFeO$_3$/h-BN-derived nanocomposites during redox cycling: (1) as prepared with 32% of Fe exsolved (~9% γFe); (2) sample (1) after consecutive oxidation (air, 500 °C) and reduction (10% H$_2$/Ar, 700 °C) (~30% of Fe total, ~13% γFe); (3) sample (2) after oxidation at 300 °C in air (~23% αFe, ~1% γFe, ~5% αFe$_2$O$_3$); (4) after reduction of sample (3) (10% H$_2$/Ar, 700 °C) (~30% αFe); (5) after oxidation of sample (4) (air, 500 °C) (~30% αFe$_2$O$_3$); (6) after reduction (10% H$_2$/Ar, 700 °C) of sample (5) (~30% of Fe total, ~16% γFe).

This is rather different from the metal exsolution from the A-site-deficient perovskites or the high-temperature hydrogen reduction of stoichiometric perovskite oxides reported previously, where the exsolved metal nanoparticles are typically clearly visible on the surfaces of oxide matrixes [10–13]. In these processes, the reducible metal cations migrate toward the outer and inner grain surfaces, which are exposed to the reducing environment, and reduce to metal, with subsequent agglomeration, grain growth, etc. The oxide matrixes, in turn, undergo shrinkage because of a loss of oxygen and metal constituents, promoting socketing of the exsolved nanoparticles [5,10,12,23]. In our ACAR-promoted exsolution process, metal reduction/exsolution is accompanied by the formation of other complex oxides of two or more elements instead of the initial perovskite oxide, even those of lower density, i.e., LaBO$_3$ vs. LaFeO$_3$. Their formation begins in the contact areas between the initial ferrite grains and the reducing reagents. These newly formed Fe-free oxides (viz. LaBO$_3$) on the top of the initial perovskite grains presumably create a diffusion barrier for the Fe species. Consequently, the exsolved metal Fe nanoparticles do not appear on top of the oxide grain surfaces, but instead localize underneath their surfaces in generated in situ voids.

Perovskite-like ferrites of alkaline-earth elements, such as Ca$_2$Fe$_2$O$_5$ and Sr$_2$Fe$_2$O$_5$, can also be used in reaction (1, 4) to produce metal nanoparticle-bearing composites. Because Ca is lighter than La and Sr, it may be more favorable for microscopic investigations in terms of the nanoparticle–matrix contrast. As stated above, the starting ferrite is the main source of oxygen for the formation of the resulting oxides in ACAR-promoted exsolution processes involving reaction (1, 4). Consequently, in the case of La ferrite, other La borates, such as LaB_3O_6, cannot be formed from the $LaFeO_3$ precursor due to a lack of oxygen. For $Ca_2Fe_2O_5$ and h-BN as starting materials, the formation of several Ca borates is possible, viz. $Ca_2B_2O_5$ and $Ca_3B_2O_6$ (without/with CaB_2O_4). The powder XRD patterns of $Ca_2Fe_2O_5$ after the ACAR-promoted exsolution reactions with h-BN and the corresponding Mössbauer spectra are shown in Figure 4. According to Figure 4(a2), the main complex oxide product is $Ca_3B_2O_6$ and, therefore, the primary reaction can be written as follow:

$$3Ca_2Fe_2O_5 + 4BN + 3H_2 \rightarrow 6Fe + 2Ca_3B_2O_6 + 2N_2 + 3H_2O \qquad (9)$$

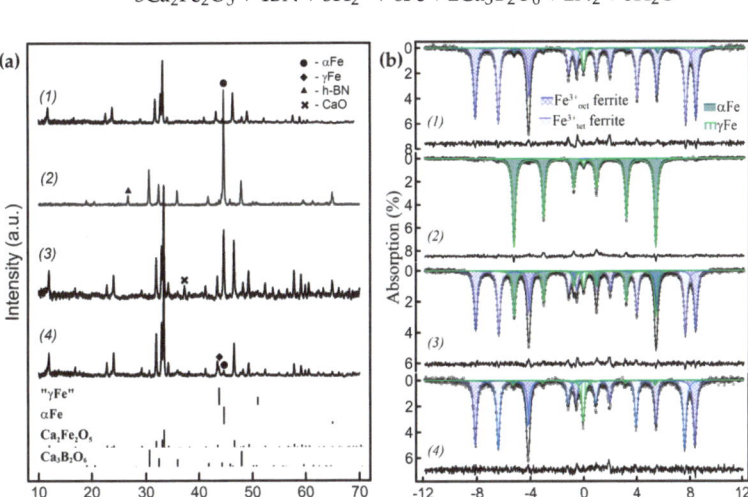

Figure 4. (**a**) Powder XRD patterns and, (**b**), corresponding Mössbauer spectra of the $Ca_2Fe_2O_5$/h-BN-derived nanocomposites with different amounts of metal Fe exsolved: (1) 7% of the total Fe amount exsolved, where 4% is in γ form; (2) 97% of Fe exsolved (2% γFe); (3) 28% of Fe exsolved (3% γFe); (4) after oxidation–reduction of sample (3), 10% of Fe exsolved (7% γFe).

Hydrogen, in addition to maintaining the low pO_2, reacts with excessive oxygen in this case. It is worth noting that no diffraction peaks corresponding to $Ca_3B_2O_6$ and/or other possible resulting oxides were observed at the low extent of reaction, indicating that the oxide products were X-ray amorphous (Figure 4(a1)). $Ca_2Fe_2O_5$ is more active in this type of reaction compared to $LaFeO_3$ and is less stable in reducing conditions as well, so minor additional peaks of CaO can be observed in the patterns of some samples depending on the preparation and reduction conditions (Figure 4(a3)). The Mössbauer spectra of the $Ca_2Fe_2O_5$-derived samples are shown in Figure 4b. The spectra are comprised of three magnetically split sextet components and a paramagnetic singlet component. The $Ca_2Fe_2O_5$ (=$CaFeO_{2.5}$) has a brownmillerite structure, which is oxygen deficient compared to ABO_3 perovskites with fully ordered oxygen vacancies at room temperature, where Fe^{3+} cations equally occupy distorted octahedral and tetrahedral oxygen polyhedra. Consequently, two of the sextets of nearly equal spectral contributions, according to their hyperfine parameters, correspond to the brownmillerite subspectrum, i.e., Fe^{3+}tet ($\delta \sim 0.19$ mm s^{-1} and $H_{hf} = 43.4$ T) and Fe^{3+}oct ($\delta \sim 0.37$ mm s^{-1} and $H_{hf} = 51.2$ T) in $Ca_2Fe_2O_5$. The third sextet ($\delta \sim 0$ mm s^{-1} and $H_{hf} = 33$ T) and a singlet ($\delta \sim -0.1$ mm s^{-1}) correspond to metal Fe in α and γ forms, respectively, i.e., to the metal Fe subspectrum (Table S2). Like in the $LaFeO_3$ case, at ~650–730 °C, the temperatures during the ACAR-promoted exsolution process were much lower than 910 °C. Both the Fe forms can be distinguished in the powder XRD patterns (Figure 4a). Similar to the previous case, the SEM images of the $Ca_2Fe_2O_5$-derived samples display smooth surfaces, without distinguishable metal Fe particles (Figure 5a). The TEM images (Figure 5b,c) show agglomerates of different contrasts, without metal Fe particles being clearly visible on the surfaces. At the same time, a small number of Fe whiskers can be observed in some samples (Figure 5d). However, as it follows from Figure 5c, the EDX mapping in several locations shows areas of Fe segregation, which, taking into account the distributions of other elements, can be identified as Fe-embedded nanoparticles. These exsolved, embedded Fe nanoparticles, identified by the EDX analysis, were about 15–25 nm in size. The EDX analysis also displays Ca-rich O-containing agglomerates (i.e., $Ca_3B_2O_6$, etc.).

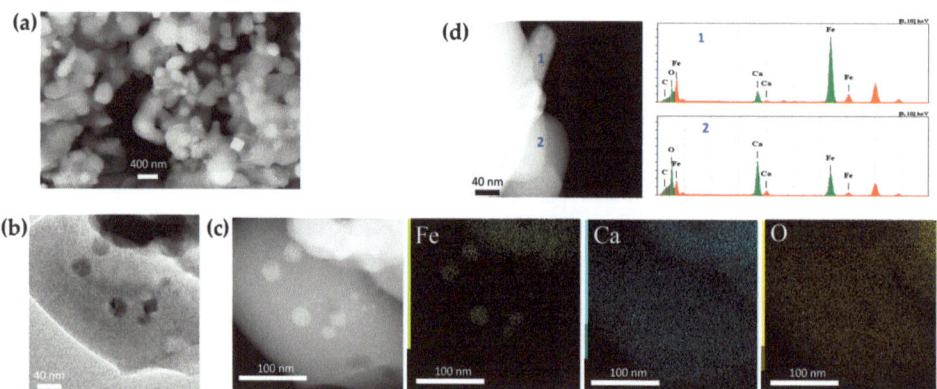

Figure 5. SEM and TEM images of the $Ca_2Fe_2O_5$/h-BN-derived nanocomposites with ~23% of metal Fe exsolved (~9% γFe): (**a**) SEM image; (**b,c**) BF TEM and HAADF-STEM images with the corresponding EDX compositional map; (**d**) HAADF-STEM image, with EDX analysis of the Fe whisker.

The data show that the $Ca_2Fe_2O_5$/h-BN system demonstrates similar behavior to the $LaFeO_3$/h-BN system in the exsolution reactions facilitated by ACARs. In both the systems, the exsolved metal Fe nanoparticles, which can be determined by EDX analysis, are mostly located within oxide matrices, consisting of initial ferrites and newly formed iron-free borates, in the generated in situ voids. Presumably, the reaction zones and, hence, the adjacent voids are somehow connected to the low pO_2 environment, otherwise, in most cases, the metal reduction reaction will not proceed to any significant extent. Metal Fe in the exsolved nanoparticles can exist in two forms, namely metastable at temperatures <910 °C γFe and stable αFe, the latter can also form whiskers in some cases. The interior particle growth in a confined space in voids appears to be a key factor, along with a nanosized dimension, for the γFe formation at temperatures well below 910 °C. The fcc close-packed crystal structure of γFe is denser than the bcc structure of $\alpha(\delta)$Fe [30,31], so that the compressive strain developed when the particles of nano-scale dimensions grow under confined conditions is conducive to the formation of γFe nanoparticles and their subsequent stabilization upon cooling. The strain-induced formation of γFe nanoparticles, smaller than ~20 Å, under confined conditions in oxide matrices (Al_2O_3, MgO), which are stable at an ambient temperature, has been reported previously [32,33]. The stabilization of γFe was also observed in iron coatings produced by arc plasma deposition on porous alumina substrates, when Fe was localized inside pores with a diameter <160 nm [34]. Note that in this case, the deposited Fe layers only covered the inner walls of the pores and did not completely fill their interior space, leaving central gaps.

The γFe/αFe ratio depends on the reaction conditions and extent: the contributions of γFe are larger in the initial stages and at the low extent of reactions. However, the nanocomposites containing only γFe have not been obtained using individual $Ca_2Fe_2O_5$ or $LaFeO_3$ ferrites. $Ca_2Fe_2O_5$ is more active during the described process than $LaFeO_3$ and reacts at lower temperatures, and the reaction can proceed almost completely, whereas for LaFeO$_3$, its extent is limited to ~50–60% (based on the Fe content in coexisting phases obtained from the Mössbauer spectra).

2.3. ACAR-Promoted Metal Exsolution Using the Substituted Ferrites

The properties of Ca$_2$Fe$_2$O$_5$ or LaFeO$_3$ ferrites can be significantly modified by substitutions. Accordingly, the effect of cation substitution on ACAR-promoted metal exsolution and γFe formation was investigated. Rare-earth and alkaline-earth ABO$_{3-\gamma}$ perovskite ferrites allow substitution in both A- and B-sites and a wide variation in oxygen content of $0 \leq \gamma \leq 0.5$ [14,15]. Figure 6b shows the powder XRD patterns of A and B double-substituted compounds, Ca$_{1.4}$Y$_{0.6}$Fe$_{1.8}$Zn$_{0.2}$O$_{5.2}$ (=Ca$_{0.7}$Y$_{0.3}$Fe$_{0.9}$Zn$_{0.1}$O$_{2.6}$), after ACAR-promoted exsolution using h-BN. Y was chosen as the lightest rare-earth 3+ cation. The Ca$_{1-x}$Y$_x$FeO$_{3-\gamma}$ solid solutions have not been studied in detail in the literature, but for our process it is important that they belong to a pseudobinary system, i.e., no additional phases other than the perovskite solid solutions coexist. The solubility ranges from both the Ca side (the brownmillerite type solid solutions with oxygen excess) and the Y side (the LaFeO$_{3-\gamma}$ type ones with mainly disordered oxygen vacancies) are not well defined and, presumably, depend on the temperature [35]. Similar to the case of unsubstituted Ca ferrite, the resulting borates can only be correctly identified at relatively large reaction extents, such as in the sample in Figure 6(b2), with ~29% of Fe exsolved, according to the Mössbauer spectrum shown in Figure 6(a2). The XRD phase analysis revealed that substitution with Y increases the number of oxide products formed, and the ACAR-promoted exsolution process in this case can be written in a simplified form as follows:

$$Ca_{0.7}Y_{0.3}Fe_{0.9}Zn_{0.1}O_{2.6} + BN + H_2 \rightarrow Fe + Ca_3B_2O_6 + Y_2O_3 + Ca_{0.7-z}Y_{0.3+z}Fe_{0.8-\beta}Zn_{0.1+\beta}O_{2.6+\gamma} + \{Zn_5B_4O_{11}\} + N_2 + H_2O \quad (10)$$

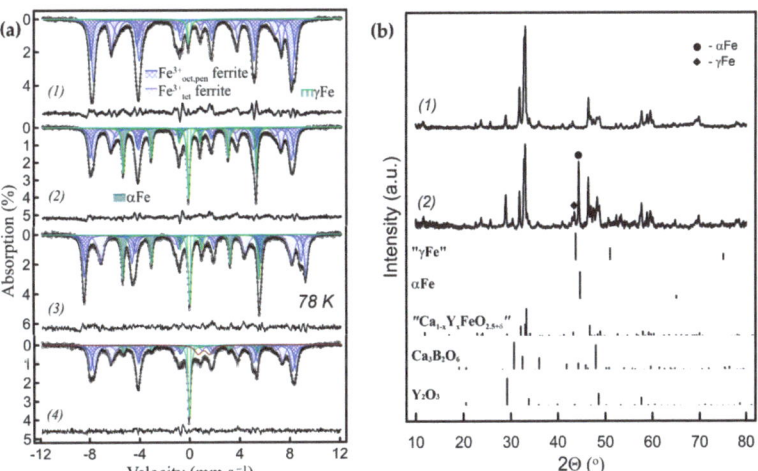

Figure 6. (a) RT Mössbauer spectra of the Ca$_{0.7}$Y$_{0.3}$Fe$_{0.9}$Zn$_{0.1}$O$_{2.6}$/h-BN-derived nanocomposites with different amounts of metal Fe exsolved. (a1) 3.5% of γFe exsolved; (a2) 29% of Fe exsolved (~9% γFe); (a3) the same sample (a2) measured at 78 K; (a4) after oxidation–reduction of sample (a2), 18.5% of Fe exsolved (14% γFe). (b) Powder XRD patterns, (b1) of the initial Ca$_{0.7}$Y$_{0.3}$Fe$_{0.9}$Zn$_{0.1}$O$_{2.6}$ ferrite and, (b2) of sample (a2).

The main oxide products are Ca$_3$B$_2$O$_6$ and Y$_2$O$_3$ (Zn$_5$B$_4$O$_{11}$ could be a minor phase). The Mössbauer spectra (Figure 6(a2,a3)) show that 29% of the Fe total content was exsolved in this sample, where γFe accounted for ~9% and αFe for 20%. The substitution manifests itself in a significant broadening of lines in the Mössbauer spectra, due to the introduction of local distortions and the disruption of the magnetic superexchange interactions. Consequently, the components of the brownmillerite subspectrum that correspond to the tetrahedral and octahedral plus pentagonal positions in the ferrite structures were fitted with combinations of several Zeeman sextets. Sextets with δ~0.34–0.37 mm s^{-1} and H_{hf}~49–51 T correspond to Fe^{3+} cations in octahedral and pentagonal positions (poorly resolved at RT), while those with δ~0.18–0.20 mm s^{-1} and H_{hf} ~40–43 T correspond to the tetrahedral ones. The two other components with narrow lines correspond to the metal Fe subspectrum, comprising of the αFe sextet with δ~0 mm s^{-1} and H_{hf} ~33 T and the γFe singlet with δ~−0.1 mm s^{-1}. The Mössbauer measurements at 78 K result in the narrower lines of the components, but the spectral contributions of the subspectra remain about the same. At a lower exsolution level of ~3–6%, all metal Fe was in the γ form (Figure 6(a1), Table S3). The TEM images of the latter samples are shown in Figure 7. Similar to the unsubstituted $Ca_2Fe_2O_5$-derived samples, it was difficult to differentiate the Fe particles among the agglomerates with different contrasts. However, the Fe nanoparticles <50 nm in size embedded in the oxide matrix were clearly identified by the EDX analysis and mapping (Figure 7a,c). The HAADF-STEM images, together with the Fourier transform imaging, confirmed that the exsolved nanoparticles are metallic γFe (Figure 7b). The EDX mapping also shows some degree of Y and O segregation on the scale of tens of nanometers, which is consistent with the XRD phase analysis.

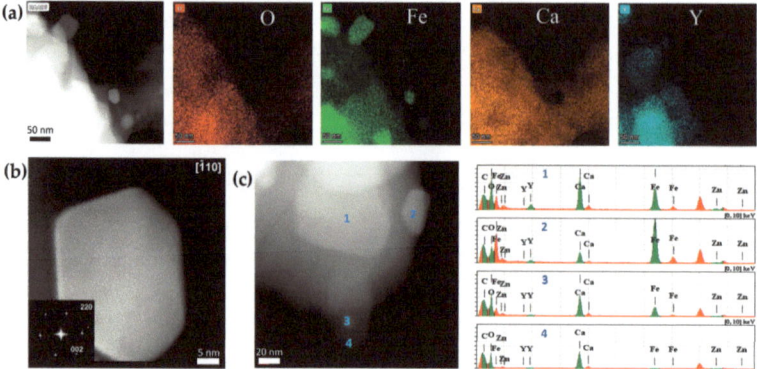

Figure 7. (a) HAADF-STEM images with the corresponding EDX compositional maps of the $Ca_{0.7}Y_{0.3}Fe_{0.9}Zn_{0.1}O_{2.6}$/h-BN-derived nanocomposites with 3.5% of γFe exsolved (Figure 6(a1)). (b) [$\bar{1}$10] HAADF-STEM image of the γFe nanoparticle, along with the Fourier transform. (c) HAADF-STEM image with EDX analysis of other $Ca_{0.7}Y_{0.3}Fe_{0.9}Zn_{0.1}O_{2.6}$/h-BN-derived nanocomposites with 6% of γFe exsolved.

Since Ca is more active in the exsolution reactions than Y, the remaining Ca–Y ferrite can also be enriched in Y despite the formation of Y_2O_3. This can be seen from the spectra of the samples with high degrees of exsolution, as shown in Figure S1 and Table S4. They reveal that the contributions of spectral components corresponding to Fe^{3+} in tetrahedral coordination decrease significantly because of the transformation of Ca ferrite-based $Ca_{1-x}Y_xFeO_{3-\gamma}$ solid solutions with x = 0.3 to Y ferrite-based solid solutions (presumably with x > 0.6 [35]).

Preferable formation of Ca-rich borates was similarly observed for the La-substituted $Ca_{1-x}La_xFeO_{3-\gamma}$ starting ferrites. This system is also pseudobinary. Within this system, the $Ca_2LaFe_3O_8$ Grenier phase, with a crystal structure intermediate between brownmillerite and perovskite types, was reported. In addition, the formation of microdomains of close compositions is possible at different values of x and γ [36–38]. The main oxide product in ACAR-promoted exsolution reactions of Ca–La ferrites with h-BN is $Ca_3B_2O_6$, like for $Ca_{0.7}La_{0.3}FeO_{2.65}$ in Figure 8(a2,a3)), while the formation of La_2O_3 was not observed. At higher exsolution degrees, double borate $Ca_3La_3(BO_3)_5$ was additionally formed (Figure 8(a3)). Since La is retained more in the $Ca_{0.7-z}La_{0.3+z}FeO_{2.65+\delta}$ solid solution, its content increases, as does the oxygen index value. At some level of La content, the resulting ferrite solid solution loses oxygen vacancy ordering, i.e., it becomes the $La_{0.3+z}Ca_{0.7-z}FeO_{3-\gamma}$ type perovskite-like solid solution. This is reflected in the Mössbauer spectra as the disappearance of Fe^{3+} tetrahedral spectral components (Figure 8(b2), Table S5). A simplified reaction is shown in (11):

$$Ca_{0.7}La_{0.3}FeO_{2.65} + BN + H_2 \rightarrow Fe + Ca_3B_2O_6 + La_{0.3+z}Ca_{0.7-z}FeO_{3-\gamma} + \{Ca_3La_3(BO_3)_5\} + N_2 + H_2O \quad (11)$$

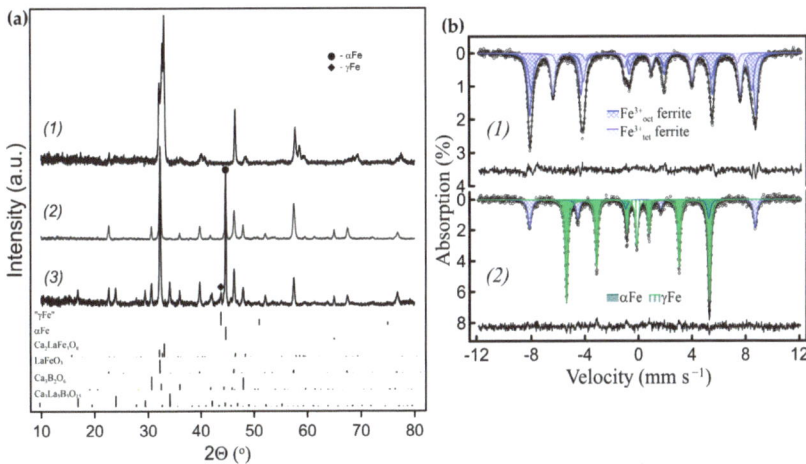

Figure 8. (a) Powder XRD patterns of the $Ca_{0.7}La_{0.3}FeO_{2.65}$/h-BN-derived nanocomposites with different amounts of metal Fe exsolved: (a1) the initial $Ca_{0.7}La_{0.3}FeO_{2.65}$ ferrite; (a2) 66% of Fe exsolved (2% γFe); (a3) 76% of Fe exsolved (8% γFe). (All Fe contributions were evaluated by Mössbauer spectroscopy). (b) Mössbauer spectra (b1) of the initial $Ca_{0.7}La_{0.3}FeO_{2.65}$ ferrite; (b2) of sample (a3).

In the case of LaFeO$_3$-based compounds, the substitution of Fe^{3+} cations by non-reducible cations can also lead to the formation of γFe exclusively. Figure 9 shows that the Zn-substituted solid solution, $LaFe_{0.8}Zn_{0.2}O_{2.9}$, reacts with h-BN, according to the simplified reaction:

$$LaFe_{0.8}Zn_{0.2}O_{2.9} + BN \rightarrow Fe + LaBO_3 + LaFe_{0.8-\beta}Zn_{0.2+\beta}O_{3-\gamma} + N_2 + \{ZnO\} \quad (12)$$

At ~8% of the total metal Fe content, all of the exsolved Fe is in the γ form (Figure 9(b1), Table S5). The TEM EDX mapping of this $LaFe_{0.8}Zn_{0.2}O_{2.9}$-derived sample with ~8% of γFe exsolved, allowed for the identification of the embedded Fe metal nanoparticles (Figure 10a). The HAADF-STEM images, together with the Fourier transform imaging, confirmed that the exsolved nanoparticles are highly twinned metal γFe nanocrystals (Figure 10b,c).

In the case of double substitution by Ca and Zn, the ACAR exsolution with h-BN will proceed as follows (Figure S2(a2,a3,b2,b3)):

$$La_{0.8}Ca_{0.2}Fe_{0.8}Zn_{0.2}O_{2.8} + BN + H_2 \rightarrow Fe + LaBO_3 + Ca_3La_3(BO_3)_5 + La_{0.8+z}Ca_{0.2-z}Fe_{0.8-\beta}Zn_{0.2+\beta}O_{3-\gamma} + \{ZnO\} + N_2 + H_2O \quad (13)$$

At low reaction extents only γFe will be exsolved (Figure S2(b2)). In both cases, Zn-containing phases may precipitate at high reaction extents, presumably in oxide (ZnO) or borate forms, but in most of our samples they did not appear in the XRD patterns.

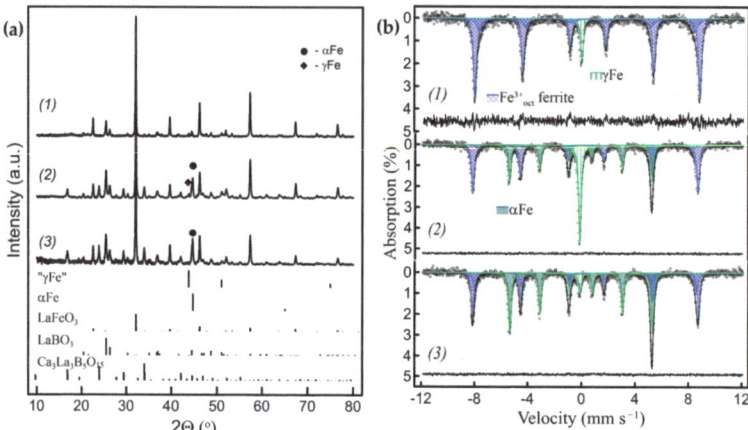

Figure 9. (a) Powder XRD patterns and (b) corresponding RT Mössbauer spectra of the nanocomposites obtained by the ACAR exsolution from Zn-substituted LaFeO$_3$: (1) from LaFe$_{0.8}$Zn$_{0.2}$O$_{2.9}$/h-BN with ~8% of γFe exsolved; (2) from La$_{0.8}$Ca$_{0.2}$Fe$_{0.8}$Zn$_{0.2}$O$_{2.8}$/h-BN with ~50% of Fe exsolved, where ~20%—γFe; (3) sample (2) after grinding in a mortar.

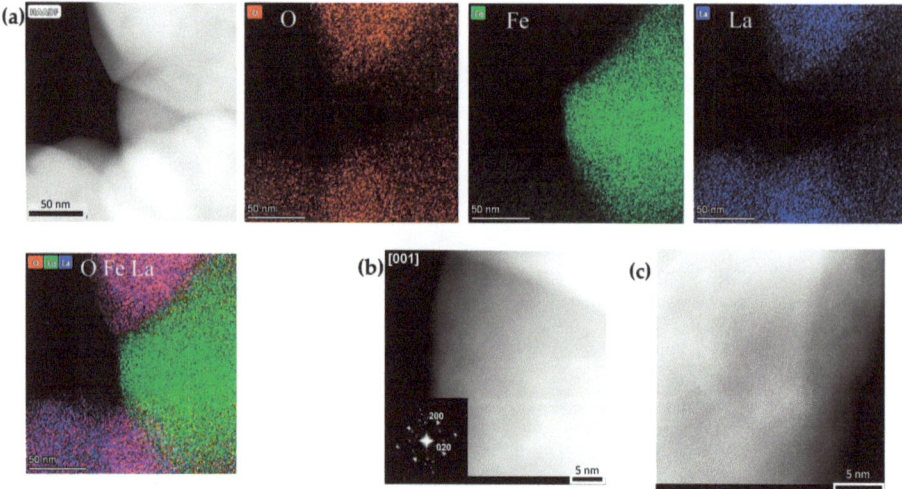

Figure 10. (a) HAADF-STEM image and corresponding EDX compositional maps of the γ-Fe nanoparticles of the LaFe$_{0.8}$Zn$_{0.2}$O$_{2.9}$/h-BN-derived nanocomposites with ~8% of γFe exsolved (Figure 9(a1,b1)). (b) [001] HAADF-STEM image of the γ-Fe nanoparticles, along with the Fourier transform. (c) HAADF-STEM image of highly twinned γ-Fe nanoparticle.

The lattice parameters of the cubic fcc cell of the exsolved γFe nanocrystals can be estimated from the room-temperature powder XRD patterns, in the range of 0.357–0.358 nm. These values correspond to those calculated for austenite solid solutions at room temperature when extrapolated to pure iron [39,40]. Since the γFe nanoparticles synthesized here at 650–750 °C are crystalline and are confined in oxide/borate matrices, they are not pyrophoric in air at room temperature. Moreover, they persist during heating/cooling cycles in a reducing atmosphere. However, a mild mechanical impact on the nanocomposites with high γFe content, by gently grinding in a mortar and pestle for 1–2 min, leads to the transformation of most of the γFe into αFe, and irreversibly so

(Figures 9(a2,a3,b2,b3) and S2(a3,a4,b3,b4), Table S6). This transition is, to some extent, analogous to transformations of retained austenite caused by mechanical deformation at low temperatures [41,42].

Note that for all the substituted ferrites investigated, the formation of Fe whiskers, which is considered undesirable during the described processes, was not observed.

2.4. Redox Behavior of the Exsolved Nanoparticles

The exsolved metal nanoparticles can be oxidized in air to iron oxides at elevated temperatures. The oxidation of metal nanoparticles is generally associated with a decrease in density and an increase in volume. As it was shown in [23], for CoNi exsolved, socketed nanoparticles, these volume changes lead to their migration from the initial sites under redox cycling conditions. In addition, redox cycling can cause the growth and coarsening of oxidized metal nanoparticles or their reintegration into perovskite matrices during oxidation [10,43,44]. For all the nanocomposites obtained herein, the embedded nanoparticles are also completely oxidized to αFe_2O_3 at 500 °C and above, in air. The γFe nanoparticles are oxidized first, and at lower temperatures of 200–300 °C, and their oxidation is accompanied by the transition of residual γFe into the α form (Figure 3(3,5), Table 1). However, the nanocomposites produced by ACAR-promoted exsolution exhibited peculiar behavior during redox cycling, or at least at the temperatures investigated. In these nanocomposites, the initial metal nanoparticles, after complete oxidation into αFe_2O_3 at 500 °C and above, can revert back to the metallic state by subsequent reduction in 10% H_2/Ar at 650–750 °C and, remarkably, the metal γFe nanoparticles can be reinstated in the γ form through such a reduction of αFe_2O_3. Moreover, the γFe fractions may increase compared to the initial content, especially when $Ca_2Fe_2O_5$ and its solid solutions have been used (Figures 3(1,2) and 4(a3,a4,b3,b4)). To our knowledge, the formation of γFe nanoparticles from Fe oxides by hydrogen reduction at temperatures below 910 °C has not been reported previously. Like the formation of the initial γFe nanoparticles, this is presumably a compressive strain-driven behavior, which can be attributed to the preservation of the localization of all the nanoparticles, i.e., initially exsolved metallic-derived oxide and restored metallic oxide, within the voids, along with their strong binding to the void walls during the redox cycles. The γFe formation can, thus, be considered as a kind of indicator of the localization of nanoparticles inside the voids during these processes. Note that at the same time, there should also be free space inside the voids close to nanoparticles, sufficient to compensate for the metal/oxide volume difference, otherwise the matrix may be destroyed.

Using the above assumption on the nanoparticle redox behavior, the reversible transformation of γFe into αFe, and vice versa, through redox cycling at different oxidation temperatures can be realized (Figure 3, Table 1). The sample of the $LaFeO_3$/h-BN based nanocomposite with ~30% of Fe exsolved was subjected to a preliminary oxidation (at 500 °C)/reduction cycle to maximize the γFe contribution (Figure 3(1,2), Table 1 (d123, d124)). To convert most of the Fe into the α form, the first stage involved oxidation at a low temperature of ~300 °C, where more chemically active γFe nanoparticles were partially oxidized/partially converted into αFe (Figure 3(3)), but the resulting oxides did not sufficiently sinter with the matrix. In the second stage, subsequent reduction at 700 °C yields αFe (Figure 3(4), see also Figure 1(a3,b4)). To regenerate γFe, the obtained αFe nanoparticles were first completely oxidized at 500 °C (Figure 3(5)), which provides sufficient sintering of the formed αFe_2O_3 particles and the matrix. Subsequent reduction, at the same temperature of 700 °C, regenerated the γFe nanoparticles (Figure 3(6)). According to the Mössbauer spectra, the Fe^0 (metal)/Fe^{3+} (in matrix) ratio remained approximately the same during redox cycling (Table 1).

The TEM images of the reduced sample with ~11% of γFe and ~6% of αFe, obtained by the ACAR-promoted exsolution reaction of $Ca_{0.7}Y_{03}(Fe_{0.9}Zn_{0.1})O_{2.6}$ with h-BN, after an oxidation–reduction cycle demonstrate features similar to uncycled (only exsolved) nanocomposites: using EDX analysis, the Fe nanoparticles can be identified in some locations, embedded in the oxide matrix, along with some Y segregation (Figure S3).

2.5. ACAR-Promoted Exsolution of FeNi$_x$ Alloys

It is well-established that the existence of a region of the γ form can be significantly extended to lower temperatures by creating an Fe alloy with certain elements, such as Ni [29]. The exsolution of NiFe alloy nanoparticles from Ni and Fe-containing perovskite-like oxides has also been reported [10,16,21], so the formation of FeNi$_x$ alloys was expected during our process as well. Ni is less chemically active and more reducible compared to Fe, so Ni additions to La, and especially Ca, ferrites significantly decrease their stability in a reducing H$_2$-containing atmosphere [27]. For this reason, the starting ferrites with low Ni content were used herein, to avoid decomposition. According to the XRD patterns shown in Figure 11a, the ACAR-promoted exsolution with h-BN from LaFe$_{0.8}$Ni$_{0.2}$O$_{2.9}$, La$_{0.8}$Ca$_{0.2}$Fe$_{0.8}$Ni$_{0.2}$O$_{2.8}$, and La$_{0.5}$Ca$_{0.5}$Fe$_{0.9}$Ni$_{0.1}$O$_{2.7}$ can be written as follows:

$$\text{LaFe}_{0.8}\text{Ni}_{0.2}\text{O}_{2.9} + \text{BN} \rightarrow \text{FeNi}_x + \text{LaBO}_3 + \text{LaFe}_{0.8+\beta}\text{Ni}_{0.2-\beta}\text{O}_{2.9+\delta} + \text{N}_2 \quad (14)$$

$$\text{La}_{0.8}\text{Ca}_{0.2}\text{Fe}_{0.8}\text{Ni}_{0.2}\text{O}_{2.8} + \text{BN} + \text{H}_2 \rightarrow \text{FeNi}_x + \text{LaBO}_3 + \text{Ca}_3\text{La}_3(\text{BO}_3)_5 + \text{La}_{0.8+z}\text{Ca}_{0.2-z}\text{Fe}_{0.8+\beta}\text{Ni}_{0.2-\beta}\text{O}_{2.8+\delta} + \text{N}_2 + \text{H}_2\text{O} \quad (15)$$

$$\text{La}_{0.5}\text{Ca}_{0.5}\text{Fe}_{0.9}\text{Ni}_{0.1}\text{O}_{2.7} + \text{BN} + \text{H}_2 \rightarrow \text{FeNi}_x + \text{Ca}_3\text{La}_3(\text{BO}_3)_5 + \text{Ca}_3\text{B}_2\text{O}_6 + \text{La}_{0.5+z}\text{Ca}_{0.5-z}\text{Fe}_{0.9+\beta}\text{Ni}_{0.9-\beta}\text{O}_{2.7+\delta} + \text{N}_2 + \text{H}_2\text{O} \quad (16)$$

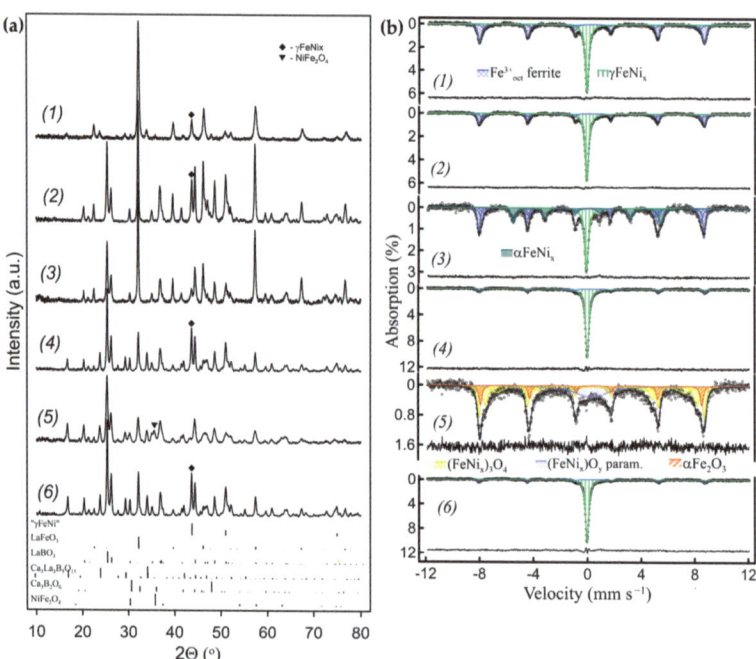

Figure 11. (a) Powder XRD patterns and (b) corresponding RT Mössbauer spectra of the nanocomposites obtained by the ACAR-promoted exsolution from Ni-substituted ferrites: (1) La$_{0.5}$Ca$_{0.5}$Fe$_{0.9}$Ni$_{0.1}$O$_{2.7}$/h-BN with ~41% of γFeNi exsolved; (2) from LaFe$_{0.8}$Ni$_{0.2}$O$_{2.9}$/h-BN with ~53% of γFeNi exsolved; (3) sample (2) after grinding in a hand mortar (~25% γFeNi, ~29% αFeNi); (4) from La$_{0.8}$Ca$_{0.2}$Fe$_{0.8}$Ni$_{0.2}$O$_{2.8}$/h-BN with ~82% of γFeNi exsolved; (5) sample (4) oxidized at 500 °C (~44%—NiFe$_2$O$_4$, ~16% αFe$_2$O$_3$, ~13% superparamagnetic phase); (6) after reduction of sample (5) (10% H$_2$/Ar, 700 °C) (~83% γFeNi).

In these reactions, the resulting borates are the same as for the Ni-free ferrites described before (Figure 11(a1,a2,a4)). Note that while it is difficult to achieve exsolution levels of more than ~50% using metal Fe for the Ni-free ferrites due to cation mobility limitations, the reaction extent can be significantly higher for Ni-containing ferrites. As follows from the Mössbauer spectra shown in Figure 11(b1,b2,b4) and Table S7, Ni additions effectively stabilize the fcc structure of the exsolved nanoparticles. Their spectra are mainly comprised of two subspectra. The first, magnetically split with broad lines, correspond to Fe^{3+} cations in ferrites. It was fitted as a set of sextets, with δ~0.37 mm s^{-1} and H_{hf}~49–53 T. The second subspectrum, which was fitted as a paramagnetic singlet with δ~−0.07 mm s^{-1}, corresponds to Fe in γFeNi$_x$ alloys. According to the Fe–Ni phase diagram, at temperatures close to ambient, the αFe-based bcc phase αFeNi$_x$ coexists with intermetallic compounds of Fe$_3$Ni and FeNi fcc types in the Fe-rich region; although, the phase boundaries at these temperatures are difficult to determine [29]. The thermodynamically stable phases of FeNi$_x$ alloys, viz. the Fe-rich bcc and the Ni-rich fcc alloys, are magnetically ordered at room temperature [45]. At elevated temperatures, there is a continuous solid solution of γFeNi$_x$ with the eutectoid temperature of ~345–400 °C at ~50 at% of Ni. At ~10–20 at% of Ni, which matches the initial ferrite stoichiometry, the transition temperature is about ~650–700 °C. Since Ni is more reducible than Fe, the alloys at low degrees of exsolution will be enriched in Ni and their transition temperatures will be even lower. These temperatures are below the temperatures of 700–750 °C at which reactions (14, 15, and 16) were carried out. When cooled to room temperature, the XRD patterns show that all the exsolved FeNi$_x$ nanoparticles retained their γ structure at room temperature (Figure 11(a1,a2,a4)). The Mössbauer subspectra corresponding to the FeNi$_x$ nanoparticles consist of singlets with δ~−0.1 mm s^{-1} (Figure 11(b1,b2,b4)), evidencing that they are paramagnetic. It is consistent with their γ form, since paramagnetic behavior at room temperature has been reported for metastable γFeNix alloys with high Fe content [46,47]. The stabilization of the most exsolved γFeNi$_x$ nanoparticles in γ form upon cooling, which can be explained by the strain developed due to their localization in voids, suggests that this is the main type of localization. Similar to γFe exsolved nanoparticles, a large part of γFeNi$_x$ nanoparticles can be transformed into ferromagnetic αFeNi$_x$ (Figure 11(b3)) by gently grinding in a mortar and pestle for 1–2 min. This treatment significantly reduced the FeNi$_x$ reflections visually, in the XRD pattern of those samples (Figure 11(a3)). The mechanical stress-induced martensite γ to α transformation in Fe-rich Fe-Ni bulk alloys at room temperature has been previously reported in the literature [48,49].

The exsolved γFeNi$_x$ nanoparticles can be completely oxidized at 500 °C and above, to the spinel solid solution Ni$_{1\pm x}$Fe$_{2\pm x}$O$_4$ (Figure 11(a5,b5)). Subsequent reduction in 10% H$_2$/Ar atmosphere at 650–700 °C will restore the γFeNi$_x$ metal nanoparticles (Figure 11(a6,b6)). It is noteworthy that the exsolution level for this sample was >80% (Table S7).

The TEM images of Ni-containing samples after the ACAR exsolution from La$_{0.5}$Ca$_{0.5}$Fe$_{0.9}$Ni$_{0.1}$O$_{2.7}$ ferrite, with ~41% of Fe exsolved (Figure 11(a1,b1)), are shown in Figure 12a–d. Similar to other samples, the images show agglomerates of different contrast (Figure 12a–c). The embedded FeNi$_x$ nanoparticles <30 nm in size, however, can be identified by EDX analysis, e.g., like in locations 2, 3, and 4 in Figure 12d. The alloy nanoparticles contained approximately about ~30 at% of Ni at this exsolution degree. Figure 12e shows the TEM images of this sample after gently grinding in a mortar with a pestle. There is not much difference compared to the unground ones, except for some rounded agglomerates of <25–30 nm located on the grain surfaces enriched in Fe and Ni, which can be identified as alloy nanoparticles (Figure 12e, loc. 2). Figure 12f shows the TEM images of the same unground sample after the oxidation (at 500 °C, air)/reduction (700 °C, 10% H$_2$/Ar) cycle. Here again, it looks similar to the original sample with embedded FeNi$_x$ nanoparticles, which were identified by EDX analysis (Figure 12f, loc. 2).

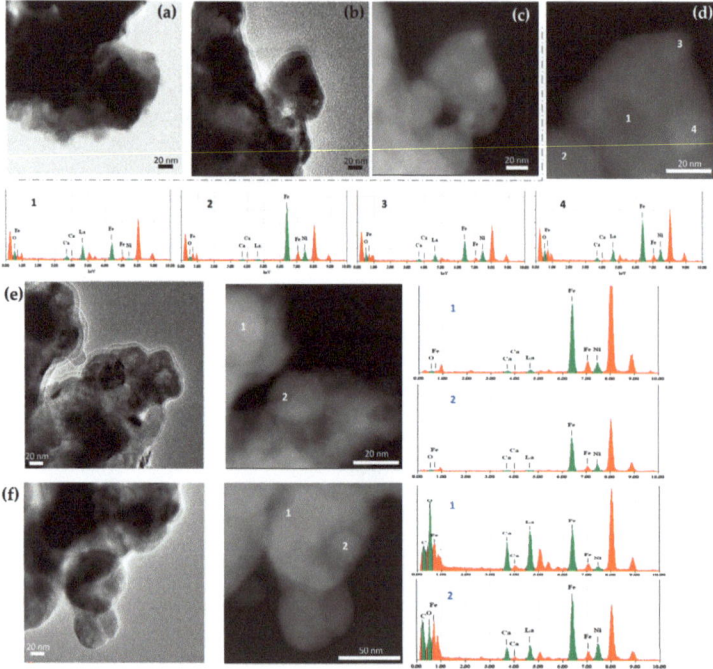

Figure 12. (**a,b**) BF TEM, (**c**) HAADF-STEM images, and (**d**) EDX elemental analysis in the selected locations of the La$_{0.5}$Ca$_{0.5}$Fe$_{0.9}$Ni$_{0.1}$O$_{2.7}$/h-BN nanocomposite with ~41% of γFeNi exsolved (Figure 11(1)) at different locations; (**e**) BF TEM and HAADF-STEM images and EDX elemental analysis in the selected locations after mortar grinding; (**f**) BF TEM and HAADF-STEM images and EDX elemental analysis of the same sample (**a**–**d**) in the selected locations of the sample after the oxidation/reduction cycle.

2.6. Separation of the Individual Nanoparticles

The reversible oxidation/reduction of metallic nanoparticles, while maintaining their localization within the in situ-created individual voids, is a remarkable feature of the nanocomposites produced by ACAR-promoted exsolution. It makes it possible to transform the initial metallic nanoparticles into various oxides, viz. Fe$_{1-x}$O, Fe$_3$O$_4$, α/γFe$_2$O$_3$ etc., while preventing their agglomeration. The components of matrices, viz. the starting and resulting ferrites, resulting borates, etc., in turn, have very diverse chemical properties depending on their composition. In particular, unsubstituted, and certain substituted, Ca ferrites are susceptible to hydrolysis and dissolve in dilute mineral acids like HCl. Ca$_3$B$_2$O$_6$ is also soluble in dilute acids. Since Fe$_3$O$_4$ (and γFe$_2$O$_3$) particles can dissolve reasonably slowly in dilute acidic aqueous solutions at RT, the crystallized and sintered Fe oxide nanoparticles, produced by the oxidation of the exsolved metal nanoparticles, can be separated from such oxide matrices by acid hydrolysis.

The TEM and SEM images of the nanoparticles separated from the Ca$_2$Fe$_2$O$_5$/h-BN and Ca$_2$FeAlO$_5$/h-BN-derived nanocomposites (~30–35% of Fe exsolved) are shown in Figures 13 and S5. Ca$_2$Fe$_{2-x}$Al$_x$O$_5$ ferrites of the brownmillerite type, including Ca$_2$FeAlO$_5$, obtained by the isovalent substitution of Fe^{3+} cations with Al^{3+}, are also prone to hydrolysis as unsubstituted Ca$_2$Fe$_2$O$_5$. They react with h-BN, similar to Ca$_2$Fe$_2$O$_5$, according to the simplified reaction (17), where Al is mainly retained in the ferrite phase (Figure S4):

$$Ca_2Fe_{2-x}Al_xO_5 + BN + H_2 \rightarrow yFe + Ca_3B_2O_6 + Ca_2Fe_{2-x-y}Al_{x+y}O_5 + N_2 + H_2O \quad (17)$$

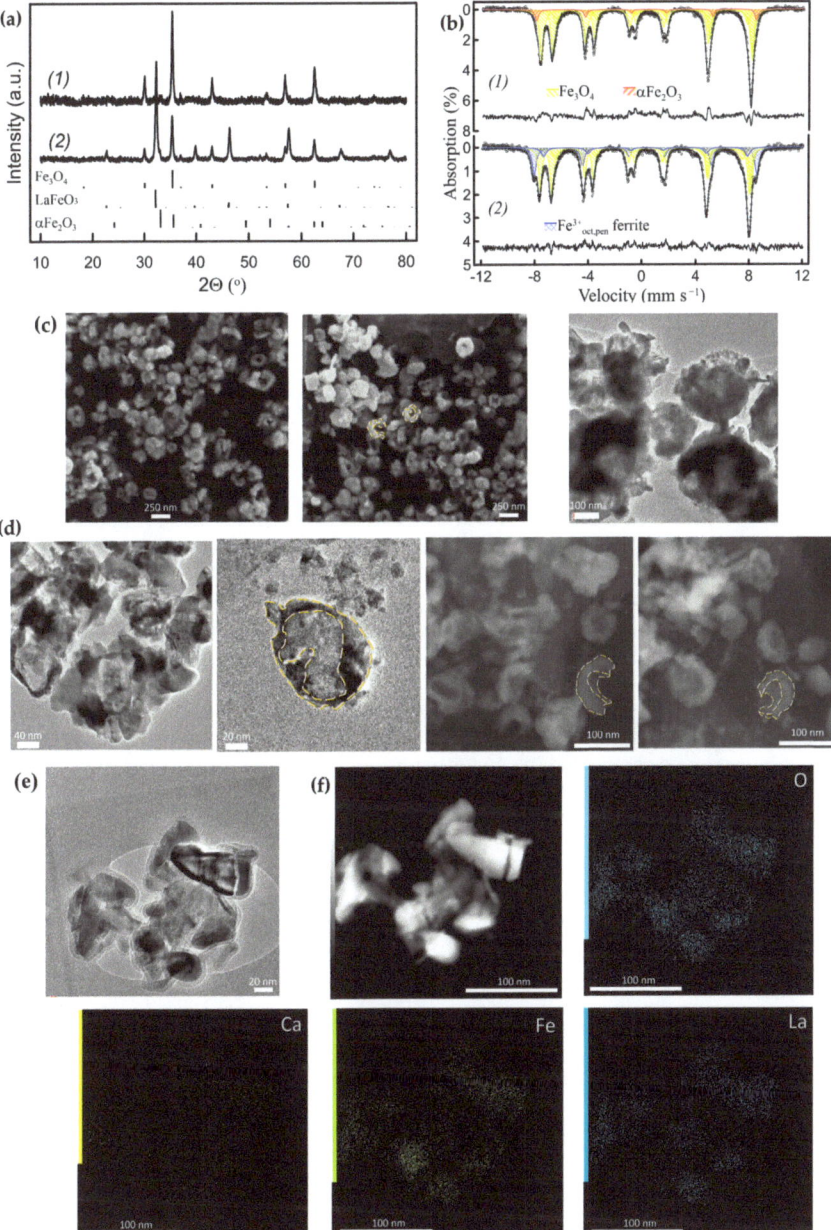

Figure 13. Separated nanoparticles: (**a**) powder XRD patterns and (**b**) corresponding RT Mössbauer spectra of the nanoparticles obtained from the Ca_2FeAlO_5/h-BN-derived nanocomposite (a1,b1) and from the $Ca_{0.8}La_{0.2}FeO_{2.6}$/h-BN derived nanocomposite (a2,b2). (**c**) SEM and BF TEM images of the Fe_3O_4 nanoparticles obtained from the $Ca_2Fe_2O_5$/h-BN-derived nanocomposite. (**d**) BF TEM and HAADF-STEM images of the Fe_3O_4 nanoparticles obtained from sample (a1,b1) (dashed lines indicate the cup shape of the particles). (**e**) BF TEM and (**f**) HAADF-STEM images with corresponding EDX compositional map of the $Fe_3O_4/La_{0.2+z}Ca_{0.8-z}FeO_{3-\gamma}$ nanoparticles obtained from sample (a2,b2).

The RT Mössbauer spectrum of the Ca_2FeAlO_5/h-BN-derived nanocomposite is shown in Figure S4(1). The subspectrum corresponding to $Ca_2Fe_{2-x}Al_xO_5$ is significantly broadened and unresolved due to the magnetic dilution by diamagnetic Al^{3+} cations. Therefore, the spectral contribution of exsolved Fe, which totaled ~35% (~5% γFe), was evaluated from the 78K Mössbauer spectra (Figure S4(2)).

Figure 13(a1,b1) shows the powder XRD patterns and Mössbauer spectra (Table S8) of the Fe_3O_4 nanoparticles separated from the Ca_2FeAlO_5/h-BN-derived nanocomposite. The transformation of metal Fe nanoparticles into Fe_3O_4 nanoparticles was carried out by their oxidation to αFe_2O_3 at 500 °C, followed by their reduction in 10% H_2/Ar flow at 460–480 °C. The matrices were dissolved in 0.5–1.5% HCl aqueous solutions at RT. The Fe_3O_4 nanoparticles were magnetically separated and washed thoroughly with distilled water. It is worth noting that the leaching process of Fe_3O_4 can be carried out with high yields. This is facilitated by the fact that the particles were produced at relatively high temperatures at both stages of synthesis, viz. at oxidation of the metallic particles to Fe_2O_3 and at their subsequent reduction to Fe_3O_4 and were, therefore, crystallized and sintered well, which should reduce their solubility in diluted acids. Similar acid leaching in diluted aqueous HCl solutions has previously been successfully utilized to extract $SrFe_{12}O_{19}$-based nanoparticles, during high-temperature glass–ceramic synthesis [50].

The HAADF-STEM and SEM images of the separated nanoparticles obtained from the $Ca_2Fe_2O_5$/h-BN and Ca_2FeAlO_5/h-BN-derived nanocomposites (~30–35% of Fe exsolved) reveal that most of them are cap-shaped (hemispherical) hollow nanostructures with sizes below ~200 nm and ~100 nm, respectively (Figures 13c,d and S5). According to Figures 13 and S5, the wall thickness of the oxide nanoparticles is of the order of a few nm, thus the wall thickness of the parent metallic nanoparticles should be of the same scale.

The separated nanoparticles provide insight into the shape and size of the parent metal nanoparticles within the matrices. These hollow cap-shaped oxide nanoparticles indicate that the exsolved metal nanoparticles are formed as an inner layer in individual voids generated in situ within the matrices. This particle shape can potentially provide free space sufficient to accommodate the volume increase during oxidation, if the hollow particles are not completely filled with other reaction products. At the same time, these shapes and locations appear to play an important role in creating the conditions necessary for the development of compressive strain in metal nanoparticles sufficient to stabilize γFe, which is somewhat analogous to the γFe formation in cylindrical pores [34].

The extracted nanoparticles were different when La-substituted $Ca_{1-x}La_xFeO_{2.5+\delta}$ ferrite was used as the starting material in the ACAR-promoted exsolution process. Figures 13e,f and S6 show the TEM images of the oxide nanoparticles extracted from the nanocomposite obtained by the reaction of $Ca_{0.8}La_{0.2}FeO_{2.6}$ with h-BN (~35% of Fe exsolved). Metallic Fe nanoparticles were converted into Fe_3O_4 nanoparticles in a similar manner as described above. According to reaction (14), the La content in the remaining perovskite oxide $Ca_{0.8-z}La_{0.2+z}FeO_{2.6+\delta}$ increases and, thus, its solubility in the dilute HCl aqueous solution decreases.

Using Mössbauer spectroscopy and XRD analyses, it was found that the extracted nanoparticles consisted mainly of spinel Fe_3O_4 and perovskite $La_{0.2+z}Ca_{0.8-z}FeO_{3-\gamma}$ phases (Figure 13(a2,b2), Table S8). The TEM with EDX analysis revealed that the $La_{0.2+z}Ca_{0.8-z}FeO_{3-\gamma}$ crystallites were segregated around Fe_3O_4 nanoparticles (Figures 13e,f and S6). Most of them were almost completely covered by the resulting La–Ca ferrite, forming open shells. The size of the Fe_3O_4 nanoparticles was estimated to be smaller (<40–50 nm) than in the previous case. The composition and microstructure of the extracted nanoparticles reflect the relative arrangement of the metal/resulting ferrite phases in the parent nanocomposites. They show that the growth of exsolved metal nanoparticles in matrices is accompanied by the formation of shells, consisting of the resulting La-rich ferrites.

2.7. Variety of ACAR-Promoted Metal Exsolution Reactions

The new approach we have developed is not limited to the aforementioned ferrites and h-BN as starting compounds. The range of suitable compounds is wider and includes, but is not limited to, for example, Sr-containing ferrites and some other nitrides like Si_3N_4, etc. In particular, Figure S7 (Table S9) illustrates the ACAR-promoted exsolution of metal Fe nanoparticles from $Sr_2Fe_2O_5$, with Si_3N_4 as a reducing agent, according to the simplified reaction (18):

$$Sr_2Fe_2O_5 + 1/3 Si_3N_4 + H_2 \rightarrow 2Fe + Sr_2SiO_4 + H_2O + 2/3 N_2 \qquad (18)$$

Even CO gas can be utilized in reactions of this type with Ca and Sr ferrites, when it is added to 10%H_2/Ar flow at ~650 °C, according to reaction (19):

$$(Sr,Ca)_2Fe_2O_5 + 2CO + H_2 \rightarrow 2Fe + 2(Sr,Ca)CO_3 + H_2O \qquad (19)$$

This reaction also leads to embedded metal α and γ Fe nanoparticles, as shown in Figure S8 [51].

The chemical activity of reducing reagents in ACAR processes is different, so that Si_3N_4 and CO are active with Ca and Sr-based ferrites, forming the corresponding silicates or carbonates, but are inactive towards La-based ferrites.

2.8. In Situ Reactions of Exsolved Nanoparticles

The exsolved metallic nanoparticles and their oxides can also be converted into other functional compounds, directly inside the matrices. For instance, embedded αFe_2O_3 nanoparticles obtained by the oxidation of exsolved metallic nanoparticles inside (La-based ferrite $La_{0.8}Ca_{0.2}Fe_{0.8}Zn_{0.2}O_{2.8}$)/h-BN-derived nanocomposites (Figure 14(1)) can be converted into embedded χ-Fe_5C_2 (Hagg carbide) nanoparticles by the reaction with gaseous CO at 300–350 °C (Figure 14(2), Table S10), according to the scheme in Figure S9. This carbide is considered to be an active phase in Fischer–Tropsch synthesis [52–54]. Its Mössbauer subspectrum at RT consists of three broad Zeeman sextets, with δ of 0.18–0.25 mm s^{-1} and H_{hf} of 10–22 T, providing reliable identification [52]. The matrix, consisting of La ferrites and La borates, does not react with CO under the above conditions and remains unchanged. The α and, more significantly, γ Fe metallic nanoparticles can then be regenerated by the oxidation of carbide to αFe_2O_3 at 500 °C, followed by reduction to metal (Figure 14(3)), indicating that the metal/oxide/carbide nanoparticles retain their localization in the voids.

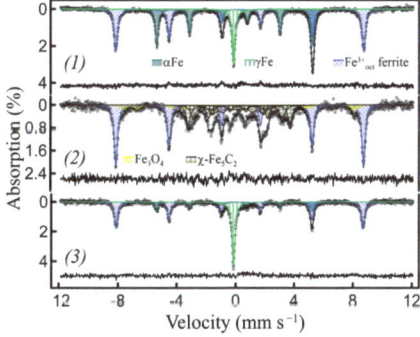

Figure 14. RT Mössbauer spectra of the $La_{0.8}Ca_{0.2}Fe_{0.8}Zn_{0.2}O_{2.8}$/h-BN-derived nanocomposites during redox cycling involving CO: (1) as prepared with ~51% of Fe exsolved (~14% γFe); (2) sample (1) after oxidation (air, 500 °C) following by the reaction with CO at 350 °C (~46% of Fe in χ-Fe_5C_2 carbide); (3) the same sample after the oxidation (air, 500 °C)–reduction (10%H_2/Ar, 700 °C) cycle (~47% of metal Fe total, ~30% γFe).

3. Materials and Methods

3.1. Materials Preparation

The initial perovskite-like compounds used for the nanocomposite syntheses were prepared using a sol–gel routine from the respective nitrate precursors and citric acid as a complexing agent. All the chemicals were of reagent grade, from commercial sources (AO Reachem, Moscow, Russia). The aqueous solution of metal nitrates in the calculated amounts were mixed with concentrated citric acid solutions in the molar ratio of citric acid to metals at ~1.1. The pH of the solution was adjusted by a dilute ammonia solution to values of ~5–7. Several samples, mostly Sr-containing ones, were synthesized using ethylenediaminetetraacetic acid (EDTA) as a complexing agent. In this case, the respective nitrate solution was dropwise added to an EDTA solution in diluted aqueous ammonia and then the pH was adjusted to ~8. The water was evaporated from the mixed solution and the resulting viscous gels were heated to ~200 °C to yield porous brownish materials. The materials were ground and then calcined in air at 500 °C (heating rate was 5 °C/min) for 4 h. The samples were then calcined in air for 8–10 h at 700–900 °C. The substituted perovskites, usually containing Fe^{4+} and/or $Fe^{(3+\delta)+}$ cations in mixed oxidation states, were additionally treated in 10% H_2/Ar gas flow at 600 °C for 1 h to reduce them to Fe^{3+}.

The nanocomposites with exsolved particles were synthesized through the solid-state method. The required amounts of perovskite powders and h-BN (Plasmotherm, Moscow, Russia) or other reducing reagents were dry grounded in an agate mortar until homogeneous for 20–30 min. The mixture powders were placed into a silica reactor in alumina crucibles and heated in 10% H_2/Ar gas flow of ~1–4 mL/min. The synthesis temperatures were ~650–700 °C and ~700–780 °C for the Ca- and La-based ferrites, respectively. For most of the samples used in this work, the exsolution reactions were carried out in a single step over ~10–16 h to reach completeness or a steady state. The experimental conditions for the synthesis of the nanocomposites are summarized in Table S1. The different extents of the exsolution reactions in this work were mainly predetermined by the ferrite/(h-BN, etc.) ratio, and the final amounts of exsolved Fe were quantified experimentally using Mössbauer spectroscopy.

3.2. Materials Characterization

The phase composition of the nanocomposites was characterized by X-ray powder diffraction (XRD), with a Huber G670 Image Plate Guinier diffractometer (CuK$_{\alpha 1}$ radiation, curved Ge monochromator, image plate detector) (Huber Diffraktionstechnik GmbH & Co., Ltd., Rimsting, Germany).

The ^{57}Fe Mössbauer spectra were recorded in terms of the transmission geometry on a commercial MS-1104EM spectrometer, using a ^{57}Co(Rh) source (13.5 mCi), equipped with a custom liquid nitrogen bath cryostat (77–320 K) (ZAO, Kordon, Russia). The fitting procedure was performed with a special original least-squares fitting software developed at Lomonosov MSU. All the isomer shift values refer to αFe at room temperature (RT).

Scanning electron microscopy (SEM) analysis was performed using a Quattro S scanning electron microscope (LaB6 field emission cathode, Thermo Fisher Scientific, Bleiswijk, The Netherlands). Transmission electron microscopy (TEM) images and energy-dispersive X-ray (EDX) spectra were obtained with a JEOL 2100 F/Cs, operated at 200 kV and equipped with a JEOL system for EDX analysis (JEOL Ltd., Tokyo, Japan), and using an aberration-corrected Titan Themis Z transmission electron microscope at 200 kV (Thermo Fisher Scientific, The Netherlands), equipped with a Super-X system for EDX analysis. For the TEM studies, the powder samples of the nanocomposites were prepared by dispersing in ethanol or heptane without additional grounding to avoid sample destruction, and then depositing a few drops of the suspension on special grids.

4. Conclusions

The novel approach to metal exsolution using $ABO_{3-\gamma}$ (B = Fe, Ni, etc.) complex perovskite oxides of reducible transition metals has been developed to produce Fe-containing metal and metal oxide nanoparticles and nanocomposites. Our strategy is based on the additional auxiliary reactions of irreducible alkaline-earth and/or rare-earth A-site cations of the parent perovskite oxides with the constituents of reducing compounds resulting in the formation of complex oxides, which facilitate the exsolution of reducible transition metals from the B-sites of perovskites. The approach was applied mainly to Ca- and La-based unsubstituted and substituted ferrites as starting compounds and h-BN as a reducing compound. The metal nanoparticles exsolved by the described approach are largely localized within oxide matrices, in individual voids generated in situ. Moreover, the metal nanoparticles of Fe and $FeNi_x$ alloys can be formed and stabilized in γ form, presumably due to induced compressive strain that develops in such locations.

The redox behavior of the exsolved nanoparticles was investigated. It was shown that they readily undergo redox cycling at moderate temperatures, where they maintain localization in the voids. The latter allows, under certain conditions, the regeneration of γFe and $\gamma FeNi_x$ from oxides during redox cycling. The nanoparticles can be chemically modified by reagents inside the matrices, while maintaining their localization.

The oxide matrices consist of unreacted initial perovskites and reaction products (borates, silicates, resulting perovskites, etc.) in various combinations and exhibit a wide variety of chemical and physical properties. They undergo significant transformation during the exsolution process, and their phase and chemical compositions depend on various factors, such as the composition of the initial ferrites and reducing reagents, the extent of reactions, and the reaction conditions, etc.

For certain Ca and Sr ferrite-based nanocomposites, a method has been developed for the separation oxide nanoparticles produced by the oxidation of the initially exsolved metal nanoparticles. The separated Fe_3O_4 nanoparticles (which can also be γFe_2O_3, etc.), obtained using Ca ferrites as starting materials, had a cup-shaped hollow shape. The separated nanoparticles obtained using La-substituted Ca ferrites were composite Fe_3O_4/(La-based resulting ferrite) layered particles. The separated nanoparticles provide insight into the shape, size, and relative arrangement of the phases of the parent metal nanoparticles within the matrices. This approach provides a new high-temperature templateless route for the synthesis of hollow and layered nanoparticles of Fe oxides and related compounds.

Our proposed strategy is applicable to a wide range of both perovskite-like oxides and reducing compounds. The developed approach provides greater flexibility in controlling the reaction extent, as well as the composition and properties of the resulting nanocomposites and nanoparticles. The approach can also be applied to the surface modification of perovskite oxide ceramics, since the starting perovskite compounds are considered stable under the reaction conditions. Our novel strategy is expected to be viable for the development of various new functional materials.

Supplementary Materials: The following supporting information can be downloaded at: https://www.mdpi.com/article/10.3390/inorganics12080223/s1, Figure S1. RT Mössbauer spectra of the $Ca_{0.7}Y_{0.3}Fe_{0.9}Zn_{0.1}O_{2.6}$/h-BN-derived nanocomposites with 57% of Fe exsolved (~11% γFe); Figure S2. (a) Powders XRD patterns of the nanocomposites obtained by the ACAR exsolution from Zn-substituted $LaFeO_3$: (a1) from $LaFe_{0.8}Zn_{0.2}O_{2.9}$/h-BN with ~8% of γFe exsolved; (a2) from $La_{0.8}Ca_{0.2}Fe_{0.8}Zn_{0.2}O_{2.8}$/h-BN with ~5% of γFe exsolved; (a3) from $La_{0.8}Ca_{0.2}Fe_{0.8}Zn_{0.2}O_{2.8}$/h-BN with ~50% of Fe exsolved, where ~20%—γFe; (a4) sample (a3) after grinding in a mortar; (a5) sample (a3) oxidized at 500 °C (all Fe contributions are evaluated by Mössbauer spectroscopy) and (b) corresponding RT Mössbauer spectra: (b1) of sample (a1); (b2) of sample (a2); (b3) of sample (a3); (b4) of sample (a4); (b5,6) of sample (a5) at RT and 78K, respectively; Figure S3. (a) BF TEM image of the $Ca_{0.7}Y_{0.3}Fe_{0.9}Zn_{0.1}O_{2.6}$/h-BN-derived nanocomposites (~18% of Fe) after the oxidation–reduction cycle; (b) HAADF-STEM images with corresponding EDX compositional maps of the sample in (a); Figure S4. The powder XRD pattern (a) and the corresponding RT Mossbauer spectrum (b) of the Ca_2FeAlO_5/h-BN-derived nanocomposite with ~35% of Fe exsolved (~5% of γFe); Figure S5. TEM images of the separated Fe_3O_4 nanoparticles obtained from the Ca_2FeAlO_5/h-BN-derived nanocomposites: (a) BF TEM; (b) HAADF-STEM images and EDX elemental analysis in the selected locations; Figure S6. TEM images of the separated Fe_3O_4/$La_{1-y}Ca_yFeO_{2.5+\delta}$ nanoparticles obtained from the $Ca_{0.8}La_{0.2}FeO_{2.6}$/h-BN-derived nanocomposite: (a) HAADF-STEM images with the corresponding EDX compositional map; (b) HAADF-STEM images and EDX elemental analysis in the selected locations; Figure S7. The powder XRD pattern (a) and the corresponding RT Mössbauer spectrum (b) of the $Sr_2Fe_2O_5$/Si_3N_4-derived nanocomposite with ~60% of Fe exsolved (~9% of γFe); Figure S8. The RT Mossbauer spectrum of nanocomposites by ACAR-promoted by CO exsolution reactions: (1) from $Sr_2Fe_2O_5$/CO with ~16% of Fe exsolved (~5% γFe); (2) from $Ca_2Fe_2O_5$/CO with ~12% of Fe exsolved (~3% γFe); Figure S9. Scheme of transformations of the exsolved Fe nanoparticles during the redox cycle involving CO treatment; Table S1. Experimental conditions for the synthesis of nanocomposites by ACAR-promoted metal exsolution; Table S2. Hyperfine parameters of the RT Mössbauer spectra of the samples in Figure 4b; Table S3. Hyperfine parameters of the RT Mössbauer spectra of the $Ca_{0.7}Y_{0.3}Fe_{0.9}Zn_{0.1}O_{2.6}$-derived nanocomposites in Figure 6a; Table S4. Hyperfine parameters of the RT Mössbauer spectra of the $Ca_{0.7}Y_{0.3}Fe_{0.9}Zn_{0.1}O_{2.6}$-derived nanocomposites in Figure S1; Table S5. Hyperfine parameters of the RT Mössbauer spectra of the samples in Figure 9b; Table S6. Hyperfine parameters of the RT Mössbauer spectra of the samples in Figure S2b; Table S7. Hyperfine parameters of the RT Mössbauer spectra of the samples in Figure 11b; Table S8. Hyperfine parameters of the RT Mössbauer spectra of the samples in Figure 13d; Table S9. Hyperfine parameters of the RT Mössbauer spectra of the samples in Figure S7b; Table S10. Hyperfine parameters of the RT Mössbauer spectra of the samples in Figure 14.

Author Contributions: Conceptualization, D.F.; Methodology D.F.; Investigation, D.F., M.R., S.M. and D.P.; Formal analysis, D.F. and D.P.; Visualization, D.F.; Writing—original draft D.F.; Writing—review and editing, D.F. and D.P. All authors have read and agreed to the published version of the manuscript.

Funding: This work was partially supported by the State Assignments of Lomonosov Moscow State University (122030200324-1).

Data Availability Statement: The data will be made available from the corresponding author on reasonable request.

Acknowledgments: The authors are grateful to Artem Abakumov for help with the TEM characterization.

Conflicts of Interest: The authors declare that there are no conflicts of interest.

References

1. Yadav, S.; Rani, N.; Saini, K. A review on transition metal oxides based nanocomposites, their synthesis techniques, different morphologies and potential applications. *IOP Conf. Ser. Mater. Sci. Eng.* **2022**, *1225*, 012004. [CrossRef]
2. Ndolomingo, M.J.; Bingwa, N.; Meijboom, R. Review of supported metal nanoparticles: Synthesis methodologies, advantages and application as catalysts. *J. Mater. Sci.* **2020**, *55*, 6195–6241. [CrossRef]
3. Gao, C.; Lyu, F.; Yin, Y. Encapsulated Metal Nanoparticles for Catalysis. *Chem. Rev.* **2021**, *121*, 834–881. [CrossRef]
4. Hong, J.; Wang, B.; Xiao, G.; Wang, N.; Zhang, Y.; Khodakov, A.Y.; Li, J. Tuning the Metal–Support Interaction and Enhancing the Stability of Titania-Supported Cobalt Fischer–Tropsch Catalysts via Carbon Nitride Coating. *ACS Catal.* **2020**, *10*, 5554–5566. [CrossRef]
5. Chen, A.; Su, Q.; Han, H.; Enriquez, E.; Jia, Q. Metal Oxide Nanocomposites: A Perspective from Strain, Defect, and Interface. *Adv. Mater.* **2019**, *31*, 1803241. [CrossRef]
6. Přech, J.; Strossi Pedrolo, D.R.; Marcilio, N.R.; Gu, B.; Peregudova, A.S.; Mazur, M.; Ordomsky, V.V.; Valtchev, V.; Khodakov, A.Y. Core–Shell Metal Zeolite Composite Catalysts for In Situ Processing of Fischer–Tropsch Hydrocarbons to Gasoline Type Fuels. *ACS Catal.* **2020**, *10*, 2544–2555. [CrossRef]
7. Yang, Q.; Liu, G.; Liu, Y. Perovskite-Type Oxides as the Catalyst Precursors for Preparing Supported Metallic Nanocatalysts: A Review. *Ind. Eng. Chem. Res.* **2018**, *57*, 1–17. [CrossRef]
8. Rane, A.V.; Kanny, K.; Abitha, V.K.; Thomas, S. Chapter 5-Methods for Synthesis of Nanoparticles and Fabrication of Nanocomposites. In *Synthesis of Inorganic Nanomaterials*; Mohan Bhagyaraj, S., Oluwafemi, O.S., Kalarikkal, N., Thomas, S., Eds.; Woodhead Publishing: Sawston, UK, 2018; pp. 121–139.
9. Kumar, A.; Dutta, S.; Kim, S.; Kwon, T.; Patil, S.S.; Kumari, N.; Jeevanandham, S.; Lee, I.S. Solid-State Reaction Synthesis of Nanoscale Materials: Strategies and Applications. *Chem. Rev.* **2022**, *122*, 12748–12863. [CrossRef]
10. Kousi, K.; Tang, C.Y.; Metcalfe, I.S.; Neagu, D. Emergence and Future of Exsolved Materials. *Small* **2021**, *17*, 27. [CrossRef]

11. Kim, J.H.; Kim, J.K.; Liu, J.; Curcio, A.; Jang, J.-S.; Kim, I.-D.; Ciucci, F.; Jung, W. Nanoparticle Ex-solution for Supported Catalysts: Materials Design, Mechanism and Future Perspectives. *ACS Nano* **2021**, *15*, 81–110. [CrossRef]
12. Sun, Z.; Hao, C.; Toan, S.; Zhang, R.; Li, H.; Wu, Y.; Liu, H.; Sun, Z. Recent advances in exsolved perovskite oxide construction: Exsolution theory, modulation, challenges, and prospects. *J. Mater. Chem. A* **2023**, *11*, 17961–17976. [CrossRef]
13. Neagu, D.; Irvine, J.T.S.; Wang, J.; Yildiz, B.; Opitz, A.K.; Fleig, J.; Wang, Y.; Liu, J.; Shen, L.; Ciucci, F.; et al. Roadmap on exsolution for energy applications. *J. Phys. Energy* **2023**, *5*, 031501. [CrossRef]
14. Tilley, R.J.D. The ABX3 Perovskite Structure. In *Perovskites*; Tilley, R.J.D., Ed.; John Wiley & Sons, Ltd.: Chichester, UK, 2016; pp. 1–41.
15. Peña, M.A.; Fierro, J.L.G. Chemical Structures and Performance of Perovskite Oxides. *Chem. Rev.* **2001**, *101*, 1981–2018. [CrossRef] [PubMed]
16. Shah, S.; Xu, M.; Pan, X.; Gilliard-Abdulaziz, K.L. Exsolution of Embedded Ni–Fe–Co Nanoparticles: Implications for Dry Reforming of Methane. *ACS Appl. Nano Mater.* **2021**, *4*, 8626–8636. [CrossRef]
17. Chen, H.; Yu, H.; Peng, F.; Yang, G.; Wang, H.; Yang, J.; Tang, Y. Autothermal reforming of ethanol for hydrogen production over perovskite LaNiO$_3$. *Chem. Eng. J.* **2010**, *160*, 333–339. [CrossRef]
18. Wang, J.; Woller, K.B.; Kumar, A.; Zhang, Z.; Zhou, H.; Waluyo, I.; Hunt, A.; LeBeau, J.M.; Yildiz, B. Ion irradiation to control size, composition and dispersion of metal nanoparticle exsolution. *Energy Environ. Sci.* **2023**, *16*, 5464–5478. [CrossRef]
19. Santaya, M.; Jiménez, C.E.; Arce, M.D.; Carbonio, E.A.; Toscani, L.M.; Garcia-Diez, R.; Knop-Gericke, A.; Mogni, L.V.; Bär, M.; Troiani, H.E. Exsolution versus particle segregation on (Ni,Co)-doped and undoped SrTi$_{0.3}$Fe$_{0.7}$O$_{3-\delta}$ perovskites: Differences and influence of the reduction path on the final system nanostructure. *Int. J. Hydrogen Energy* **2023**, *48*, 38842–38853. [CrossRef]
20. Amaya-Duenas, D.M.; Chen, G.X.; Weidenkaff, A.; Sata, N.; Han, F.; Biswas, I.; Costa, R.; Friedrich, K.A. A-site deficient chromite with in situ Ni exsolution as a fuel electrode for solid oxide cells (SOCs). *J. Mater. Chem. A* **2021**, *9*, 5685–5701. [CrossRef]
21. Santaya, M.; Jiménez, C.E.; Troiani, H.E.; Carbonio, E.A.; Arce, M.D.; Toscani, L.M.; Garcia-Diez, R.; Wilks, R.G.; Knop-Gericke, A.; Bär, M.; et al. Tracking the nanoparticle exsolution/reoxidation processes of Ni-doped SrTi$_{0.3}$Fe$_{0.7}$O$_{3-\delta}$ electrodes for intermediate temperature symmetric solid oxide fuel cells. *J. Mater. Chem. A* **2022**, *10*, 15554–15568. [CrossRef]
22. Oh, D.; Colombo, F.; Nodari, L.; Kim, J.H.; Kim, J.K.; Lee, S.; Kim, S.; Kim, S.; Lim, D.-K.; Seo, J.; et al. Rocking chair-like movement of ex-solved nanoparticles on the Ni-Co doped La$_{0.6}$Ca$_{0.4}$FeO$_{3-\delta}$ oxygen carrier during chemical looping reforming coupled with CO$_2$ splitting. *Appl. Catal. B* **2023**, *332*, 122745. [CrossRef]
23. Neagu, D.; Papaioannou, E.I.; Ramli, W.K.W.; Miller, D.N.; Murdoch, B.J.; Ménard, H.; Umar, A.; Barlow, A.J.; Cumpson, P.J.; Irvine, J.T.S.; et al. Demonstration of chemistry at a point through restructuring and catalytic activation at anchored nanoparticles. *Nat. Commun.* **2017**, *8*, 1855. [CrossRef] [PubMed]
24. Weber, M.L.; Wilhelm, M.; Jin, L.; Breuer, U.; Dittmann, R.; Waser, R.; Guillon, O.; Lenser, C.; Gunkel, F. Exsolution of Embedded Nanoparticles in Defect Engineered Perovskite Layers. *ACS Nano* **2021**, *15*, 4546–4560. [CrossRef] [PubMed]
25. Wang, J.; Syed, K.; Ning, S.; Waluyo, I.; Hunt, A.; Crumlin, E.J.; Opitz, A.K.; Ross, C.A.; Bowman, W.J.; Yildiz, B. Exsolution Synthesis of Nanocomposite Perovskites with Tunable Electrical and Magnetic Properties. *Adv. Funct. Mater.* **2022**, *32*, 2108005. [CrossRef]
26. Syed, K.; Wang, J.; Yildiz, B.; Bowman, W.J. Bulk and surface exsolution produces a variety of Fe-rich and Fe-depleted ellipsoidal nanostructures in La$_{0.6}$Sr$_{0.4}$FeO$_3$ thin films. *Nanoscale* **2022**, *14*, 663–674. [CrossRef] [PubMed]
27. Nakamura, T.; Petzow, G.; Gauckler, L.J. Stability of the perovskite phase LaBO$_3$ (B = V, Cr, Mn, Fe, Co, Ni) in reducing atmosphere I. Experimental results. *Mater. Res. Bull.* **1979**, *14*, 649–659. [CrossRef]
28. Dreyer, M.; Krebs, M.; Najafishirtari, S.; Rabe, A.; Friedel Ortega, K.; Behrens, M. The Effect of Co Incorporation on the CO Oxidation Activity of LaFe$_{1-x}$Co$_x$O$_3$ Perovskites. *Catalysts* **2021**, *11*, 550. [CrossRef]
29. Ohnuma, I.; Shimenouchi, S.; Omori, T.; Ishida, K.; Kainuma, R. Experimental determination and thermodynamic evaluation of low-temperature phase equilibria in the Fe–Ni binary system. *Calphad* **2019**, *67*, 101677. [CrossRef]
30. Bundy, F.P. Pressure—Temperature Phase Diagram of Iron to 200 kbar, 900 °C. *J. Appl. Phys.* **2004**, *36*, 616–620. [CrossRef]
31. Bhadeshia, H.; Honeycombe, R. (Eds.) Chapter 1-Iron and Its Interstitial Solutions. In *Steels: Microstructure and Properties*, 4th ed.; Butterworth-Heinemann: Oxford, UK, 2017; pp. 1–22.
32. Shalimov, A.; Potzger, K.; Geiger, D.; Lichte, H.; Talut, G.; Misiuk, A.; Reuther, H.; Stromberg, F.; Zhou, S.; Baehtz, C.; et al. Fe nanoparticles embedded in MgO crystals. *J. Appl. Phys.* **2009**, *105*, 064906. [CrossRef]
33. Yuan, C.L.; Hu, C.; Mei, Y.X.; Hong, A.J.; Yang, Y.; Xu, K.; Yu, T.; Luo, X.F. Strain-induced FCC Fe nanocrystals confined in Al$_2$O$_3$ matrix. *J. Alloys Compd.* **2017**, *727*, 1100–1104. [CrossRef]
34. Yamada, Y.; Tanabe, K.; Nishida, N.; Kobayashi, Y. Iron films deposited on porous alumina substrates. *Hyperfine Interact.* **2016**, *237*, 9. [CrossRef]
35. Fu, B.; Huebner, W.; Trubelja, M.F.; Stubican, V.S. (Y$_{1-x}$Ca$_x$)FeO$_3$: A Potential Cathode Material for Solid Oxide Fuel Cells. *ECS Proc. Vol.* **1993**, *1993*, 276. [CrossRef]
36. Grenier, J.-C.; Fournès, L.; Pouchard, M.; Hagenmuller, P.; Komornicki, S. Mössbauer resonance studies on the Ca$_2$Fe$_2$O$_5$-LaFeO$_3$ system. *Mater. Res. Bull.* **1982**, *17*, 55–61. [CrossRef]
37. Vallet-Regí, M.; González-Calbet, J.; Alario-Franco, M.A.; Grenier, J.-C.; Hagenmuller, P. Structural intergrowth in the Ca$_x$La$_{1-x}$FeO$_{3-x/2}$ system ($0 \leq x \leq 1$): An electron microscopy study. *J. Solid State Chem.* **1984**, *55*, 251–261. [CrossRef]

38. Price, P.M.; Browning, N.D.; Butt, D.P. Microdomain Formation, Oxidation, and Cation Ordering in LaCa$_2$Fe$_3$O$_{8+y}$. *J. Am. Ceram. Soc.* **2015**, *98*, 2248–2254. [CrossRef]
39. Rammo, N.N.; Abdulah, O.G. A model for the prediction of lattice parameters of iron–carbon austenite and martensite. *J. Alloys Compd.* **2006**, *420*, 117–120. [CrossRef]
40. Reed, R.P.; Schramm, R.E. Lattice Parameters of Martensite and Austenite in Fe–Ni Alloys. *J. Appl. Phys.* **2003**, *40*, 3453–3458. [CrossRef]
41. Tavares, S.S.M.; Pardal, J.M.; da Silva, M.J.G.; Abreu, H.F.G.; da Silva, M.R. Deformation induced martensitic transformation in a 201 modified austenitic stainless steel. *Mater. Charact.* **2009**, *60*, 907–911. [CrossRef]
42. Sohrabi, M.J.; Naghizadeh, M.; Mirzadeh, H. Deformation-induced martensite in austenitic stainless steels: A review. *Arch. Civ. Mech. Eng.* **2020**, *20*, 124. [CrossRef]
43. Lai, K.-Y.; Manthiram, A. Evolution of Exsolved Nanoparticles on a Perovskite Oxide Surface during a Redox Process. *Chem. Mater.* **2018**, *30*, 2838–2847. [CrossRef]
44. Weber, M.L.; Sohn, Y.J.; Dittmann, R.; Waser, R.; Menzler, N.H.; Guillon, O.; Lenser, C.; Nems, S.; Gunkel, F. Reversibility limitations of metal exsolution reactions in niobium and nickel co-doped strontium titanate. *J. Mater. Chem. A* **2023**, *11*, 17718–17727. [CrossRef]
45. Xiong, W.; Zhang, H.; Vitos, L.; Selleby, M. Magnetic phase diagram of the Fe–Ni system. *Acta Mater.* **2011**, *59*, 521–530. [CrossRef]
46. Cherdyntsev, V.V.; Pustov, L.Y.; Kaloshkin, S.D.; Tomilin, I.A.; Shelekhov, E.V.; Estrin, E.I.; Baldokhin, Y.V. Phase transformations in powder iron-nickel alloys produced by mechanical alloying. *Phys. Met. Metallogr.* **2009**, *107*, 466–477. [CrossRef]
47. Kumar, S.; Roy, K.; Maity, K.; Sinha, T.P.; Banerjee, D.; Das, K.C.; Bhattacharya, R. Superparamagnetic Behavior of Fe–Ni Alloys at Low Ni Concentration. *Phys. Status Solidi* **1998**, *167*, 175–181. [CrossRef]
48. Sato, H.; Nishiura, T.; Moritani, T.; Watanabe, Y. Atypical phase transformation behavior of Fe-33%Ni alloys induced by shot peening. *Surf. Coat. Technol.* **2023**, *462*, 129470. [CrossRef]
49. Guimarães, J.R.C.; Rios, P.R. The mechanical-induced martensite transformation in Fe–Ni–C alloys. *Acta Mater.* **2015**, *84*, 436–442. [CrossRef]
50. Trusov, L.A.; Sleptsova, A.E.; Duan, J.; Gorbachev, E.A.; Kozlyakova, E.S.; Anokhin, E.O.; Eliseev, A.A.; Karpov, M.A.; Vasiliev, A.V.; Brylev, O.A.; et al. Glass-Ceramic Synthesis of Cr-Substituted Strontium Hexaferrite Nanoparticles with Enhanced Coercivity. *Nanomaterials* **2021**, *11*, 924. [CrossRef] [PubMed]
51. Filimonov, D.S.; Pokholok, K.V.; Rozova, M.G. Fe/perovskite nanocomposites as a result of partial decompositon of substituted strontium ferrites. In Proceedings of the 22th International Meeting Order, Disorder and Properties of Oxides, Rostov-on-Don, Russia, 4–9 September 2019; pp. 185–187.
52. Gu, B.; Ordomsky, V.V.; Bahri, M.; Ersen, O.; Chernavskii, P.A.; Filimonov, D.; Khodakov, A.Y. Effects of the promotion with bismuth and lead on direct synthesis of light olefins from syngas over carbon nanotube supported iron catalysts. *Appl. Catal. B* **2018**, *234*, 153–166. [CrossRef]
53. Chernavskii, P.A.; Kazantsev, R.V.; Pankina, G.V.; Pankratov, D.A.; Maksimov, S.V.; Eliseev, O.L. Unusual Effect of Support Carbonization on the Structure and Performance of Fe/MgAl$_2$O$_4$ Fischer–Tropsch Catalyst. *Energy Technol.* **2021**, *9*, 2000877. [CrossRef]
54. Opeyemi Otun, K.; Yao, Y.; Liu, X.; Hildebrandt, D. Synthesis, structure, and performance of carbide phases in Fischer-Tropsch synthesis: A critical review. *Fuel* **2021**, *296*, 120689. [CrossRef]

Disclaimer/Publisher's Note: The statements, opinions and data contained in all publications are solely those of the individual author(s) and contributor(s) and not of MDPI and/or the editor(s). MDPI and/or the editor(s) disclaim responsibility for any injury to people or property resulting from any ideas, methods, instructions or products referred to in the content.

Article

Mesoporous Titania Nanoparticles for a High-End Valorization of *Vitis vinifera* Grape Marc Extracts

Anil Abduraman [1], Ana-Maria Brezoiu [1], Rodica Tatia [2], Andreea-Iulia Iorgu [1,†], Mihaela Deaconu [1], Raul-Augustin Mitran [3], Cristian Matei [1] and Daniela Berger [1,*]

1. Faculty of Chemical Engineering and Biotechnologies, National University of Science and Technology Politehnica Bucharest, 011061 Bucharest, Romania; anil.abduraman@stud.chimie.upb.ro (A.A.); ana_maria.brezoiu@upb.ro (A.-M.B.); mihaela.deaconu@upb.ro (M.D.); cristian.matei@upb.ro (C.M.)
2. National Institute of R&D for Biological Sciences, 060031 Bucharest, Romania
3. "Ilie Murgulescu" Institute of Physical Chemistry, Romanian Academy, 202 Splaiul Independentei, 060021 Bucharest, Romania; raul.mitran@gmail.com
* Correspondence: daniela.berger@upb.ro
† Current address: Department of Chemistry and Manchester Institute of Biotechnology, The University of Manchester, Manchester M1 7DN, UK.

Abstract: Mesoporous titania nanoparticles (NPs) can be used for encapsulation polyphenols, with applications in the food industry, cosmetics, or biomedicine. TiO_2 NPs were synthesized using the sol-gel method combined with solvothermal treatment. TiO_2 NPs were characterized through X-ray diffraction, FTIR spectroscopy, the N_2 adsorption method, scanning and transmission electron microscopy, and thermal analysis. The sample prepared using Pluronic F127 presented a higher surface area and less agglomerated NPs than the samples synthesized with Pluronic P123. Grape marc (GM), a by-product from wine production, can be exploited for preparing extracts with good antioxidant properties. In this regard, we prepared hydroethanolic and ethanolic GM extracts from two cultivars, Feteasca Neagra (FN) and Pinot Noir. The extract components were determined by spectrometric analyses and HPLC. The extract with the highest radical scavenging activity, the hydroethanolic FN extract, was encapsulated in titania (FN@TiO_2) and compared with SBA-15 silica support. Both resulting materials showed biocompatibility on the NCTC fibroblast cell line in a 50–300 µg/mL concentration range after 48 h of incubation and even better radical scavenging potential than the free extract. Although titania has a lower capacity to host polyphenols than SBA-15, the FN@TiO_2 sample shows better cytocompatibility (up to 700 µmg/mL), and therefore, it could be used for skin-care products.

Keywords: mesoporous titania; grape marc extract; mesoporous silica; extract encapsulation; antioxidant properties; phenolic compounds; biocompatibility

1. Introduction

An effective carrier for biologically active molecules should meet the following conditions: a high specific surface area and pore volume, tunable pore size, good biocompatibility and no toxicity, and the possibility to adjust the interactions between biologically active molecules and the support. The most used inorganic matrix for the encapsulation of biologically active compounds is mesoporous silica, which presents outstanding porosity (up to 1200 m^2/g specific surface area and 1.5 cm^3/g pore volume), good biocompatibility depending on particle size and shape, surface properties, concentration, etc. [1–3]. Silica nanoparticles are generally assessed as safe and biocompatible by the US Food and Drug Administration (FDA) and have been employed as additives in cosmetics and foods [4].

Among functional inorganic materials, nanostructured titania is of particular interest due to its remarkable features: good chemical and thermal stability, low cost, biocompatibility, resistance to photochemical erosion, and excellent optical and electrical properties.

Titania is used intensively in environmental applications to remove pollutants from both air and water [5]. Titania, which has already been successfully applied as an implant material or in cosmetic formulations, can also be employed as a carrier for biologically active compounds [6]. For example, in 2012, mesoporous titania was tested as a carrier for ibuprofen [7].

Titania exhibits photocatalytic properties under UV irradiation but has a lower adsorption capacity than mesoporous silica, so by combining the properties of both silica and titania, titania–silica composites have been successfully applied as photocatalysts for environment purification [8], as well as for biomedical applications [9]. Titania nanoparticles (NPs) are poorly soluble in biological fluids and thus pose challenges when used in biomedical applications. However, titania NPs have been applied in photodynamic therapy for the tumors' treatment [10,11]. There are reports that showed that titania NPs help to neutralize bacterial or fungal strains [12].

Unlike silica, mesoporous titania exhibits lower porosity values (up to 250 m^2/g specific surface area and 0.4 cm^3/g pore volume) and a lower concentration of surface OH groups that can be involved in functionalization reactions, but it could interact with biologically active substances through donor–acceptor bonds.

Typically, mesoporous titania is produced by the soft-templated sol-gel method based on the hydrolysis and condensation reactions involving the cooperative assembly of titanium precursor, usually, titanium isopropoxide or titanium butoxide, with the structure directing agent in the presence of a complexation agent that slows down the rate of hydrolysis and condensation reactions. The sol-gel process of Ti precursors differs from that of silicon alkoxides because of their higher chemical reactivity, resulting from the lower electronegativity of titanium and its ability to spontaneously enhance the coordination number with water molecules, with the hydrolysis rate of titanium alkoxide being five times faster than silicates, hindering the cooperative assembly of inorganic species with surfactant molecules [13].

Red *Vitis vinifera* L. grape marc (GM) resulting from winemaking is an affordable source of polyphenols, including anthocyanins, flavonoids, stilbenes, flavan-3-ols, and phenolic acids, which have antioxidant, anti-inflammatory, antidiabetic, and cardioprotective properties, antiproliferative effects [14], and potent broad virucidal activity [15]. For instance, Balea et al. [16] reported that Feteasca Neagra extract prepared in 70% ethanol from fermented grape pomace exhibited antiproliferative activity on A549 lung carcinoma, MDA-MB-231 human breast adenocarcinoma, and B164A5 murine melanoma, with the best results being obtained for the last cell line.

There are studies on phenolic compounds in wines. For example, malvidin glycosides are formed during wine maturation [17]. Goldberg et al. [18] highlighted that climate influences the concentration of quercetin, as wines originating from areas with a warmer climate and greater exposure to sunlight had a higher concentration of quercetin. The presence of p-coumaric acid in wines seems to be more random than that of quercetin, with this phenolic acid being important because it acts as a precursor of flavonols, flavan-3-ols, and trihydroxy stilbenes [17].

Lately, the demand for natural extracts with antioxidant properties for cosmetics and nutraceuticals is of growing interest and the use of an abundant waste such as grape marc from wine production can contribute to the sustainability of this field [19]. Wasilewski et al. [20] reported a shower gel formulation enriched with compounds extracted from grape pomace, which was safe for utilization as natural cosmetics.

The recovery of biologically active substances from GM is done by extraction, with the yield and chemical profile depending on the type of solvent, the extraction technique and its parameters, and the quality of the grape pomace, which in turn is strongly influenced by the climate, etc. [21–23].

Herein, we report the synthesis of a series of mesoporous titania NPs by combining the sol-gel process and solvothermal treatment using two metal precursors, titanium(IV) isopropoxide and titanium(IV) butoxide, and two nonionic surfactants, the triblock copolymer EO20-PO70-EO20 (Pluronic P123) or EO99-PO70-EO99 (Pluronic F127), in various molar ratios, 2-propanol or n-butanol, as the solvent and acetic acid was used to delay the hydrolysis rate of the metallic alkoxide. Titania NPs with the best features were used to encapsulate a hydroethanolic extract prepared from red grape marc, and the resulting material was compared with the material obtained by incorporating the same extract in mesoporous SBA-15 silica. We also evaluated the chemical profile of several ethanolic or hydroethanolic extracts prepared from the red grape marc from two varieties: Feteasca Neagra and Pinot Noir.

2. Results

2.1. Obtaining and Characterization of Mesoporous Matrices

A series of mesoporous titania nanoparticles were synthesized by the sol-gel method assisted by solvothermal treatment using as a template agent either the triblock copolymer, EO20-PO70-EO20 (Pluronic P123) or the nonionic surfactant EO99-PO70-EO99 (Pluronic F127). The scheme showing the steps of obtaining the titania samples can be seen in Figure 1, while Table 1 lists the titanium precursor, the solvent, the molar ratio between the titanium precursor and template agent used in the synthesis of titania NPs, and how the samples are denoted. Titania samples obtained after solvothermal treatment and purified through Soxhlet extraction (samples labeled Sn_E) or calcined at 400 °C (samples labeled Sn_C) were characterized by wide-angle powder X-ray diffraction (XRD), FTIR spectroscopy, thermal analysis, scanning electron microscopy, transmission electron microscopy, and porosity evaluation based on N_2 adsorption–desorption isotherms.

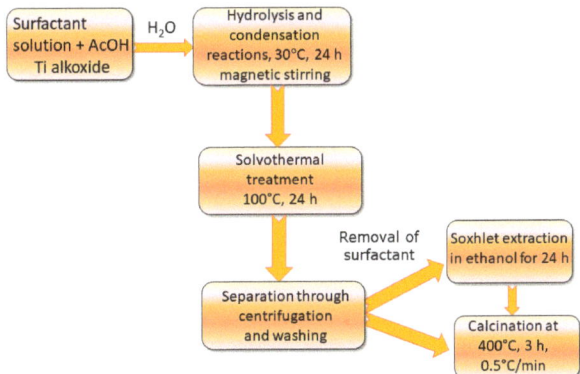

Figure 1. Main steps of the titania samples preparation.

Table 1. Synthesis conditions for mesoporous titania obtaining.

Sample	Solvent	Titania Precursor	Template Agent	Ti Precursor/Template Agent Molar Ratio
S1	n-butanol	Ti(OnBu)$_4$	P123	1/0.017
S2	n-butanol	Ti(OnBu)$_4$	P123	1/0.034
S3	2-propanol	Ti(OiPr)$_4$	P123	1/0.017
S4	2-propanol	Ti(OiPr)$_4$	P123	1/0.034
S5	2-propanol	Ti(OiPr)$_4$	F127	1/0.017
S6	n-butanol	Ti(OnBu)$_4$	F127	1/0.017

2.1.1. X-ray Diffraction

The XRD patterns of the titania samples demonstrated the formation of an anatase crystalline phase, having tetragonal symmetry for all samples irrespective of the structure directing agent, titanium precursor, or quantity of the surfactant used in the synthesis. The thermal treatment determined the preservation of the anatase phase (JCPDS no. 21-1272) and an increase in its crystallinity, with a crystallite size of $D_{101} = 8$ nm, calculated using

Rigaku PDXL software version 1.8 from the most intense diffraction peak at 2θ = 25.27° for uncalcined materials (Figure 2A), and D_{101} = 10 nm for the samples obtained at 400 °C (Figure 2B).

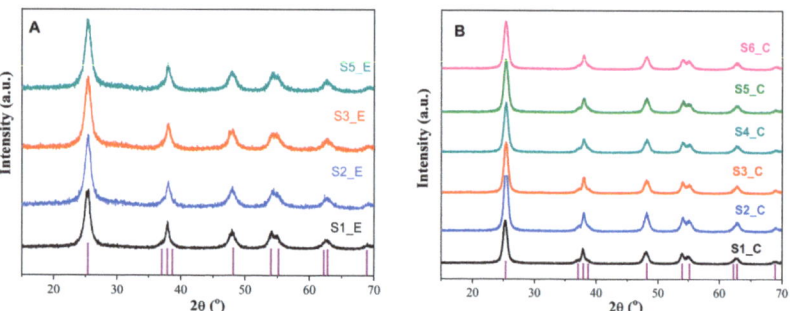

Figure 2. XRD patterns of titania samples obtained after solvothermal treatment and purified by Soxhlet extraction (**A**) and for TiO$_2$ samples calcined at 400 °C, 3 h. (**B**) The JCPDS no. 21-1272 for anatase phase is shown as reference.

2.1.2. Morphology of Mesoporous Inorganic Matrices

The morphology of titania samples was investigated by scanning and transmission electron microscopy (Figure 3). The use of titanium(IV) butoxide as a precursor yielded more agglomerated nanoparticles (Figure 3A,B,D) than in the case of titanium(IV) isopropoxide (Figure 3C,E). TEM investigation demonstrated the crystalline nature of TiO$_2$ NPs and the interparticle pores formation (Figure 3F,G). Titania samples present polyhedral nanoparticles with around 10 nm dimension, which are agglomerated. The particle sizes of the calcined S3_C (Figure 3F) and S5_C (Figure 3G) samples observed on the TEM investigation were consistent, with the crystallite size found based on XRD analyses. Mesoporous SBA-15 silica presents rod-type particles with an average diameter of 447 nm and a length/diameter ratio in the range of 2.04–2.20 (Figure 3H).

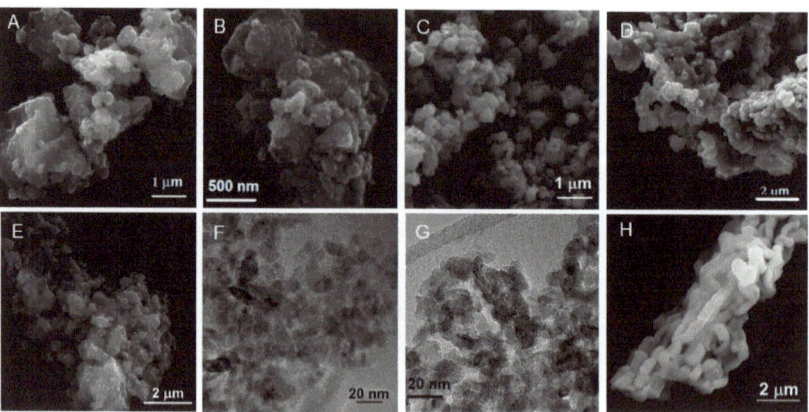

Figure 3. SEM micrographs for the following samples: S1_E (**A**) and S6-E (**B**) purified by Soxhlet extraction, S5_C (**C**), S2_C (**D**) and S4_C (**E**) titania samples obtained at 400 °C, TEM images of S3_C (**F**) and S5_C (**G**) calcined samples, and SEM image of SBA-15 silica (**H**).

2.1.3. FTIR Spectroscopy

In the FTIR spectra of the titania samples isolated after solvothermal treatment (S3 and S5), one can notice the characteristic bands of methyl and methylene groups in the range of 2850–2930 cm^{-1}, as well as the ether bridge vibrations ranging in the

1410–1530 cm^{-1} domain that demonstrate the presence of the polymer on the surface of titanium dioxide nanoparticles (Figure 4). The extraction process in ethyl alcohol for 24 h did not completely remove the structure directing agent, as its characteristic vibrations were still present but less intense in the FTIR spectra of the samples purified by extraction, S1_E and S3_E (Figure 4). The calcining step at 400 °C caused the removal of the template agent regardless of the polymer type (Figure 4—spectrum of S3_C sample). In the FTIR spectra of all TiO$_2$ samples, irrespective of the stage in which the samples were analyzed, one can notice the specific bands of Ti–O and Ti–O–Ti bonds at 677 cm^{-1} and 466 cm^{-1}, respectively [24], as well as the stretching modes of vibration of hydroxyl groups with the lowest transmittance value at 3420 cm^{-1} and the bending band of adsorbed water molecules at 1634 cm^{-1}. The stretching vibrations of -OH groups are less intense in the case of the calcined S3_C sample than in the spectra of the other NPs isolated after solvothermal treatment, S3 and S5, or after Soxhlet extraction, S1_E and S3_E (Figure 4).

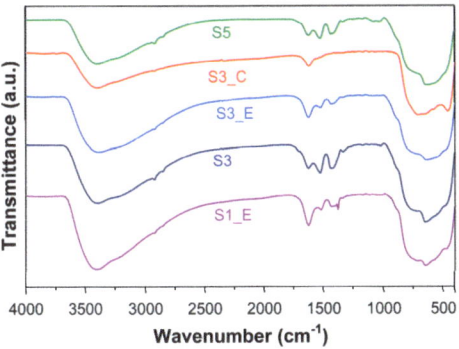

Figure 4. FTIR spectra of the following samples: S1_E and S3_E obtained after Soxhlet extraction, S3 and S5 isolated after solvothermal treatment, and S3_C calcined at 400 °C.

2.1.4. Thermal Analysis

The copolymers decompose in steps up to 400 °C. The extraction process was not very efficient, a content of about 10% (wt.) copolymer remained in the materials purified by Soxhlet extraction (Figure 5A). In agreement with FTIR spectra, the DTA-TG analysis showed that the thermal treatment at 400 °C for 3 h completely removed the surfactant, as no effect was recorded on the DTA traces of the calcined samples (Figure 5B).

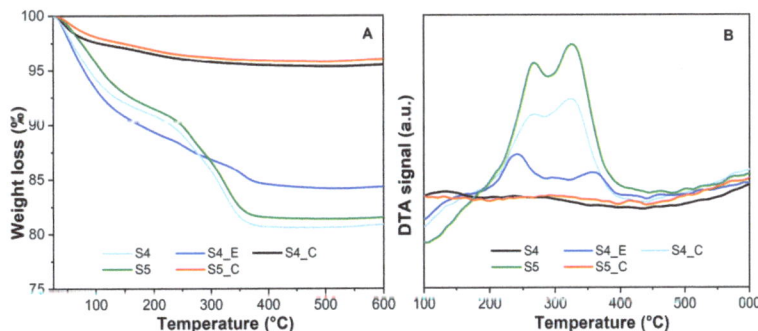

Figure 5. (**A**) TG analyses performed in synthetic air for S4 and S5 samples isolated after solvothermal treatment, S4_E purified by Soxhlet extraction and S4_C and S5_C calcined at 400 °C, 3 h. The TG curves for calcined samples were included to evaluate the content of hydroxyl groups attached to the titania surface that was subtracted when the content of the polymer was determined. (**B**) Corresponding DTA curves.

2.1.5. Evaluation of the Porosity for Mesoporous Inorganic Matrices

The textural properties of the titania samples were determined from nitrogen adsorption–desorption isotherms, recorded at liquid nitrogen temperature, that are type IV with hysteresis at $P/P_0 > 0.5$, characteristic for mesoporous materials with wormlike pores. The porosity of the titania samples is mainly due to interparticle pores, evidenced by a sharp increase in the volume of adsorbed gas in the region of high P/P_0 values (Figures 6A,B and S1). Table 2 gathered the parameters determined from nitrogen adsorption–desorption isotherms: specific surface area, S_{BET}, determined by applying BET theory, total pore volume measured at relative pressure, $P/P_0 = 0.985$, and average pore diameter, d_{BJH}, calculated from the desorption branch of the isotherms with BJH method. The uncalcined titania samples have a higher specific surface area, S_{BET}, than the calcined materials (312 and 274 m^2/g for S2_E and S4_E, respectively, and 154 m^2/g in the case of S4_C), probably because of the contribution of remaining copolymer on titania nanoparticles surface (Table 2). When Ti(OnBu)$_4$ was used as the precursor in the synthesis, this resulted in higher values of average pore size (6.3 nm for S6_C in comparison with 5.4 nm for S5_C for which Ti(OiPr)$_4$ was used), although the specific surface area and total pore volume did not change significantly (153 m^2/g and 0.33 cm^3/g for S5_C and 153 m^2/g and 0.31 cm^3/g for S6_C-Table 2). The use of the copolymer F127 instead of Pluronic P123 for titania nanoparticles synthesis did not have an important contribution to the porosity of the resulting materials. A higher copolymer quantity introduced in the synthesis led to a diminution of the pore size (7.4 nm and 5.8 nm for S3_C and S4_C, respectively, Table 2). SBA-15 silica had higher porosity than all titania NPs with the following textural features: $S_{BET} = 984$ m^2/g, V = 1.31 cm^3/g, and the average diameter of the mesopores computed from the desorption branch of isotherm, $d_{BJH} = 6.3$ nm.

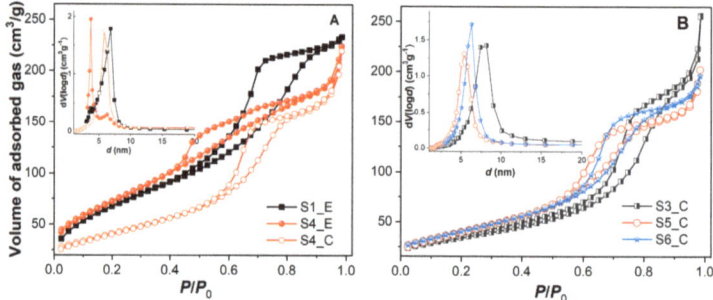

Figure 6. Nitrogen adsorption–desorption isotherms recorded at liquid nitrogen temperature of S1_E and S4_E samples purified by Soxhlet extraction and S4_C (**A**), S3_C, S5_C, and S6_C calcined at 400 °C (**B**). The corresponding pore distribution curves calculated with BJH model from desorption branch of the isotherms are inserted in both figures. All samples were outgassed in vacuum at 120 °C for 17 h.

Table 2. The textural parameters computed from N$_2$ adsorption–desorption isotherms for mesoporous inorganic matrices.

Sample	S_{BET} (m^2/g)	V (cm^3/g)	d_{BJH} (nm)
S1_E	258	0.34	6.8
S2_E	312	0.43	3.7
S2_C	142	0.42	6.8
S3_C	126	0.42	7.4
S4_E	274	0.30	3.7
S4_C	154	0.36	5.8
S5_C	153	0.33	5.4
S6_C	153	0.31	6.3
SBA-15	984	1.31	6.3

2.2. Obtaining and Characterization of Red GM Extracts

The polyphenols extraction was carried out using *Vitis vinifera* L. red grape marc (GM) from two cultivars, Feteasca Neagra (FN) and Pinot Noir (PN), from the Research Station for Viticulture Murfatlar (Constanta County, Romania). The extracts were obtained by conventional technique at reflux using FN and PN grape pomace samples collected after wine production in 2018 or 2019 and ethanol or 50% ethanol aqueous solution as solvent.

For all extracts, the following data were determined by UV-vis methods: the total polyphenols index (TP) expressed as gallic acid (GA) equivalents based on the chemical reaction between phenolic substances and Folic Ciocâlteu reagent, the total flavonoids index (TF) expressed as quercetin (Q) equivalents, based on the chemical reaction between flavonoids and the aqueous solution of $AlCl_3$, and the total anthocyanin pigments (TA) expressed as cyanidin-glucoside (CG) equivalents using the method described by Lee et al. [25]. All these data are presented in Table 3. The radical scavenging activity (RSA) was expressed as Trolox (T) equivalents using two assays: DPPH and ABTS.

Table 3. Solvents used for extraction, yield values, and spectrometric data expressed per gram of dry extract for prepared extracts.

Extract	Year, Solvent	Yield (%)	TP (mgGA/g)	TF (mgQ/g)	TA (mg CG/g)
FN	2019, 50% ethanol	13.2	295.46 ± 1.33	14.82 ± 0.51	7.35 ± 1.89
FN(E)	2018, ethanol	0.4	71.02 ± 0.22	12.31 ± 0.64	20.42 ± 4.26
FN(E-W)	2018, 50% ethanol	6.5	147.71 ± 8.6	5.06 ± 0.12	3.56 ± 0.24
PN	2019, 50% ethanol	11.8	138.49 ± 3.49	11.94 ± 0.03	4.29 ± 0.00
PN(E)	2018, ethanol	1.3	47.09 ± 0.99	13.34 ± 4.11	2.86 ± 0.00

All spectrometric measurements were performed in triplicate.

Upon first glance at the data in Table 3, one can observe that the type of red GM and the harvest year strongly influenced the quantity of phenolic compounds in the extracts. The extract richest in polyphenols is the hydroethanolic FN extract prepared from Feteasca Neagra GM collected in 2019. Both GMs used in this study from 2018 were very poor in phytocompounds, probably because the climate has altered the quality of these wastes. For all extracts prepared from GMs collected in 2018, we obtained very low yields (in the range of 0.4–6.5%) and TP index values (47.09–147.71 mgGA/g extract). Regarding the solvent, the 50% ethanol aqueous solution was more effective in the recovery of polyphenols (147.71 for FN(E-W) vs. 71.02 mgGA/g extract for FN(E) extract) than absolute ethanol, which was better at extracting flavonoids (12.31 and 5.06 mgQ/g extract for FN(E) and FN(E-W), respectively, or 13.34 and 11.94 mgQ/g extract for PN(E) and PN, respectively).

Figure 7 shows the radical scavenging activity (RSA) expressed as Trolox equivalents (TE) using two assays: DPPH and ABTS. The FN sample prepared from GM collected in 2019 has the highest value for antioxidant activity assessed by both DPPH and ABTS assays, (566.00 ± 13.08 and 572.37 ± 5.78 mgTE/g extract, respectively), followed by the other hydroethanolic FN extract obtained from GM collected in 2018 (316.92 ± 21.6 and 267.11 ± 9.04 mgTE/g extract, respectively). Both ethanolic extracts prepared from GMs collected in 2018 had weak radical scavenging activity (35.18 ± 4.88 and 47.89 ± 7.12 mgTE/g extract for FN(E) and PN(E), respectively), which could be correlated with the low amount of polyphenols, though the FN(E) sample has the highest index of TA (20.42 ± 4.26 mgCG/g extract)—Table 3.

Phenolic substances present in the extracts were quantified using reverse-phase high-performance liquid chromatography and the data are listed in Table 4. The chromatograms can be seen in Figure S2. We identified in all the extracts the following phenolic acids: gallic, protocatechuic, vanillic, and syringic acids from the class of flavonoids, rutin and quercetin, and catechin (except for the FN(E) sample) and (-) epicatechin (except the extracts prepared in absolute ethanol) from the flavan-3-ol class. The only extract in which delphinidin (0.593 ± 0.010 mg/g extract) and the esterified phenolic acid—caftaric acid (0.050 mg/g extract)—were quantified was FN(E-W), while *p*-coumaric acid was found only in the FN(E) sample. Being poorly soluble in water but soluble in ethanol, *trans*-resveratrol was quantified in higher amounts in the ethanolic extracts: FN(E) and PN(E) (Table 4).

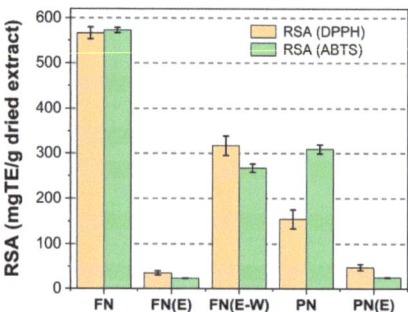

Figure 7. Radical scavenging activity expressed as Trolox equivalents per mass of the dried extract using DPPH and ABTS methods.

Table 4. Chemical profile of prepared extracts determined by HPLC-PDA analysis.

Phenolic Substance	Concentration (mg/g Dried Extract)				
	FN	FN(E)	FN(E-W)	PN	PN(E)
Gallic Acid	1.150 ± 0.004	0.594 ± 0.001	0.437 ± 0.000	0.238 ± 0.008	0.723 ± 0.000
Protocatechuic acid	0.541 ± 0.001	0.714 ± 0.000	0.078 ± 0.002	0.367 ± 0.001	1.748 ± 0.001
Catechin	0.087 ± 0.001	nd	5.978 ± 0.000	1.164 ± 0.002	0.504 ± 0.001
Caftaric acid	nd	nd	0.050 ± 0.000	nd	nd
Vanillic acid	1.091 ± 0.003	0.826 ± 0.000	0. 334 ± 0.007	1.424 ± 0.004	0.309 ± 0.000
Syringic acid	2.492 ± 0.004	1.490 ± 0.000	0.674 ± 0.000	2.043 ± 0.006	1.869 ± 0.003
p-Coumaric acid	nd	0.083 ± 0.000	nd	nd	0.035 ± 0.001
(-) Epicatechin	2.755 ± 0.006	nd	4.757 ± 0.001	0.644 ± 0.002	nd
Delphinidin	nd	nd	0.593 ± 0.010	nd	nd
Rutin	0.582 ± 0.001	0.191 ± 0.001	0.702 ± 0.001	0.427 ± 0.000	0.139 ± 0.000
trans-Resveratrol	0.023 ± 0.001	0.082 ± 0.000	nd	nd	0.080 ± 0.000
Quercetin	0.160 ± 0.001	0.370 ± 0.000	0.133 ± 0.000	0.073 ± 0.001	0.280 ± 0.000

nd—not determined.

2.3. Characterization of Materials Containing FN Extract

The hydroethanolic FN extract with the best antioxidant potential (572.37 mgTrolox/g extract) was selected to be encapsulated into mesoporous SBA-15 silica with an ordered pore framework with hexagonal symmetry determined by small-angle X-ray diffraction (Figure S3) and in S5_C titania nanoparticles through the impregnation method, followed by solvent evaporation under low pressure (3 mbar) according to the procedure described elsewhere [26].

The resulting materials containing FN extract were analyzed by TG-DTA analysis to evaluate the amount of phenolic compounds incorporated in mesoporous inorganic matrices. The main feature influencing the quantity of phenolic compounds that can be encapsulated in the mesopores of inorganic matrices is the total pore volume. The extract amount from the considered inorganic supports was computed based on the weight loss for materials containing extract, taking into account the weight loss of hydroxyl groups during the support heating, and the removal of moisture, which corresponds to the first endothermic event of the DTA curve for each sample subjected to the analysis (Figure 8). As expected, the quantity of encapsulated phenolic substances was lower in the case of FN@TiO$_2$, 20% (wt) (Figure 8A) than for FN@SBA-15 material, 39% (wt) (Figure 8B). Another observation from the thermal analyses is that the temperature at which the polyphenol decomposition rate reached the maximum, 312 °C, was higher in the case of FN@SBA-15 (Figure 8B, black dash-dot curve) than 281 °C for FN@TiO$_2$ (Figure 8A, red dash-dot line), which could be explained by the bigger exposure of biologically active compounds to oxidation in the case of their embedding in mesoporous titania since its mesopores are interparticle pores, and the polyphenols were on the titania NPs' surfaces, unlike in the case of the SBA-15 matrix, which has cylindrical channels of mesopores.

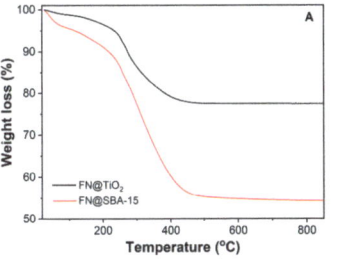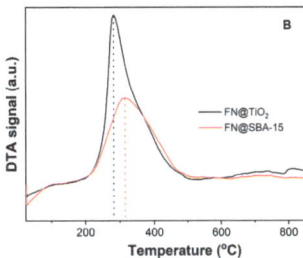

Figure 8. TG analyses of FN@TiO$_2$ and FN@SBA-15 samples (**A**) and the corresponding DTA curves (**B**) performed in air flow.

The sharp decrease in the porosity of materials with encapsulated extract demonstrated the presence of phytocompounds in the mesopores of the inorganic matrices (Figure S4). In the case of FN@SBA-15 material, the values of S_{BET} and V decreased from 984 m^2/g and 1.31 cm^3/g, respectively, for SBA-15, to 209 m^2/g and 0.36 cm^3/g, respectively. In the case of FN@TiO$_2$, practically all pores were filled with FN extract (V = 0.01 cm^3/g), though the extract quantity was lower than in the case of FN@SBA-15.

FTIR spectroscopy highlighted the presence of phenolic components in the spectra of the FN extract and FN@SBA-15 and FN@TiO$_2$ samples (Figure 9A) through the presence of the asymmetric stretching vibrations of C=O bonds belonging to the carboxylic groups at 1726 cm^{-1}, asymmetric and symmetric stretching bands of C–H bonds of methylene groups at 2929 cm^{-1} and 2843 cm^{-1}, respectively. The stretching vibration of C–O bonds overlapped with the deformation band of O–H groups linked on aromatic rings from 1391 cm^{-1} and the vibration mode of aromatic C–H bonds at 1100 cm^{-1}, the last band being superimposed the very intense asymmetrical stretching band of the Si–O–Si bonds of the SBA-15 matrix in the case of FN@SBA-15 material. In all the spectra, including that of mesoporous inorganic matrices, the broad band with the minimum transmittance at 3470 cm^{-1} due to hydroxyl groups, which belong either to the phenolic substances or the inorganic matrices, and the bending vibration of adsorbed water at 1627 cm^{-1} can be observed [27].

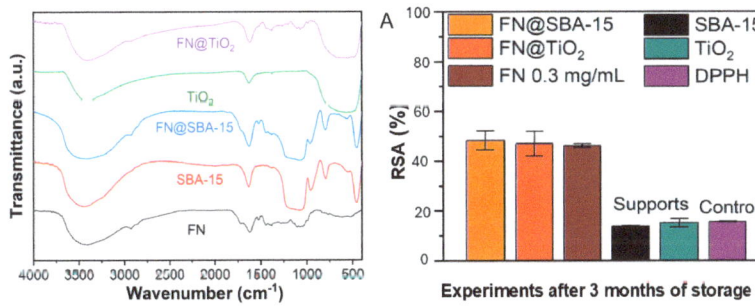

Figure 9. FTIR spectra of FN extract, SBA-15 matrix, FN@SBA-15 material, TiO$_2$ support, and FN@TiO$_2$ sample (**A**). Radical scavenging activity assessed by DPPH assay for FN@TiO$_2$ and FN@SBA-15 in comparison with FN extract alone and corresponding inorganic matrices, TiO$_2$ and SBA-15, in the same quantities as in the FN-loaded supports (**B**).

Spectrometric determinations of radical scavenging activity (DPPH method) for materials containing FN extract showed the preservation of this property over time (Figure 9B). The antioxidant potential was also determined for the FN extract after three months of storage in dark conditions, in a refrigerator, and for mesoporous inorganic matrices, which showed no contribution to the radical scavenging potential of FN@TiO$_2$ and FN@SBA-15.

2.4. Assessment of Cytocompatibility of the Polyphenolic Extract, Materials Containing Extract, and Corresponding Mesoporous Supports

The cytocompatibility of FN and PN extracts, FN extract encapsulated in SBA-15 and TiO$_2$, and the corresponding inorganic materials was assessed on NCTC clone L929 murine fibroblasts cell line at 24 h and 48 h incubation times using MTT assay. All samples were tested in triplicate.

Both extracts, FN and PN, are biocompatible at concentrations lower than 300 µg/mL, either after 24 h or 48 h of incubation. PN extract showed lower cytotoxicity than FN extract at 700 µg/mL concentration, and in the case of FN extract, cell viability decreased to 35.86% at 24 h and 19.47% after 48 h, while for PN extract, cytotoxic effects were observed only after 48 h of incubation (69.17% cell viability). Cell viability decreased sharply at a treatment dose of 1000 µg/mL for both FN and PN extracts to 15.64% and 47.71%, respectively after 24 h and 7.59% and 15.64%, respectively, after 48 h of incubation (Figure 10A).

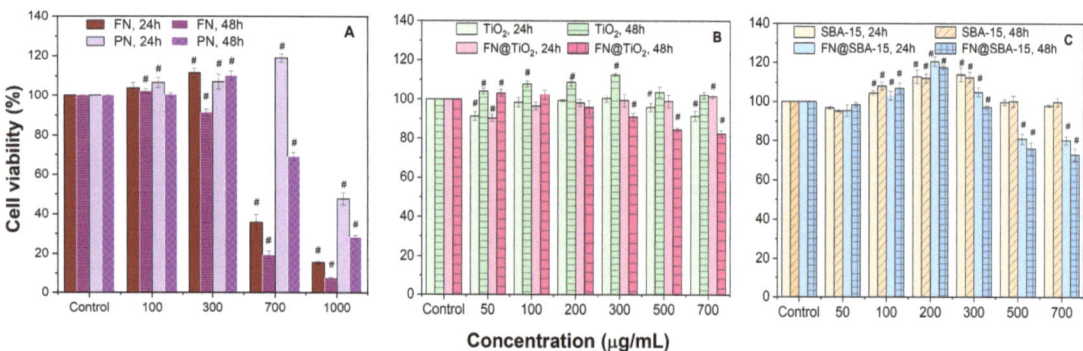

Figure 10. Cell viability of normal L929 cells for the free hydroethanolic PN and FN grape pomace extracts (**A**) and FN extract encapsulated in TiO$_2$ (**B**) and SBA-15 matrices compared to the corresponding supports, TiO$_2$ and SBA-15, assessed using MTT assay (**C**). Results are presented as average value of three replicates ± standard deviation (n = 3). # ($p < 0.05$) shows significant differences between the fibroblasts incubated with samples in comparison to control. Treatments with samples of cells are not considered toxic when cell viability is higher than 80%.

The FN extract encapsulated in TiO$_2$ NPs, FN@TiO$_2$, showed biocompatible behavior for all tested concentrations, between 50 µg/mL and 700 µg/mL, with cell viability values in the range of 104.22–82.89%. TiO$_2$ NPs also showed no cytotoxic effect on NCTC fibroblasts at concentrations up to 700 µg/mL (Figure 10B). FN@SBA-15 showed cytocompatibility at concentrations in the range of 50–300 µg/mL after both incubation periods. At high concentrations, 500 and 700 µg/mL of FN@SBA-15, the cell viability decreased to 75.95% and 72.91%, respectively. SBA-15 support presents no cytotoxicity at the tested concentrations (Figure 10C).

3. Discussion

Antioxidants are very important for cellular health because they help neutralize free radicals and thus can prevent various inflammation-based diseases. In recent years, many interdisciplinary research groups have studied their biological effects on human health, their recovery from various sources, including plant waste [28–30], and how to preserve or enhance their effectiveness by developing new approaches, such as including antioxidants in various formulations with improved features and benefits [26,31,32].

Grape pomace, a by-product from wine production, is a source of antioxidants and antimicrobial agents, especially against foodborne bacteria strains, which can be used in various fields, such as fertilizers, the food industry, cosmetics, and supplements, after their recovery by solid–liquid extraction [33–35]. Grape pomace extracts can improve the antibacterial activity of antibiotics because they inhibit biofilm formation and could be used in the treatment of resistant-biofilm-related infections [35]. Until now, this valuable source of antioxidants and antimicrobials has not been very well exploited.

We valorized fermented dried grape marc from two cultivars, Feteasca Neagra and Pinot Noir from Constanta County, Romania, collected in two years, by preparing extracts through conventional extraction technique in non-toxic solvents: 50% ethanol aqueous solution and absolute ethanol for further application in cosmetics. We observed that the quantity of phenolic compounds is very different depending on the variety and the harvest year of the grapes. The extracts obtained from Feteasca Neagra were generally richer in polyphenols and had more total anthocyanins expressed per mass of dried extract than those prepared from Pinot Noir GM, results that are in agreement with the data reported by Balea et al. [16].

Thus, the richest extract in bioactive compounds obtained in this study was the hydroethanolic FN extract with 39.00 ± 0.17 mg GA/g GM, followed by hydroethanolic PN extract with 16.34 ± 0.41 mg GA/g GM, with the TP index values being in the domain of total phenolic content reported by Constantin et al. [36] for the extracts prepared in water at 28 °C from Feteasca Neagra grapes, Galati County, Romania (0.77–83.62 mg/g dried mass). They reported the following distribution of phenolic compounds: 3%, 1%, and 96% in grape pulp, skins, and seeds, respectively. This study and previous results demonstrated that the climatic conditions and cultivar strongly influence the content of phenolic compounds in not only wines but also in fermented GM, and thus, if exploited, the quality of extracts. For instance, we reported for a polyphenolic extract prepared in the same way as the PN sample in this study, but using GM collected in 2020, values for TP of 43.16 mgGA/g GM and RSA of 689.09 mgTE/g extract, which are 2.6 and 2.2 times higher, respectively, than those of the hydroethanolic PN extract from reference [26]. If we compare the ethanolic FN(E) extract with our previous results from an extract obtained from the same GM variety but collected in 2017, the TP index and antioxidant potential are 3.9 and 3.4 times lower for the FN(E) sample than for the FN extract (GM from 2017) [23]. Previously, we also prepared and characterized Cabernet Sauvignon extract for which the TP index was 265.21 ± 4.97 mg GA/g extract and the RSA value was 344 mgTE/g extract (DPPH assay), with both values being lower than those of FN extract (295.46 mg GA/g extract and 566.00 mgTE/g extract, respectively) discussed here.

With respect to the TA index expressed as CG equivalents per mass of dry extract, all the extracts prepared from Feteasca Neagra had higher TA index values, especially the ethanolic extract (20.42 for FN(E) vs. 7.35 and 3.56 mgCG/g for FN and FN(E-W), respectively) than extracts obtained from Pinot Noir (2.86 and 4.29 mgCG/g for PN(E) and PN, respectively). Also, for the Pinot Noir extracts, for the preparation of which we used GM collected in 2020, we determined a lower value of the TA index, 4.52 mgCG/g extract, [23] than for Feteasca Neagra extracts.

Nevertheless, bioactive substances derived from plants are susceptible to degradation. Hence, one approach to enhance their chemical stability and thus their shelf-life is to encapsulate bioactive compounds in various carriers. In this regard, Castro et al. [37] proposed formulations for a grape seed extract using microdispersion based on soy lecithin and pectin with a very good yield of the entrapment and controlled delivery of the extract polyphenols. Raschip et al. [38] reported the embedding of Feteasca Neagra and Merlot extracts in ice-templated 3D xanthan–PVA composites. They showed that the polymeric films containing Feteasca Neagra extract presented a high antioxidant potential, while the Merlot extract embedded in xanthan–PVA composites exhibited a better antibacterial potential against Gram-positive or Gram-negative bacterial strains than the samples containing FN extract or samples without extract. By microencapsulation through the freeze-drying method of grape pomace extracts in sodium alginate with gum Arabic coated with gelatin, an improved in vitro bioaccessibility of polyphenols was demonstrated [39].

Reports on polyphenols, like quercetin- or resveratrol-based delivery systems for cancer therapy have emerged in recent years [40–42]. For example, functional lipid–polymer nanoparticles with high biocompatibility have been developed as carriers for bioactive substances from *Curcuma zedoaria* and *Platycodon grandiflorum* to treat breast cancer metastases [43], in addition to quercetin-loaded lipid nanoparticles with improved release properties for anticancer therapy, thus mitigating the side-effects of chemotherapy [40].

Among carriers, mesoporous inorganic materials, especially silica, have been used for embedding various extracts, the resulting extract-loaded materials exhibiting improved stability, radical scavenging potential or anti-inflammatory properties, and desirable biocompatibility [23,44,45]. Usually, mesoporous silica nanoparticles increase the solubility of poorly water-soluble bioactive compounds due to their nanoconfinement in mesopores in amorphous state, improving their bioavailability [46].

In our previous papers, we showed the improved properties of the extracts when loaded on supports based on mesoporous MCM-41 silica, depending on the surface functionalization [47,48]. For example, we reported that the hydroethanolic Pinot Noir extract (GM collected in 2020) presented better anti-inflammatory properties when encapsulated in fucoidan-coated aminopropyl functionalized MCM-41 silica than that of corresponding free extract or reference drugs [26]. We carried out a stability study of polyphenolic extract from wild bilberries free and encapsulated in mesoporous silica. During the accelerated degradation study, the free extract lost a part of the polyphenols, while when it was incorporated in mesoporous silica, better stability in time was observed [48].

Herein, for the first time, we tested mesoporous SBA-15 silica as a carrier for phenolic compounds, which has a larger pore volume and wider pore diameter than MCM-41. We demonstrated that SBA-15 silica has desirable cytocompatibility, being suitable for use as a matrix for the incorporation of bioactive compounds (with the FN@SBA-15 sample having a high content of polyphenols, 39% wt.). The FN@SBA-15 formulation showed non-toxic effects up to 300 μg/mL on NCTC fibroblast cells (Figure 10C). Also, the encapsulation of FN extract in SBA-15 support did not alter the radical scavenging properties of the extract (Figure 9B). Also, liposomal formulations of sea buckthorn and grape pomace extracts showed an enhanced antioxidant activity of extracts due to the protective effect of liposomes against acidic degradation [49].

The use of titania NPs is approved by the Food and Drug Administration and the EU Commission in cosmetics and sunscreen formulations [11,50]. The challenge in using TiO_2 NPs for biomedical applications is related to their poor solubility in water and biological fluids, but titania can be used for topical applications.

Previously, we reported the use of titania NPs as support for common sage and wild thyme extracts [51]. TiO_2 NPs, used to incorporate these extracts, were obtained by sol-gel method, followed by an aging step of the reaction mixture at reflux using Pluronic P123 as a template agent. Finally, titania was calcined at 450 °C. In this study, we report the synthesis of a series of titania nanoparticles synthesized through the sol-gel technique combined with a solvothermal treatment, showing the parameters that influence the textural features of NPs.

We demonstrated that both Pluronic P123 and Pluronic F127 can be used for mesopore formation. By using this method, we obtained smaller pore diameters ranging from 3.7 to 7.4 nm versus 10.5 nm and higher S_{BET} in the range of 126–312 m^2/g (Table 2) compared to 115 m^2/g [51], 112 m^2/g for a sample also prepared in the presence of Pluronic F127 in ethanol [52], or 39 m^2/g [53]. The use of Pluronic F127 led to less aggregated TiO_2 NPs, the formed spherical agglomerates having nanometric size (100–400 nm). In this study, we selected the titania sample obtained in the presence of F127 copolymer, which was thermally treated at 400 °C/3 h to encapsulate the hydroethanolic FN sample, the resulting sample being denoted as FN@TiO_2. The biocompatibility of titania support and FN@TiO_2

was assessed on NCTC fibroblasts, and the results demonstrated no cytotoxic effects of either titania NPs or the FN@TiO$_2$ sample for all tested concentrations, in the range of 50–700 μg/mL after 24 h and 48 h incubation time periods (Figure 10B). As in the case of FN@SBA-15 sample, titania nanoparticles containing FN extract preserved the radical scavenging capacity of the extract (Figure 9B). Our findings are in agreement with the literature data that showed a good in vitro biocompatibility of titania nanostructures on various cell lines [54–56].

Regarding the morphology of titania NPs prepared in the presence of both Pluronic F127 and P123, small crystals of 10 nm average size with polyhedral shape were obtained. Samsudin et al. observed that Pluronic F127 acts as a crystallographic controlling agent favoring the formation of {0 0 1} anatase facet that finally led to truncated octahedral bipyramidal particles, while the {1 0 1} facets led to the formation of almost spherical particles [53]. During TEM investigation, a similar particles shape was observed of titania NPs synthesized in the presence of either Pluronic F127 or Pluronic P123, so both triblock copolymers probably favor the formation of the {0 0 1} anatase facet.

4. Materials and Methods

All chemicals utilized for the red GM polyphenolic extracts preparation and analyses, as well as for the synthesis of the mesoporous inorganic matrices, are provided in the Supplementary Materials.

4.1. Obtaining of Mesoporous Inorganic Matrices

The first step of the synthesis of titania nanoparticles consisted of dissolving 1.25 g of the structure directing agent (Pluronic P123 or Pluronic F127) in 50 mL of solvent (2-propanol or n-butanol) at 40 °C, under magnetic stirring. Then, 1.5 mL of glacial acetic acid were added dropwise to slow down the hydrolysis reaction of the titanium precursor. A volume of Ti(IV) isopropoxide or Ti(IV) n-butoxide was poured into the reaction mixture kept at 40 °C, under magnetic stirring. After 1 h, in the resulting solution, 1 mL of deionized water was dropped to promote the hydrolysis and condensation reactions of titanium alkoxide. The reaction mixture was maintained under magnetic stirring at 40 °C for 24 h. Then, the resulting mixture was transferred into a reactor under autogenous pressure for a solvothermal treatment carried out at 100 °C for 24 h. After cooling, the solid was separated by centrifugation, washed with corresponding solvent, water, and ethanol, and then dried at 60 °C. The removal of the structure directing agent was carried out by Soxhlet extraction in ethanol for 24 h. Then, a part of each sample was calcined at 400 °C for 3 h with a heating rate of 0.5 °C/min.

Mesoporous SBA-15 silica was synthesized by an established procedure [57] starting with 2.5 g of Pluronic P123 dissolved at room temperature in 92.8 mL aqueous solution of hydrochloric acid (prepared by adding 13.8 mL HCl 37% (wt) in 79 mL deionized water), and then tetraethyl orthosilicate (5.9 mL) was added to the solution of the structure directing agent. The resulting mixture was stirred at 35 °C for 24 h and then was solvothermally treated in statical conditions under autogenous pressure for another 24 h at 100 °C. The solid was filtered off, washed 3 times with 25 mL ethanol and 3 times with 50 mL deionized water, and dried at room temperature overnight. Finally, SBA-15 silica was thermally treated at 550 °C for 5 h.

4.2. Materials Characterization

The materials were investigated through X-ray diffraction in the range of 2θ = 10–70° and at a scanning rate of 2°/min and 0.01° step (Miniflex 2, Rigaku Holdings Corporation, Tokyo, Japan), FTIR spectroscopy using 64 scans, a resolution of 2 cm^{-1} in 4000–400 cm^{-1} range (KBr pellets technique; Bruker Tensor 27, Bruker Corporation Optik GmbH, Bremen, Germany), thermogravimetric analysis coupled with differential thermal analysis carried out in air flow (50 mL/min), with a heating rate of 10°/min in the temperature range of 20–850 °C (DTA-TG, Mettler Toledo GA/SDTA851e, Greifensee, Switzerland), scanning electron microscopy (Tescan Vega 3 LMH microscope, Brno, Czech Republic), and transmission electron microscopy (FEI TECNAI F30 G2 S-TWIN, Hillsboro, OR, USA), as well as N_2 adsorption–desorption isotherms recorded at liquid nitrogen temperature after outgassing of the samples at 120 °C for 17 h (Quantachrome Autosorb iQ$_2$, Quantachrome Instruments, Boynton Beach, FL, USA).

4.3. Extracts Preparation

The grape marc (GM) samples were collected after fermentation stage during winemaking process. The fermented GM was dried in fresh air, in a thin layer, at ambient temperature on metallic mesh. Every 24 h, the GM was aerated by turning to facilitate the evaporation of water and ethylic alcohol and to avoid the development of bacteria and fungi because of the moisture. The dry GM was grounded in a food processor to increase the contact surface between vegetal waste and solvent.

For all extracts, a maceration step of 18 h was carried out at room temperature, under magnetic stirring before the three steps of 60 min, performed by refluxing the mixture containing red GM (3 g) and solvent (18 mL), adding on each stage a new solvent volume, and keeping the same ratio between the vegetal waste mass and solvent volume (1/6 g/mL). All three fractions obtained after extraction were put together, and then the solvent was completely evaporated using a DLAB RE100-Pro rotary evaporator (DLAB SCIENTIFIC Co., Ltd., Beijing, China).

4.4. Characterization of Polyphenolic Extracts

Total polyphenols index, total flavonoids content, total amount of anthocyanins, and radical scavenging potential (DPPH and ABTS assays) were determined using UV-vis spectroscopy (Shimadzu UV-1800 spectrophotometer, Shimadzu Corporation, Kyoto, Japan), the methods being described in reference [23]. The components of the extracts from twenty-three standard substances (see Supplementary Materials) were quantified by reverse-phase HPLC-PDA (Shimadzu Nexera X2 with SPD-M30A detector) operating in the wavelength range of 250–600 nm and using a Nucleoshell® C18 column 4.6 × 100 mm (2.7 μm) (Macherey-Nagel GmbH & Co. KG, Düren, Germany). The details of the HPLC method were provided elsewhere [23].

4.5. Cytocompatibility Evaluation

The cytocompatibility of FN and PN extracts, as well as of nanoparticles containing FN extract compared to that of mesoporous inorganic matrices in which FN extract was incorporated, was evaluated on NCTC clone L929 murine fibroblasts obtained from the European Collection of Authenticated Cell Cultures. The samples were sterilized under UV irradiation for 2 h, and then stock suspensions/solutions of 1 mg/mL were prepared in culture medium, which contained 10% fetal bovine serum and 1% antibiotics. The stock solutions/suspensions were incubated at 37 °C for 24 h in humid atmosphere containing 5% carbon dioxide. After 24 h, the stock suspensions were dispersed in ultrasounds for at least 1 h, and the extract solutions were filtered off through 0.22 μm Millipore membrane. Mouse NCTC fibroblast cells were seeded in sterile 96-well culture plates (4×10^4 cells/mL) in culture medium. Fibroblasts were treated with the tested samples and incubated for 24 h and 48 h. The in vitro testing of biocompatibility of the samples was carried out at the following concentrations: 100, 300, 700, and 1000 μg/mL for the free FN and PN extracts and 50, 100, 200, 300, 500, and 700 μg/mL for encapsulated FN extract in TiO_2 NPs and SBA-15 silica (FN@TiO_2 and FN@SBA-15) and corresponding supports (TiO_2 and SBA-15). The determination of cell viability (MTT assay, Sigma-Aldrich, Merck Company, Darmstadt, Germany) was performed after 24 h and 48 h incubation periods of the cells with samples. The experiments were performed according to the procedure described in reference [58]. The spectrophotometric determinations were performed on a Berthold Mithras LB 940 Multimode Plate Reader (Berthold Technologies GmbH & Co., Bad Wildbad, Germany), at 570 nm wavelength. Untreated cells were considered the control with 100% cellular viability, based on which were calculated the cell viability for the samples.

Experiments were performed in triplicate, and the results were presented as mean ± standard deviation. Student's t-test was performed for the statistical analysis, using two-tailed distribution and two-sample equal variance, in Microsoft 365 Excel software. Statistical differences were considered for $p < 0.05$.

5. Conclusions

We report the parameters that influenced the morphology of mesoporous titania NPs synthesized by sol-gel method combined with solvothermal treatment. The use of Pluronic F127 as a porogenic agent in the synthesis showed the formation of less agglomerated TiO_2 nanoparticles based on SEM and TEM investigation with higher S_{BET} and lower pore diameter (153 m^2/g vs. 126 m^2/g and 5.4 nm vs. 7.4 nm, respectively) than in the case of Pluronic P123.

We determined the chemical profiles of several extracts prepared from two types of red GM (Feteasca Neagra and Pinot Noir), and we could conclude that it depended on the cultivar and solvent. The extract with the best antioxidant properties, hydroethanolic FN extract prepared from GM collected in 2019, was encapsulated in TiO_2 NPs and compared with the resulting material obtained by the incorporation of FN extract in SBA-15 silica support.

We have successfully obtained and characterized materials containing FN extract, FN@TiO_2, and FN@SBA-15, with desirable cytocompatibility on NCTC clone L929 murine fibroblasts in the concentration range of 50–300 µg/mL after 24 h and 48 h of incubation time and radical scavenging properties.

A long-term stability study for the FN extract alone and encapsulated in titania nanoparticles must be performed to obtain data regarding shelf-life. Also, further in vivo studies should be carried out for the safe use of the proposed extract-loaded nanoparticles in skincare products.

Supplementary Materials: The following supporting information can be downloaded at: https://www.mdpi.com/article/10.3390/inorganics12100263/s1. Figure S1. N_2 adsorption–desorption isotherms for S2_E and S2_C samples recorded at 77 K. Inset the corresponding pore size distribution curves calculated with BJH model from desorption branch; Figure S2. Chromatograms for the following extracts: hydroethanolic FN extract (A), ethanolic FN(E) extract (B), hydroethanolic PN extract (C), and ethanolic PN(E) extract; Figure S3. Small-angle X-ray diffraction of mesoporous SBA-15 silica; Figure S4. N_2 adsorption–desorption isotherms for FN@SBA-15 sample (a) and SBA-15 silica (b).

Author Contributions: Conceptualization, D.B.; methodology, A.-M.B., R.T., A.-I.I., M.D., R.-A.M., C.M. and D.B.; validation, A.A.; A.-M.B., R.T., A.-I.I., M.D., R.-A.M., C.M. and D.B.; formal analysis, A.A., A.-M.B., R.T., A.-I.I., M.D., R.-A.M. and C.M.; investigation, A.A., A.-M.B., R.T., A.-I.I., M.D., R.-A.M. and C.M.; resources, D.B.; R.T., R.-A.M.; data curation, A.A., A.-M.B.; M.D. and C.M.; writing—original draft preparation, A.A.; writing—review and editing, D.B.; supervision, D.B.; project administration, D.B.; funding acquisition, D.B. All authors have read and agreed to the published version of the manuscript.

Funding: This research was funded by UEFISCDI (Romania) project PCCDI no. 85/2018 and project PCE no. 117/2022.

Institutional Review Board Statement: The NCTC clone 929 mouse fibroblast cell line was provided by The European Collection of Authenticated Cell Cultures (ECACC).

Data Availability Statement: The raw data of this paper will be made available by the authors on request.

Conflicts of Interest: The authors declare no conflicts of interest.

References

1. Manzano, M.; Vallet-Regí, M. Mesoporous Silica Nanoparticles for Drug Delivery. *Adv. Funct. Mater.* **2019**, *30*, 1902634. [CrossRef]
2. Castillo, R.R.; Lozano, D.; González, B.; Manzano, M.; Izquierdo-Barba, I.; Vallet-Regí, M. Advances in mesoporous silica nanoparticles for targeted stimuli-responsive drug delivery: An update. *Expert Opin. Drug Deliv.* **2019**, *16*, 415. [CrossRef] [PubMed]
3. Janjua, T.I.; Cao, Y.; Yu, C.; Popat, A. Clinical translation of silica nanoparticles. *Nat. Rev. Mater.* **2021**, *6*, 1072–1074. [CrossRef] [PubMed]
4. Farjadian, F.; Roointan, A.; Mohammadi-Samani, S.; Hosseini, M. Mesoporous Silica Nanoparticles: Synthesis, Pharmaceutical Applications, Biodistribution, and Biosafety Assessment. *Chem. Eng. J.* **2019**, *359*, 684–705. [CrossRef]
5. Mao, T.; Zha, J.; Hu, Y.; Chen, Q.; Zhang, J.; Luo, X. Research Progress of TiO_2 Modification and Photodegradation of Organic Pollutants. *Inorganics* **2024**, *12*, 178. [CrossRef]
6. Aw, M.S.; Addai-Mensah, J.; Losic, D. A multi-drug delivery system with sequential release using titania. *Chem. Commun.* **2012**, *48*, 3348–3350. [CrossRef]
7. Ghenadi, E.; Nichele, V.; Signoretto, M.; Cerrato, G. Structure-Directing Agents for the Synthesis of TiO_2-Based Drug-Delivery Systems. *Chem. Eur. J.* **2012**, *18*, 10653–10660.
8. Inada, M.; Enomoto, N.; Hojo, J. Fabrication and structural analysis of mesoporous silica–titania for environmental purification. *Microporous Mesoporous Mater.* **2013**, *182*, 173–177. [CrossRef]
9. Georgescu, D.; Brezoiu, A.M.; Mitran, R.A.; Berger, D.; Matei, C.; Negreanu-Pîrjol, B.S. Mesostructured silica–titania composites for improved oxytetracycline delivery systems. *Comptes Rendus Chim.* **2017**, *20*, 1017–1025. [CrossRef]
10. Behnam, M.A.; Emami, F.; Sobhani, Z.; Dehghanian, A.R. The application of titanium dioxide (TiO_2) nanoparticles in the photo-thermal therapy of melanoma cancer model. *Iran. J. Basic Med. Sci.* **2018**, *21*, 1133–1139.
11. Mansoor, A.; Khurshid, Z.; Khan, M.T.; Mansoor, E.; Butt, F.A.; Jamal, A.; Palma, P.J. Medical and Dental Applications of Titania Nanoparticles: An Overview. *Nanomaterials* **2022**, *12*, 3670. [CrossRef] [PubMed]
12. Kubacka, A.; Diez, M.; Rojo, D. Understanding the antimicrobial mechanism of TiO_2-based nanocomposite films in a pathogenic bacterium. *Sci. Rep.* **2014**, *4*, 4134. [CrossRef]
13. Zhang, R.; Elzatahry, A.A.; Al-Deyab, S.S.; Zhao, D. Mesoporous titania: From synthesis to application. *Nano Today* **2012**, *7*, 344–366. [CrossRef]
14. Drosou, C.; Kyriakopoulou, K.; Bimpilas, A.; Tsimogiannis, D.; Krokida, M. A comparative study on different extraction techniques to recover red grape pomace polyphenols from vinification byproducts. *Ind. Crops Prod.* **2015**, *75*, 141–149. [CrossRef]
15. Iacono, E.; Di Marzo, C.; Di Stasi, M.; Cioni, E.; Gambineri, F.; Luminare, A.G.; De Leo, M.; Braca, A.; Quaranta, P.; Lai, M.; et al. Broad-spectrum virucidal activity of a hydroalcoholic extract of grape pomace. *Bioresour. Technol. Rep.* **2024**, *25*, 101745. [CrossRef]
16. Balea, S.S.; Pârvu, A.E.; Pârvu, M.; Vlase, L.; Dehelean, C.A.; Pop, T.I. Antioxidant, Anti-Inflammatory and Antiproliferative Effects of the *Vitis vinifera* L. var. Feteasca Neagra and Pinot Noir Pomace Extracts. *Front. Pharmacol.* **2020**, *11*, 990. [CrossRef]
17. Robards, K.; Prenzler, P.D.; Tucker, G.; Swatsitang, P.; Glover, W. Phenolic compounds and their role in oxidative processes in fruits. *Food Chem.* **1999**, *66*, 401–436. [CrossRef]
18. Goldberg, D.M.; Tsang, E.; Karumanchiri, A.; Soleas, G.J. Quercetin and p-coumaric acid concentrations in commercial wines. *Am. J. Enol. Vitic.* **1998**, *49*, 142–151. [CrossRef]
19. Hoss, I.; Rajha, H.N.; El Khoury, R.; Youssef, S.; Manca, M.L.; Manconi, M.; Louka, N.; Maroun, R.G. Valorization of Wine-Making By-Products' Extracts in Cosmetics. *Cosmetics* **2021**, *8*, 109. [CrossRef]
20. Wasilewski, T.; Hordyjewicz-Baran, Z.; Zarebska, M.; Stanek, N.; Zajszły-Turko, E.; Tomaka, M.; Bujak, T.; Nizioł-Łukaszewska, Z. Sustainable Green Processing of Grape Pomace Using Micellar Extraction for the Production of Value-Added Hygiene Cosmetics. *Molecules* **2022**, *27*, 2444. [CrossRef]
21. Moro, K.I.B.; Bender, A.B.B.; da Silva, L.P.; Penna, N.G. Green Extraction Methods and Microencap-sulation Technologies of Phenolic Compounds from Grape Pomace: A Review. *Food Bioprocess Technol.* **2021**, *14*, 1407–1431. [CrossRef]
22. Okur, P.S.; Okur, I. Recent Advances in the Extraction of Phenolic Compounds from Food Wastes by Emerging Technologies. *Food Bioprocess Technol.* **2024**. [CrossRef]
23. Brezoiu, A.-M.; Matei, C.; Deaconu, M.; Stanciuc, A.-M.; Trifan, A.; Gaspar-Pintiliescu, A.; Berger, D. Polyphenols extract from grape pomace. Characterization and valorisation through encapsulation into mesoporous silica-type matrices. *Food Chem. Toxicol.* **2019**, *133*, 110787. [CrossRef] [PubMed]
24. Nithya, N.; Bhoopathi, G.; Magesh, G.; Balasundaram, O.N. Synthesis and characterization of yttrium doped titania nanoparticles for gas sensing activity. *Mater. Sci. Semicond. Process.* **2019**, *99*, 14–22. [CrossRef]
25. Lee, J.; Durst, R.W.; Wrolstad, R.E. Determination of Total Monomeric Anthocyanin Pigment Content of Fruit Juices, Beverages, Natural colorants and Wines by the pH- differential method: Collaborative Study. *J. AOAC Int.* **2005**, *88*, 1269–1278. [CrossRef]
26. Deaconu, M.; Abduraman, A.; Brezoiu, A.-M.; Sedky, N.K.; Ioniță, S.; Matei, C.; Ziko, L.; Berger, D. Anti-Inflammatory, Antidiabetic, and Antioxidant Properties of Extracts Prepared from Pinot Noir Grape Marc, Free and Incorporated in Porous Silica-Based Supports. *Molecules* **2024**, *29*, 3122. [CrossRef] [PubMed]
27. Socrates, G. *Infrared and Raman Characteristic Group Frequencies*, 3rd ed.; John Wiley & Sons Ltd.: Chichester, UK, 2001.

28. Wu, Y.; Gao, H.; Wang, Y.; Peng, Z.; Guo, Z.; Ma, Y.; Zhang, R.; Zhang, M.; Wu, Q.; Xiao, J.; et al. Effects of Different Extraction Methods on Contents, Profiles, and Antioxidant Abilities of Free and Bound Phenolics of Sargassum Polycystum from the South China Sea. *J. Food Sci.* **2022**, *87*, 968–981. [CrossRef]
29. Pandey, K.B.; Rizvi, S.I. Role of red grape polyphenols as antidiabetic agents. *Integr. Med. Res.* **2014**, *3*, 119–125. [CrossRef]
30. Denny, C.; Lazarini, J.G.; Franchin, M.; Melo, P.S.; Pereira, G.E.; Massarioli, A.P.; Moreno, I.A.M.; Paschoal, J.A.R.; Alencar, S.M.; Rosalen, P.L. Bioprospection of Petit Verdot grape pomace as a source of anti-inflammatory compounds. *J. Funct. Foods* **2014**, *8*, 292–300. [CrossRef]
31. Szewczyk, A.; Brzezinska-Rojek, J.; Osko, J.; Majda, D.; Prokopowicz, M.; Grembecka, M. Antioxidant-Loaded Mesoporous Silica—An Evaluation of the Physicochemical Properties. *Antioxidants* **2022**, *11*, 1417. [CrossRef]
32. Tsali, A.; Goula, A.M. Valorization of grape pomace: Encapsulation and storage stability of its phenolic extract. *Powder Technol.* **2018**, *340*, 194–207. [CrossRef]
33. Machado, T.O.X.; Portugal, I.; de A. C. Kodel, H.; Droppa-Almeida, D.; Dos Santos Lima, M.; Fathi, F.; Oliveira, B.P.P.; de Albuquerque-Júnior, R.L.C.; Claudio Dariva, C.; Souto, E.B. Therapeutic potential of grape pomace extracts: A review of scientific evidence. *Food Biosci.* **2024**, *60*, 104210.
34. Almanza-Oliveros, A.; Bautista-Hernández, I.; Castro-López, C.; Aguilar-Zárate, P.; Meza-Carranco, Z.; Rojas, R.; Michel, M.R.; Martínez-Ávila, G.C.G. GrapePomace—Advances in Its Bioactivity, Health Benefits, and Food Applications. *Foods* **2024**, *13*, 580. [CrossRef]
35. Sateriale, D.; Forgione, G.; Di Rosario, M.; Pagliuca, C.; Colicchio, R.; Salvatore, P.; Paolucci, M.; Pagliarulo, C. Vine-Winery Byproducts as Precious Resource of Natural Antimicrobials: In Vitro Antibacterial and Antibiofilm Activity of Grape Pomace Extracts against Foodborne Pathogens. *Microorganisms* **2024**, *12*, 437. [CrossRef]
36. Constantin, O.E.; Skrt, M.; Poklar Ulrih, N.; Râpeanu, G. Anthocyanins profile, total phenolics and antioxidant activity of two Romanian red grape varieties: Fetească neagră and Băbească neagră (Vitis vinifera). *Chem. Pap.* **2015**, *69*, 1573–1581. [CrossRef]
37. Castro, M.L.; Azevedo-Silva, J.; Valente, D.; Machado, A.; Ribeiro, T.; Ferreira, J.P.; Pintado, M.; Ramos, O.L.; Borges, S.; Baptista-Silva, S. Elevating Skincare Science: Grape Seed Extract Encapsulation for Dermatological Care. *Molecules* **2024**, *29*, 3717. [CrossRef]
38. Raschip, I.E.; Fifere, N.; Dinu, M.V. A Comparative Analysis on the Effect of Variety of Grape Pomace Extracts on the Ice-Templated 3D Cryogel Features. *Gels* **2021**, *7*, 76. [CrossRef] [PubMed]
39. Martinović, J.; Ambrus, R.; Planinić, M.; Šelo, G.; Klarić, A.-M.; Perković, G.; Bucić-Kojić, A. Microencapsulation of Grape Pomace Extracts with Alginate-Based Coatings by Freeze-Drying: Release Kinetics and In Vitro Bioaccessibility Assessment of Phenolic Compounds. *Gels* **2024**, *10*, 353. [CrossRef] [PubMed]
40. Ariraman, S.; Seetharaman, A.; Babunagappan, K.V.; Sudhakar, S. Quercetin-loaded nanoarchaeosomes for breast cancer therapy: A ROS mediated cell death mechanism. *Mater. Adv.* **2024**, *5*, 6944–6956. [CrossRef]
41. Zu, Y.; Zhang, Y.; Wang, W.; Zhao, X.; Han, X.; Wang, K.; Ge, Y. Preparation and in vitro/in vivo evaluation of resvera-trol-loaded carboxymethyl chitosan nanoparticles. *Drug Deliv.* **2016**, *23*, 981–991. [CrossRef]
42. Vieira, I.R.S.; Tessaro, L.; Lima, A.K.O.; Velloso, I.P.S.; Conte-Junior, C.A. Recent Progress in Nanotechnology Improving the Therapeutic Potential of Polyphenols for Cancer. *Nutrients* **2023**, *15*, 3136. [CrossRef]
43. Shi, J.; Zhang, R.; Wang, Y.; Sun, Y.; Gu, X.; An, Y.; Chai, X.; Wang, X.; Wang, Z.; Lyu, Y.; et al. Herb-Nanoparticle Hybrid System for Improved Oral Delivery Efficiency to Alleviate Breast Cancer Lung Metastasis. *Int. J. Nanomed.* **2024**, *19*, 7927–7944. [CrossRef] [PubMed]
44. Budiman, A.; Rusdin, A.; Wardhana, Y.W.; Puluhulawa, L.E.; Cindana Mo'o, F.R.; Thomas, N.; Gazzali, A.M.; Aulifa, D.L. Exploring the Transformative Potential of Functionalized Mesoporous Silica in Enhancing Antioxidant Activity: A Comprehensive Review. *Antioxidants* **2024**, *13*, 936. [CrossRef]
45. Deaconu, M.; Prelipcean, A.M.; Brezoiu, A.M.; Mitran, R.A.; Seciu-Grama, A.M.; Matei, C.; Berger, D. Design of scaffolds based on zinc-modified marine collagen and bilberry leaves extract-loaded silica nanoparticles as wound dressings. *Int. J. Nanomed.* **2024**, *19*, 7673–7689. [CrossRef] [PubMed]
46. Trendafilova, I.; Popova, M. Porous Silica Nanomaterials as Carriers of Biologically Active Natural Polyphenols: Effect of Structure and SurfaceModification. *Pharmaceutics* **2024**, *16*, 1004. [CrossRef]
47. Brezoiu, A.-M.; Bajenaru, L.; Berger, D.; Mitran, R.-A.; Deaconu, M.; Lincu, D.; Guzun, A.S.; Matei, C.; Moisescu, M.G.; Negreanu-Pirjol, T. Effect of Nanoconfinement of Polyphenolic Extract from Grape Pomace into Functionalized Mesoporous Silica on its Biocompatibility and Radical Scavenging Activity. *Antioxidants* **2020**, *9*, 696. [CrossRef]
48. Brezoiu, A.-M.; Deaconu, M.; Mitran, R.-A.; Sedky, N.K.; Schiets, F.; Marote, P.; Voicu, I.-S.; Matei, C.; Ziko, L.; Berger, D. The Antioxidant and Anti-Inflammatory Properties of Wild Bilberry Fruit Extracts Embedded in Meso-porous Silica Type Supports: A Stability Study. *Antioxidants* **2024**, *13*, 250. [CrossRef]
49. Popovici, V.; Boldianu, A.-B.; Pintea, A.; Caraus, V.; Ghendov-Mosanu, A.; Subotin, I.; Druta, R.; Sturza, R. In Vitro Antioxidant Activity of Liposomal Formulations of Sea Buckthorn and Grape Pomace. *Foods* **2024**, *13*, 2478. [CrossRef] [PubMed]
50. Bousiakou, L.G.; Dobson, P.J.; Jurkin, T.; Marić, I.; Aldossary, O.; Ivanda, M. Optical, structural and semiconducting properties of Mn doped TiO_2 nanoparticles for cosmetic applications. *J. King Saud Univ.-Sci.* **2022**, *34*, 101818. [CrossRef]

51. Brezoiu, A.M.; Prundeanu, M.; Berger, D.; Deaconu, M.; Matei, M.; Oprea, O.; Vasile, E.; Negreanu-Pîrjol, T.; Mun-tean, D.; Danciu, C. Properties of *Salvia officinalis* L. and *Thymus serpyllum* L. Extracts Free and Embedded into Mesopores of Silica and Titania Nanomaterials. *Nanomaterials* **2020**, *10*, 820. [CrossRef]
52. Shi, Y.; Chen, S.; Sun, H.; Shu, Y.; Quan, X. Low-temperature selective catalytic reduction of NOx with NH_3 over hierarchically macro-mesoporous Mn/TiO_2. *Catal. Commun.* **2013**, *42*, 10–13. [CrossRef]
53. Samsudin, E.M.; Abd Hamid, S.B.; Juan, J.C.; Basirun, W.J. Influence of triblock copolymer (pluronic F127) on enhancing the physico-chemical properties and photocatalytic response of mesoporous TiO_2. *Appl. Surf. Sci.* **2015**, *355*, 959–968. [CrossRef]
54. Wang, Y.; Zhang, F.; Chen, S.; Osaka, A.; Chen, W. Facile synthesis, characterization, and in vitro biocompatibility of free-standing titania hollow microtubes. *Int. J. Appl. Ceram. Technol.* **2024**, *21*, 3897–3905. [CrossRef]
55. Torres-Romero, A.; Cajero-Juárez, M.; Nuñez-Anita, R.E.; Contreras-García, M.E. Ceria-Doped Titania Nanoparticles as Drug Delivery System. *J. Nanosci. Nanotechnol.* **2020**, *20*, 3971–3980. [CrossRef] [PubMed]
56. Sahare, P.; Alvarez, P.G.; Yanez, J.M.; Luna-Bárcenas, G.; Chakraborty, S.; Paul, S.; Estevez, M. Engineered titania nanomaterials in advanced clinical applications. *Beilstein J. Nanotechnol.* **2022**, *13*, 201–218. [CrossRef] [PubMed]
57. Zhao, D.; Feng, J.; Huo, Q.; Melosh, N.; Fredrickson, G.H.; Chmelka, B.F.; Stucky, G.D. Triblock Copolymer Syntheses of Mesoporous Silica with Periodic 50 to 300 Angstrom Pores. *Science* **1998**, *279*, 548. [CrossRef]
58. Gavrila, A.I.; Zalaru, C.M.; Tatia, R.; Seciu-Grama, A.-M.; Negrea, C.L.; Calinescu, I.; Chipurici, P.; Trifan, A.; Popa, I. Green Ex-traction Techniques of Phytochemicals from *Hedera helix* L. and In Vitro Characterization of the Extracts. *Plants* **2023**, *12*, 3908.

Disclaimer/Publisher's Note: The statements, opinions and data contained in all publications are solely those of the individual author(s) and contributor(s) and not of MDPI and/or the editor(s). MDPI and/or the editor(s) disclaim responsibility for any injury to people or property resulting from any ideas, methods, instructions or products referred to in the content.

Article

Precipitative Coating of Calcium Phosphate on Microporous Silica–Titania Hybrid Particles in Simulated Body Fluid

Reo Kimura [1], Kota Shiba [2,*], Kanata Fujiwara [1], Yanni Zhou [1], Iori Yamada [1,3] and Motohiro Tagaya [1,*]

1. Department of Materials Science and Bioengineering Technology, Nagaoka University of Technology, Kamitomioka 1603-1, Nagaoka 940-2188, Japan
2. Center for Functional Sensor & Actuator, Research Center for Functional Materials, National Institute for Materials Science (NIMS), 1-1, Namiki, Tsukuba 305-0044, Japan
3. Research Fellow of the Japan Society for the Promotion of Science (DC), 5-3-1 Koji-machi, Chiyoda-ku, Tokyo 102-0083, Japan
* Correspondence: shiba.kota@nims.go.jp (K.S.); tagaya@mst.nagaokaut.ac.jp (M.T.)

Abstract: Titania and silica have been recognized as potential drug delivery system (DDS) carriers. For this application, controllable biocompatibility and the suppression of the initial burst are required, which can be provided by a calcium phosphate (CP) coating. However, it is difficult to control the morphology of a CP coating on the surface of carrier particles owing to the homogeneous nucleation of CP. In this study, we report the development of a CP-coating method that homogeneously corresponds to the shapes of silica–titania (SiTi) porous nanoparticles. We also demonstrate that controlled surface roughness of CP coatings could be achieved in SBF using SiTi nanoparticles with a well-defined spherical shape, a uniform size, and a tunable nanoporous structure. The precipitation of CP was performed on mono-dispersed porous SiTi nanoparticles with different Si/Ti molar ratios and pore sizes. The pore size distribution was found to significantly affect the CP coating in SBF immersion; the surfaces of the nanoparticles with bimodal pore sizes of 0.7 and 1.1–1.2 nm became rough after CP precipitation, while those with a unimodal pore size of 0.7 nm remained smooth, indicating that these two pore sizes serve as different nucleation sites that lead to different surface morphologies.

Keywords: bioceramic nanoparticles; simulated body fluid; nanopore; CP precipitative coating; silica–titania nanohybrid

Citation: Kimura, R.; Shiba, K.; Fujiwara, K.; Zhou, Y.; Yamada, I.; Tagaya, M. Precipitative Coating of Calcium Phosphate on Microporous Silica–Titania Hybrid Particles in Simulated Body Fluid. *Inorganics* **2023**, *11*, 235. https://doi.org/10.3390/inorganics11060235

Academic Editors: Roberto Nisticò and Silvia Mostoni

Received: 11 April 2023
Revised: 23 May 2023
Accepted: 24 May 2023
Published: 28 May 2023

Copyright: © 2023 by the authors. Licensee MDPI, Basel, Switzerland. This article is an open access article distributed under the terms and conditions of the Creative Commons Attribution (CC BY) license (https://creativecommons.org/licenses/by/4.0/).

1. Introduction

Various nanomaterials composed of bioinert ceramics have been synthesized for use as artificial joints, implants, and drug delivery system (DDS) carriers [1,2]. In these applications, a DDS carrier needs to fulfill many requisites, including not only the inherent biocompatibility of bioceramic-based materials but also other properties including being of a uniform shape, size, and size distribution and possessing high affinity for aqueous media in order to form a stable suspension [3–5]. A sol–gel method based on the hydrolysis and condensation of a metal alkoxide has been used to synthesize a variety of biocompatible metal oxide nanoparticles with controlled morphologies, for which titania and silica have been extensively studied [6–8]. For example, amorphous titania in nanotube form has been investigated with respect to its use as a DDS carrier [9,10]. Since the surface of titania exhibits a Zeta potential of −18 mV [11] in water (at pH 7.4), serious aggregate formation can occur depending on the experimental conditions. By contrast, amorphous silica shows a higher Zeta potential of −60 mV in water (at pH 7.4) [12], allowing for the formation of a relatively stable suspension [13,14]. Although mixed oxide nanoparticles composed of silica and titania (SiTi nanoparticles) [15] have been explored as another potential option, their use also poses the problems such as biotoxicity due to the release of a silicate ion elusion into the biological solution [16] and difficulty in controlling drug release owing to the initial burst [17,18]. Therefore, the surfaces of SiTi nanoparticles need to be properly designed before being used for DDS applications.

Calcium phosphate (CP) coating has been developed as a technique to improve the osteoconductivity of the surfaces of titanium implants [19–21]. The CP coating is thought to suppress the initial burst of drug molecules [22], allowing them to be released gradually over several weeks. This is due to the fact that the CP coating itself can act as a reservoir for drug molecules, which slowly dissolve and diffuse over time as the coating degrades. The general CP-coating methods are electrochemical deposition, sputtering, and plasma spraying, which are performed under unphysiological conditions such as at high temperatures to provide different chemical and crystalline states with respect to the bone hydroxyapatite [23–25], leading to lower bioactivity in vivo. The biomimetic method has attracted attention due to its potential benefits. In this approach, CP is precipitated on particles in simulated body fluid (SBF) under conditions that mimic the biological environment of a living body [26,27]. Biomimetic CP synthesized under these conditions has been found to be more bioactive than CP synthesized under higher-temperature conditions [28–30]. However, biomimetic CP has generally only been coated on flat substrates [31–35]. In the case of nanoparticles, uniform nucleation occurs at different positions from the particle surfaces due to their high curvature and lower ability to induce heterogeneous nucleation. Therefore, a coating technique that adapts to the shapes of nanoparticles and provides a uniform coating has not been developed.

In this study, we demonstrate that controlled surface roughness of CP coatings can be achieved in SBF by using SiTi nanoparticles with a well-defined spherical shape, a uniform size, and a tunable nanoporous structure (Scheme 1). We synthesized SiTi nanoparticles using a microfluidic approach [15], which allowed us to design their size and shape so that they were suitable for DDS, and used them as a scaffold for the CP coating. The SiTi nanoparticles serve two critical functions in the SBF: they (1) provide CP nucleation sites that promote the substitution of phosphate ions with silicate ions and (2) create nanopores that induce the selective adsorption of hydrated ions in SBF. As described, the silicate ions elute readily into biological fluids [36,37] and can be replaced by phosphate ions [38], facilitating Ca^{2+} ion adsorption and subsequent CP nucleation. Moreover, the hydrated ions in SBF, including Na^+, K^+, and Ca^{2+}, can be adsorbed into the nanopores in their hydrated states and their sizes differ from each other. We prove that SiTi nanoparticles with tunable nanostructures can effectively function as an ion (molecular) sieve [39,40] that enables the selective adsorption of Ca^{2+} from SBF, leading to the formation of CP coatings. We also discuss the effect of nanopore sizes on the surface roughness of CP coatings.

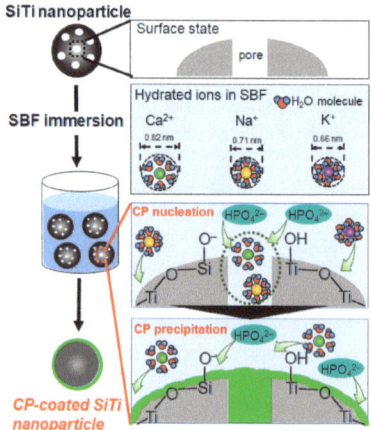

Scheme 1. Illustration of the CP precipitation process of the SiTi nanoparticles via immersion in SBF.

2. Results and Discussion

2.1. Synthesis Result of XSiTi Nanoparticles

The Si/Ti molar ratios were measured via XRF and the values were X, and the sample was named as XSiTi (X = 0, 0.1, 0.7, and 1.2). The FE-SEM images and size distributions of the XSiTi nanoparticles are shown in Figure 1. All the SiTi nanoparticles exhibited spherical and mono-dispersed states. The diameter of the particles was around 150–200 nm, which is considered a size that does not induce cytotoxicity [41,42].

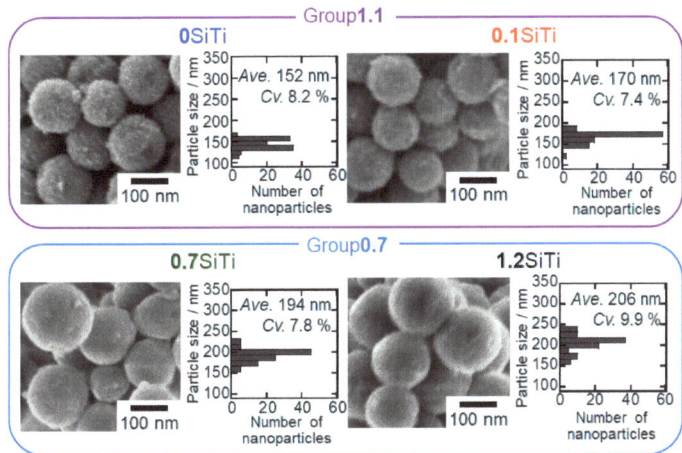

Figure 1. FE-SEM images and particle size distributions of the SiTi nanoparticles.

The XRD patterns (Figure S1) of the XSiTi nanoparticles indicated that all the nanoparticles were amorphous. Comparing the properties of these particles (e.g., particle shape, particle size, and CV value) with those of previously reported particles [15], we confirmed that they were identical and that equivalent particles were synthesized.

Figure 2 shows the N_2 adsorption and desorption isotherms of the SiTi nanoparticles. In the results regarding the specific surface area calculated using the αs-plot (Figure 2a), it is evident that the surface area increased with the increase in the Si/Ti molar ratio. According to the nanopore size distributions based on the MP (micropore) method in the results regarding the XSiTi nanoparticles (Figure 2b,d), 0SiTi and 0.1SiTi exhibited bimodal nanopore sizes of 0.7 and 1.1~1.2 nm, and 0.7SiTi and 1.2SiTi exhibited only monomodal pores of 0.7 nm. The different nanopore sizes occurred due to the increase in the Si/Ti molar ratio. We propose that the hydrated ions of SBF were potentially diffused and adsorbed into the nanopores (Figure 2c,e). We suggest that nanopores 0SiTi and 0.1SiTi, with pore sizes of 1.1~1.2 nm, enable the diffusion and adsorption of the hydrated Ca^{2+}, Na^+, and K^+ ions, while 0.7SiTi and 1.2SiTi, with a pore size of 0.7 nm, only allows for the diffusion and adsorption of the hydrated Na^+ and K^+ ions. According to the nanopore size distribution, 0SiTi and 0.1SiTi were defined as Group1.1, while 0.7SiTi and 1.2SiTi were defined as Group0.7. By comparing these nanopore diameters with those of previously reported particles [15], we confirmed that they are identical and that comparable particles had been synthesized.

2.2. Results Regarding the SBF-Immersed SiTi Nanoparticles

The chemical element (Ca, Na, and K) amounts adsorbed by the SiTi nanoparticles through immersion in SBF were evaluated via XRF (Figure 3). The adsorbed elements increased with an increased immersion time. Referring to the results regarding the change in the amount of Ca on the nanoparticles (Figure 3a), the amount in Group1.1 was clearly larger than that in Group0.7 at the initial stage, indicating that Ca was preferentially adsorbed on the nanoparticles in Group**1.1**. By observing the amount changes of Na and K adsorbed on the nanoparticles (Figure 3b,c), it is evident that those of Group**0.7** were significantly larger than those in Group**1.1** at the initial stage, indicating that Na and K were preferentially adsorbed on the nanoparticles in Group**0.7**. These differences are thought to be due to the difference in the nanopore sizes between Group**1.1** and Group**0.7**. The diameters of the hydrated ions that could diffuse and be adsorbed inside the nanopores were determined as shown in Figure 2. The adsorption of Ca in Group**1.1** reached equilibrium within 1 day of immersion, while the other ions in Group**0.7** did not reach equilibrium even after 3 days. In addition, most of the adsorption of Na and K for Group**0.7** reached equilibrium within 1 day of immersion, whereas that for Group**1.1** did not reach equilibrium until 3 days.

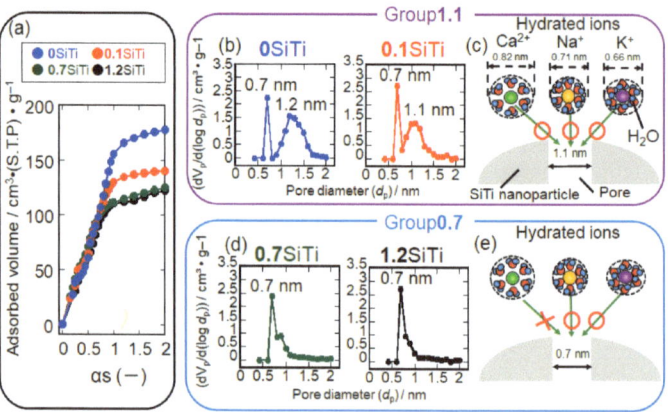

Figure 2. (a) N_2 adsorption (●) and desorption (○) isotherms of the SiTi nanoparticles; (b,d) the MP pore size distributions; and (c,e) illustrations of the hydrated ion interactions with the micropores. The specific surface areas of 0SiTi, 0.1SiTi, 0.7SiTi, and 1.2SiTi were 382, 443, 466, and 570 $m^2 \cdot g^{-1}$, respectively.

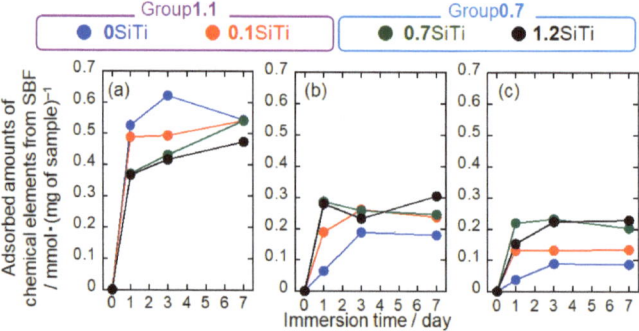

Figure 3. Adsorbed amount changes of the chemical elements of (a) Ca, (b) Na, and (c) K on the SiTi nanoparticles from SBF with immersion time.

The nucleation sites pertaining to the CP precipitation of SiTi nanoparticles immersed in SBF are discussed in Figure 4. Figure 4a shows the FT-IR spectra of the change in the absorbance band due to the OH group of the nanoparticles. The band intensity of Group**1.1** did not change after immersion, whereas Group**0.7** showed an increase in band intensity. The result of the change in the Si/Ti molar ratio of the nanoparticles after immersion is shown in Figure 4b. Group**1.1** did not change in terms of its Si/Ti molar ratio, whereas Group**0.7** showed a significant decrease, suggesting that the Si component was eluted from Group0.7 into SBF. Regarding the changes in the average particle sizes of the nanoparticles following immersion (Figure 4c), all the nanoparticles showed a decrease in the size, and a significant decrease was observed in Group0.7. Group1.1 containing lower Si-content did not show a change, while Group0.7 with higher Si-content showed a change, indicating that the Si component's elution can induce CP precipitation. The mechanism behind the CP nucleation in the precipitation on Group0.7 is suggested in Figure 4d. Group0.7 preferentially absorbed the hydrated Na^+ and K^+ ions inside the nanopores. The Si-components in Group0.7 were eluted as the silicate ions outside the nanopores, and the phosphate ions interacting with the H_2O component in SBF were adsorbed into the eluted sites [38]. The intensity of the OH group of Group0.7 increased through immersion in SBF due to the subsequent adsorption of the hydrated Ca^{2+} ions and the consequent promotion of CP nucleation. Therefore, the outside nanopore surfaces are considered the CP nucleation sites for Group0.7.

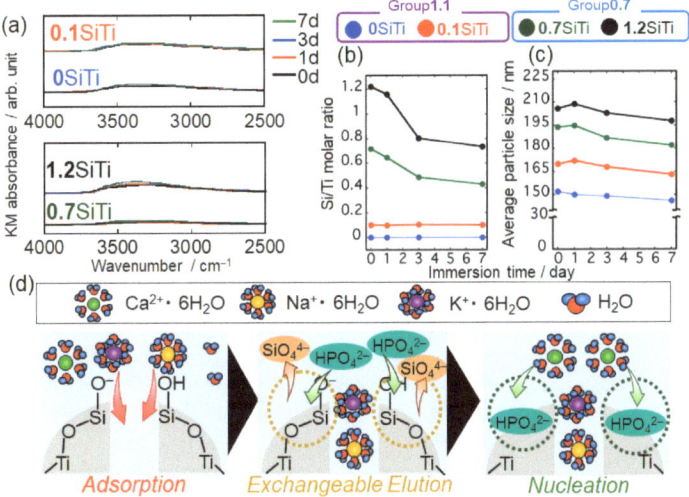

Figure 4. (a) FT-IR spectra, (b) Si/Ti molar ratio, and (c) average particle size changes of the SiTi nanoparticles with immersion time in SBF, and (d) illustrations of the possible interfacial reactions of Group0.7 with the ions in SBF. Xd represents the immersion time of X days (X = 0, 1, 3, and 7).

In Figure 5, the characteristics of phosphate ion adsorption for the CP nucleation sites on Group1.1 and Group0.7 are evaluated and discussed. According to the changes of the absorption band generated by phosphate ions after immersion (Figure 5a), Group1.1 and Group0.7 showed increases in the absorbance bands of the stretching vibrations due to Ti–P–O [43], P–O, and P–OH bonds [44] at 1100, 1039–997, and 866–842 cm^{-1} following immersion. For Group0.7, the bands produced by Si–O–Si [45] and Si–OH [46] at 1039–997 and 866–842 cm^{-1} were also included in the spectra, and the shapes were different from those of Group1.1. The amount changes in the adsorbed phosphorous components of Group1.1 and Group0.7 showed an increase in the amount after immersion (Figure 5b). In particular, Group1.1 reached the adsorption equilibrium after approximately 1 day of immersion, whereas Group0.7 showed a slower adsorption rate, suggesting that the CP precipitate emerged at a relatively earlier stage in Group1.1 compared to that of Group0.7. Figure 5c shows the possible illustrations of the nucleation sites of Group1.1 and Group0.7. In Group1.1, the hydrated Ca^{2+} ions in addition to the Na^+ and K^+ ions were preferentially diffused and absorbed inside the nanopores, which serve as sites for CP nucleation. In Group0.7, the hydrated Na^+ and K^+ ions were preferentially diffused and absorbed inside the nanopores, and the phosphate ions were replaced with the sites where the silicate ions were eluted, suggesting that the outside of the nanopores serve as CP nucleation sites.

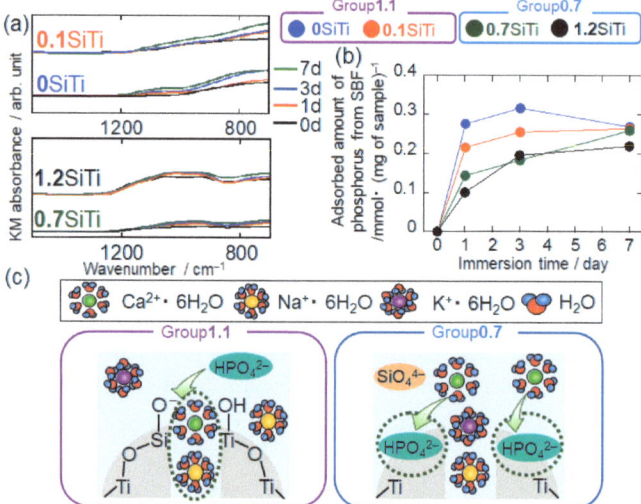

Figure 5. (a) FT-IR spectral changes of the SiTi nanoparticles with immersion time in SBF and (b) the adsorbed amount changes of phosphorus from SBF. (c) Illustrations of the possible calcium phosphate nucleation processes of Group**1.1** and Group**0.7**.

According to the FE-SEM images and particle size distributions of the SiTi nanoparticles (Figure 6), even after immersion for 7 days, the particles still exhibited spherical shapes and mono-dispersed states, indicating a preserved particle size of approximately 150–200 nm. In particular, Group**1.1** exhibited rough surfaces, whereas Group**0.7** retained smooth surfaces. These results show that CP was roughly precipitated on Group**1.1** but was smoothly precipitated on Group**0.7**, indicating that a smooth CP coating was achieved using Group**0.7**.

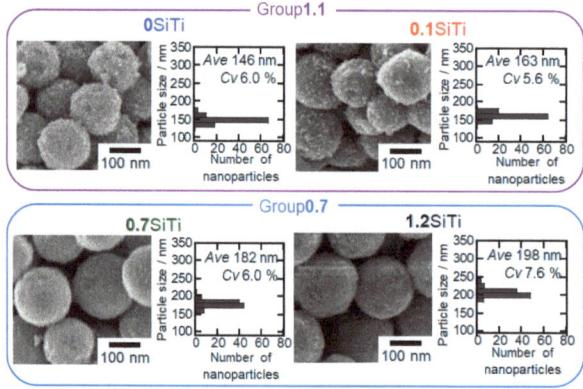

Figure 6. FE-SEM images and particle size distributions of the SiTi nanoparticles after immersion in SBF for 7 days.

The elemental mapping results and TEM images of 0.1SiTi and 1.2SiTi after their immersion are shown in Figure 7. The particle images (i.e., BF: STEM HAADF images) and shapes (i.e., locations) of the chemical elements were similar between Group**1.1** and Group**0.7** (Figure 7a,d), indicating that a homogeneous CP precipitation on the surfaces could be achieved by immersing the nanoparticles in SBF. The Ca signal for 0.1SiTi (Group**1.1**) was weaker than that for 1.2SiTi (Group**0.7**), which is possibly due to the different CP nucleation mechanisms between Group**1.1** and Group**0.7** (as shown in Figure 5c). In Group**1.1**, the hydrated Ca^{2+} ions in addition to the Na^+ and K^+ ions were preferentially diffused and adsorbed inside the nanopores, which served as CP nucleation sites. The results suggested that the number of nucleation sites in Group**1.1** is smaller than that in Group**0.7**, indicating a lower amount of the CP precipitation. The contrast of 0.1SiTi (Group**1.1**) was different from that of 1.2SiTi (Group**0.7**), indicating the presence of rough surfaces due to the CP precipitation of Group**1.1** (Figure 7b,c,e,f).

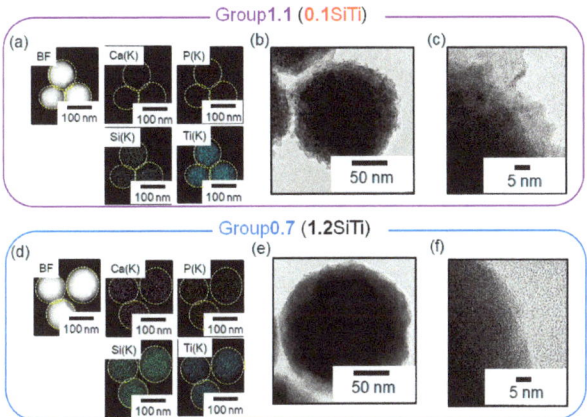

Figure 7. (**a**,**d**) STEM and EDS elemental mapping (Ca, P, Si, and Ti) images of the SiTi nanoparticles after immersion in SBF for 7 days. The detected energies for Ca(K), P(K), Si(K), and Ti(K) were 3.69, 2.01, 1.74, and 4.52 keV, respectively. The dotted yellow circles indicate areas where chemical elements are present. (**b**,**c**,**e**,**f**) TEM images of the SiTi nanoparticles after immersion in SBF for 7days.

The αs plots of the SiTi nanoparticles after immersion in SBF are shown in Figure S2. The changes in the specific surface area of the nanoparticles, which were determined based on the aforementioned results, are shown in Figure 8. Group**1.1** and Group**0.7** showed a decrease in the specific surface area following immersion, indicating the adsorption of the ions inside the nanopores. In particular, Group**1.1** showed a faster rate of decrease in surface area compared with Group**0.7** since nitrogen (N_2) molecules could not enter the nanopores where CP had effectively precipitated inside. Based on Figure 8, it can be observed that Group**1.1** presents a higher rate of reduction in external surface area determined from the αs-plot, indicating that the effective precipitation of CP was due to pore blockage. As a result, Group**1.1** exhibits a higher concentration of adsorbed phosphorous (i.e., phosphate ions). The lower reduction in the specific surface area in Group**0.7** suggests lesser pore blockage through calcium ion adsorption. Since the slight reduction in the surface area is attributed to Na^+ and K^+ ions, it can be inferred that this reduction in surface area is less significant in the present paper. Therefore, the distribution of pores in Group **1.1** is considered a random array shape. Moreover, the peaks of Group**1.1** with bimodal distributions decreased after immersion (Figure S3). The nanopores of 0SiTi at 1.2 nm decreased to 1.0 nm, while the nanopores of 0.1SiTi at 1.1 nm decreased to 0.9 nm. Figure S4 shows the N_2 adsorption and desorption isotherms during immersion. According to a previous report [47], the isotherm type of Group**1.1** was type IV before immersion, which changed to type I after immersion. The isotherm of Group**0.7** remained type I after SBF immersion, indicating that the nanopore structures in Group**1.1** were preserved upon their immersion. Regarding pore size distribution, Group**1.1** shows bimodal shapes in Figure S3. It was suggested that pore blockage in the 1.1 nm sized particles of Group**1.1** would occur, whereas the pores at 0.7 nm remained unblocked, thereby maintaining microporous structures. After immersion, the adsorption isotherm of Group**1.1** in SBF was changed such that is similar in shape to that of Group**0.7** (Figure S4).

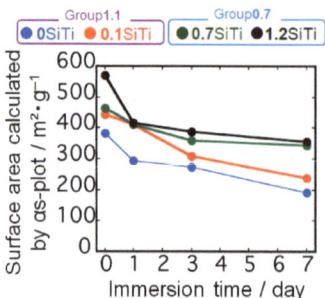

Figure 8. Specific surface area changes of the SiTi nanoparticles with immersion time in SBF. The reduction percentages in the surface areas of 0SiTi, 0.1SiTi, 0.7SiTi, and 1.2SiTi were 50, 46, 26, and 38%.

Accoring to the XRD pattern results, after their immersion in SBF for 14 days, Group**1.1** and Group**0.7** remained amorphous (Figure S5). All of the nanoparticles showed an amorphous calcium phosphate (ACP) halo peak at 2θ = 30°, indicating the precipitation of ACP on their surfaces.

2.3. Mechanism of ACP Precipitation on XSiTi Nanoparticles after Immersion

Based on the above results and discussion, the mechanisms of ACP precipitation in Group**1.1** and Group**0.7** are shown in Scheme 2. Regarding Group**1.1**, the Ca^{2+} ions were diffused and adsorbed inside the nanopores after immersion within one day. The Ca^{2+} ions inside the nanopores reacted with the phosphate ions in SBF, and the nanopores became the ACP nucleation sites, leading to rough ACP precipitation. For Group**0.7**, only the Na^+ and K^+ ions were diffused and adsorbed inside the nanopores after immersion for one day. The phosphate ions exchanged with the eluted silicate ions outside the nanopores and became the ACP nucleation sites. Therefore, it was determined that the ACP precipitation state was smooth without changing the surface morphology of Group**0.7**.

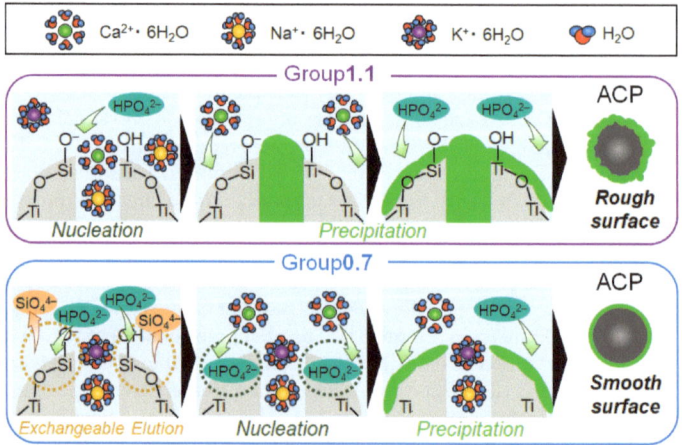

Scheme 2. Illustration of the precipitation processes of the SiTi nanoparticles in this study.

3. Materials and Methods

3.1. Chemicals

TEOS ($C_8H_{20}O_4Si$: CAS No. 78-10-4) and TTIP ($C_{12}H_{28}O_4Ti$: CAS No. 546-68-9) were purchased from Tokyo Chemical Industry Co., Ltd. 2-Propanol (IPA, CAS No.

67-63-0), hydrochloric acid (HCl, 1 N, CAS No. 7647-01-0), ethanol (EtOH, 99.5 vol %, CAS No. 64-17-5), tris-hydroxymethylaminomethane (Tris, $C_4H_{11}NO_3$, CAS No. 77-86-1), sodium chloride (NaCl, CAS No. 7647-14-5), potassium chloride (KCl, CAS No. 7447-40-7), dipotassium hydrogenphosphate (K_2HPO_4, CAS No. 7758-11-4), magnesium chloride hexahydrate ($MgCl_2 \cdot 6H_2O$, CAS No. 7791-18-6), calcium chloride ($CaCl_2$, CAS No. 10043-52-4), and sodium sulfate (Na_2SO_4, CAS No. 7757-82-6) were purchased from FUJIFILM Wako Pure Chemical Co., Ltd. Sodium hydrogen carbonate ($NaHCO_3$, CAS No. 144-55-8) was purchased from Nacalai Tesque Co., Ltd. Octadecylamine (ODA, $CH_3(CH_2)_{17}NH_2$, CAS No. 124-30-1) was purchased from Sigma Aldrich Co., Ltd. All reagents are unpurified.

3.2. Synthesis

3.2.1. Synthesis of SiTi Nanoparticles

In a previous report [15], SiTi nanoparticles were synthesized via microfluidic synthesis. Initially, three solutions (A–C) were prepared. Volumes of 1.63 mL of TTIP and 0, 0.187, 1.705, and 15.34 mL of TEOS were added to solution A to form Si/Ti molar ratios of 0, 0.15, 1.4, and 12, respectively, and 43.30, 43.11, 41.60, and 27.96 mL volumes of IPA were added according to the ratios. Solution B was prepared by mixing 44.60 mL of IPA and 0.277 mL of ultrapure water. Solution C was prepared by mixing 236.1 mL of IPA, 3.00 mL of ultrapure water, and 0.205 g of ODA. Solutions A and B were then mixed and reacted in a microreactor to generate nucleation via a sol–gel process, and the reaction solution was dropped into Solution C at a flow rate of 60 mL/min at 1000 rpm and left to grow the particles for 24 h under the room temperature. The liquid portion was removed via centrifugation, washed with ethanol and ultrapure water, and then dried at 60 °C for 24 h to obtain the SiTi nanoparticles with ODA (SiTi-ODA). Next, 10.2 mL of 1 N HCl and 150 mL of ethanol were added into 1 g of the dried SiTi-ODA, and the mixture was stirred at 700 rpm for 3 h at room temperature to remove ODA through solvent extraction. The solid phase was then removed via centrifugation and washed once with ethanol and once with ultrapure water. The particles were dried at 60 °C for 24 h to obtain nanoporous SiTi nanoparticles.

3.2.2. Immersion of SiTi Nanoparticles into SBF

The 1.0 SBF (Na^+, 142 mM; K^+, 5.0 mM; Mg^{2+}, 1.5 mM; Ca^{2+}, 2.5 mM; Cl^-, 148.8 mM; HCO_3^-, 4.2 mM; HPO_4^{2-}, 1.0 mM; SO_4^{2-}, 0.5 mM; and Tris, 50 mM) was prepared according to the method provided in a previous report [48], and the pH value was adjusted to 7.4 with HCl. Then, 0.5 SBF and 1.5 SBF were prepared at 0.5 and 1.5 times the inorganic ion concentrations of 1.0 SBF. After the XSiTi nanoparticles were added to 0.5 SBF, the pH value was adjusted to 8.60 with Tris and kept at 37 °C for 1 day. The particles were then immersed in 1.5 SBF for 7 days. The solid phase was removed via centrifugation and dried at 37 °C for 24 h to obtain CP-coated SiTi nanoparticles.

3.3. Characterization

The morphologies were observed on a carbonblack-coated Cu grid using a field emission scanning electron microscope (FE-SEM: HITACHI Co., Ltd., SU-8230) at an accelerating voltage of 200 kV; the vertical size, side size, and particle size distributions of the SiTi nanoparticles' shapes were calculated by counting 150 particles, and their average (Ave.) and coefficient of variation (Cv.) values were also calculated. Size distributions of the SiTi nanoparticle images obtained through FE-SEM were calculated by randomly selecting 150 particles.

X-ray diffraction (XRD) patterns were obtained using a powder X-ray diffractometer (Rigaku Co., Ltd., Smart Lab) with CuKα radiation (λ = 0.15418 nm), a voltage of 40 kV, and a current of 200 mA.

Specific surface area and pore size distribution determined via N_2 adsorption and desorption isotherms were measured at -196 °C with a BELSORP-Mini II instrument

(Microtrac BEL Co., Ltd.) to estimate the total surface areas. Prior to measurement, 100 mg of each sample was degassed and pretreated at 80 °C under a vacuum. The following methods were used to analyze the nanopores. The specific surface area was evaluated using the αs-plots [49], and the pore size distribution was determined using micropore analysis (MP). Furthermore, t-plots were used to calculate the specific surface area inside the pores and the adsorbed layers' thickness [50,51]; then, the pore volume was obtained. In this study, the Harkins−Jula equation representing the standard t-curve was used to investigate the standard isotherm. This curve is one of the most commonly used MP methods. Pore size was defined as dp, which was plotted against dVp/dlog dp to show the pore size distribution.

Elemental composition was evaluated using an X-ray fluorescence analyzer (XRF: ZSX Primus II, Rigaku Co., Ltd.). XRF analysis was performed on sample powders in the state of pellets, which were pressurized and molded without dilution. The fundamental parameter method was conducted using software for semi-quantitative analysis (EZ Scan Program, Rigaku Co., Ltd.). Specifically, the amount of each element (Ca, Na, K, and P) adsorbed from SBF was detected and then evaluated in terms of mmol·(mg of sample)$^{-1}$ on a semi-quantitative basis.

Infrared absorption spectra were measured using a Fourier transform infrared spectrometer (FT-IR: FT/IR-4600, Japan Spectroscopic Co., Ltd.) operating in the wavenumber range 4000–500 cm^{-1} with a KBr background, 128 accumulation times, and a spectral resolution of 4 cm^{-1}. FT-IR spectra were measured using KBr powder, and all weights were determined with 49 mg of KBr and 1 mg of sample. All the spectra were recorded after subtracting the background spectrum of KBr.

Transmission electron microscopy (TEM) was performed using a JEOL JEM-2100F transmission electron microscope. Scanning TEM high-angle annular dark-field (STEM−HAADF) images and elemental mapping energy-dispersive X-ray (EDX) spectroscopy images were recorded using a JEM-2100F and a JED-2300 instrument (EX-24200M1G2T, JEOL Ltd.) at an accelerating voltage of 200 kV. The sample suspension was dropped onto a Cu grid (a high-resolution carbon substrate on STEM 100CuP grids, Okenshoji Co., Ltd.), and the grids were dried under vacuum for a few days before each measurement. STEM and EDS elemental mapping (Ca, P, Si, and Ti) images of the SiTi nanoparticles were taken after the nanoparticles' immersion in SBF for 7 days. The detected energies for Ca(K), P(K), Si(K), and Ti(K) were 3.69, 2.01, 1.74, and 4.52 keV, respectively.

4. Conclusions

We established a CP-coating method that homogeneously corresponds to the shapes of SiTi nanoparticles. CP precipitation was performed on mono-dispersed nanoporous SiTi nanoparticles with different Si/Ti molar ratios and pore sizes. The pore size distribution was found to significantly affect the CP coating in SBF immersion; the surfaces of the nanoparticles with bimodal pore sizes of 0.7 and 1.1~1.2 nm became rough after CP precipitation, while those with unimodal pore sizes of 0.7 nm remained smooth, indicating that these two pore sizes work as different nucleation sites that lead to different surface morphologies. These CP-coated SiTi nanoparticles could improve osteoconductivity while retaining the properties of SiTi nanoparticles, which we believe may be suitable for use in the DDS carriers in the future.

Supplementary Materials: The following supporting information can be downloaded at: https://www.mdpi.com/article/10.3390/inorganics11060235/s1, Scheme S1: XRD patterns of the SiTi nanoparticles.; Figure S2: αs-plots of the SiTi nanoparticles with the immersion time Xd (X days, X=0, 1, 3, 7) in SBF; Figure S3: The MP pore size distribution of SiTi nanoparticles with the immersion time in SBF; Figure S4: N$_2$ adsorption (close marks) and desorption (open marks) isotherms of the SiTi nanoparticles with the immersion time in SBF; Figure S5: XRD patterns of the SiTi nanoparticles at the immersion time in SBF for 14 days.

Author Contributions: Conceptualization, R.K. and M.T.; methodology, R.K., K.F. and M.T.; software, R.K.; validation, K.F. and M.T.; formal analysis, I.Y. and K.S.; investigation, R.K.; resources, M.T.; data curation, Y.Z.; writing—original draft preparation, R.K. and Y.Z.; writing—review and editing, K.S., Y.Z., I.Y. and M.T.; supervision, K.S. and M.T.; project administration, K. S. and M.T. All authors have read and agreed to the published version of the manuscript.

Funding: This research was funded by the Japan Society for the Promotion of Science (JSPS) KAKENHI (Grant-in-Aid for Challenging Exploratory Research, Grant 22K18916).

Institutional Review Board Statement: Not applicable.

Informed Consent Statement: Not applicable.

Data Availability Statement: Not applicable.

Acknowledgments: This study was partially supported by a grant from the Japan Society for the Promotion of Science (JSPS) KAKENHI (Grant-in-Aid for Challenging Exploratory Research, Grant 22K18916). Additionally, a portion of this work was supported by NIMS Electron Microscopy Analysis Station, Nanostructural Characterization Group. The authors also thank the Analysis and Instrumentation Center at the Nagaoka University of Technology for providing their facilities.

Conflicts of Interest: The authors declare no conflict of interest.

References

1. Lau, M.; Giri, K.; Garcia-Bennett, A.E. Antioxidant Properties of Probucol Released from Mesoporous Silica. *Eur. J. Pharm. Sci.* **2019**, *138*, 105038. [CrossRef] [PubMed]
2. Marchi, J.; Ussui, V.; Delfino, C.S.; Bressiani, A.H.A.; Marques, M.M. Analysis in Vitro of the Cytotoxicity of Potential Implant Materials. I: Zirconia-Titania Sintered Ceramics. *J. Biomed. Mater. Res. B. Appl. Biomater.* **2010**, *94*, 305–311. [CrossRef] [PubMed]
3. Tanaka, T.; Shimazu, R.; Nagai, H.; Tada, M.; Nakagawa, T.; Sandhu, A.; Handa, H.; Abe, M. Preparation of Spherical and Uniform-Sized Ferrite Nanoparticles with Diameters between 50 and 150 Nm for Biomedical Applications. *J. Magn. Magn. Mater.* **2009**, *321*, 1417–1420. [CrossRef]
4. Rosenholm, J.M.; Peuhu, E.; Eriksson, J.E.; Sahlgren, C.; Lindén, M. Targeted Intracellular Delivery of Hydrophobic Agents Using Mesoporous Hybrid Silica Nanoparticles as Carrier Systems. *Nano. Lett.* **2009**, *9*, 3308–3311. [CrossRef] [PubMed]
5. Kataoka, T.; Shiba, K.; Nagata, S.; Yamada, I.; Chai, Y.; Tagaya, M. Preparation of Monodispersed Nanoporous Eu(III)/Titania Loaded with Ibuprofen: Optimum Loading, Luminescence, and Sustained Release. *Inorg. Chem.* **2021**, *60*, 8765–8776. [CrossRef] [PubMed]
6. Chiavaioli, F.; Biswas, P.; Trono, C.; Jana, S.; Bandyopadhyay, S.; Basumallick, N.; Giannetti, A.; Tombelli, S.; Bera, S.; Mallick, A.; et al. Sol-Gel-Based Titania-Silica Thin Film Overlay for Long Period Fiber Grating-Based Biosensors. *Anal. Chem.* **2015**, *87*, 12024–12031. [CrossRef]
7. Arcos, D.; Vallet-Regí, M. Sol-Gel Silica-Based Biomaterials and Bone Tissue Regeneration. *Acta. Biomater.* **2010**, *6*, 2874–2888. [CrossRef]
8. Li, Z.; Hou, B.; Xu, Y.; Wu, D.; Sun, Y.; Hu, W.; Deng, F. Comparative Study of Sol-Gel-Hydrothermal and Sol-Gel Synthesis of Titania-Silica Composite Nanoparticles. *J. Solid. State. Chem.* **2005**, *178*, 1395–1405. [CrossRef]
9. Aw, M.S.; Addai-Mensah, J.; Losic, D. A Multi-Drug Delivery System with Sequential Release Using Titania Nanotube Arrays. *Chem. Comm.* **2012**, *48*, 3348–3350. [CrossRef]
10. Aw, M.S.; Losic, D. Ultrasound Enhanced Release of Therapeutics from Drug-Releasing Implants Based on Titania Nanotube Arrays. *Int. J. Pharm.* **2013**, *443*, 154–162. [CrossRef]
11. Kasar, S.; Kumar, S.; Kar, A.S.; Godbole, S.V.; Tomar, B.S. Sorption of Eu(III) by Amorphous Titania, Anatase and Rutile: Denticity Difference in Surface Complexes. *Col. Surf. A. Physicochem. Eng. Asp.* **2013**, *434*, 72–77. [CrossRef]
12. Ekka, B.; Sahu, M.K.; Patel, R.K.; Dash, P. Titania Coated Silica Nanocomposite Prepared via Encapsulation Method for the Degradation of Safranin-O Dye from Aqueous Solution: Optimization Using Statistical Design. *Water. Resour. Ind.* **2019**, *22*, 100071. [CrossRef]
13. Deng, Y.; Qi, D.; Deng, C.; Zhang, X.; Zhao, D. Superparamagnetic High-Magnetization Microspheres with an $Fe_3O_4@SiO_2$ Core and Perpendicularly Aligned Mesoporous SiO_2 Shell for Removal of Microcystins. *J. Am. Chem. Soc.* **2008**, *130*, 28–29. [CrossRef] [PubMed]
14. Jana, N.R.; Earhart, C.; Ying, J.Y. Synthesis of Water-Soluble and Functionalized Nanoparticles by Silica Coating. *Chem. Mater.* **2007**, *19*, 5074–5082. [CrossRef]
15. Shiba, K.; Sato, S.; Ogawa, M. Preparation of Well-Defined Titania-Silica Spherical Particles. *J. Mater. Chem.* **2012**, *22*, 9963–9969. [CrossRef]
16. He, Q.; Shi, J.; Zhu, M.; Chen, Y.; Chen, F. The Three-Stage in Vitro Degradation Behavior of Mesoporous Silica in Simulated Body Fluid. *Microporous. Mesoporous. Mater.* **2010**, *131*, 314–320. [CrossRef]

17. Raula, J.; Eerikäinen, H.; Peltonen, L.; Hirvonen, J.; Kauppinen, E. Aerosol-Processed Polymeric Drug Nanoparticles for Sustained and Triggered Drug Release. *J. Control. Release* **2010**, *148*, e52–e53. [CrossRef]
18. Zhao, C.; Zhuang, X.; He, P.; Xiao, C.; He, C.; Sun, J.; Chen, X.; Jing, X. Synthesis of Biodegradable Thermo- and PH-Responsive Hydrogels for Controlled Drug Release. *Polymer* **2009**, *50*, 4308–4316. [CrossRef]
19. Paital, S.R.; Dahotre, N.B. Wettability and Kinetics of Hydroxyapatite Precipitation on a Laser-Textured Ca-P Bioceramic Coating. *Acta. Biomater.* **2009**, *5*, 2763–2772. [CrossRef]
20. Barfeie, A.; Wilson, J.; Rees, J. Implant Surface Characteristics and Their Effect on Osseointegration. *Br Dent J* **2015**, *218*. [CrossRef]
21. Shadanbaz, S.; Dias, G.J. Calcium Phosphate Coatings on Magnesium Alloys for Biomedical Applications: A Review. *Acta. Biomater.* **2012**, *8*, 20–30. [CrossRef] [PubMed]
22. Liu, Y.; De Groot, K.; Hunziker, E.B. Osteoinductive Implants: The Mise-En-Scène for Drug-Bearing Biomimetic Coatings. *Ann. Biomed. Eng.* **2004**, *32*, 398–406. [CrossRef] [PubMed]
23. Wolke, J.G.C.; Van Der Waerden, J.P.C.M.; Schaeken, H.G.; Jansen, J.A. In Vivo Dissolution Behavior of Various RF Magnetron-Sputtered Ca-P Coatings on Roughened Titanium Implants. *Biomater.* **2003**, *24*, 2623–2629. [CrossRef] [PubMed]
24. Leeuwenburgh, S.; Wolke, J.; Schoonman, J.; Jansen, J. Electrostatic Spray Deposition (ESD) of Calcium Phosphate Coatings. *J. Biomed. Mater. Res. A* **2003**, *66*, 330–334. [CrossRef]
25. Radin, S.R.; Ducheyne, P. Plasma Spraying Induced Changes of Calcium Phosphate Ceramic Characteristics and the Effect on in Vitro Stability. *J. Mater. Sci. Mater. Med.* **1992**, *3*, 33–42. [CrossRef]
26. Shang, Y.; Yamada, S.; Chai, Y.; Tagaya, M. Synthesis of Spherical Phosphorus-Containing Mesoporous Silica for Improving Their Reaction Behavior in Simulated Body Fluid. *Key. Eng. Mater.* **2018**, *782 KEM*, 59–64. [CrossRef]
27. Kataoka, T.; Shiba, K.; Wang, L.Y.; Yamada, S.; Tagaya, M. Hybrid Preparation of Terbium(III)-Doped Mesoporous Silica Particles with Calcium Phosphates. *RSC Adv.* **2017**, *7*, 19479–19485. [CrossRef]
28. Lin, X.; De Groot, K.; Wang, D.; Hu, Q.; Wismeijer, D.; Liu, Y. A Review Paper on Biomimetic Calcium Phosphate Coatings. *Open Biomed. Eng. J.* **2015**, *9*, 56–64. [CrossRef]
29. Li, F.; Feng, Q.L.; Cui, F.Z.; Li, H.D.; Schubert, H. A Simple Biomimetic Method for Calcium Phosphate Coating. *Surf. Coat. Technol.* **2002**, *154*, 88–93. [CrossRef]
30. Leeuwenburgh, S.; Layrolle, P.; Barrre, F.; De Bruijn, J.; Schoonman, J.; Van Blitterswijk, C.A.; De Groot, K. Osteoclastic Resorption of Biomimetic Calcium Phosphate Coatings in Vitro. *J. Biomed. Mater. Res.* **2001**, *56*, 208–215. [CrossRef]
31. Cao, J.; Lian, R.; Jiang, X.; Liu, X. Formation of Porous Apatite Layer after Immersion in SBF of Fluorine-Hydroxyapatite Coatings by Pulsed Laser Deposition Improved in Vitro Cell Proliferation. *ACS. Appl. Bio. Mater.* **2020**, *3*, 3698–3706. [CrossRef]
32. Tanahashi, M.; Matsuda, T. Surface Functional Group Dependence on Apatite Formation on Self- Assembled Monolayers in a Simulated Body Fluid. *J. Biomed. Mater. Res.* **1997**, *34*, 305–315. [CrossRef]
33. Gu, Y.W.; Khor, K.A.; Cheang, P. In Vitro Studies of Plasma-Sprayed Hydroxyapatite/Ti-6Al-4V Composite Coatings in Simulated Body Fluid (SBF). *Biomater.* **2003**, *24*, 1603–1611. [CrossRef] [PubMed]
34. Cui, W.; Beniash, E.; Gawalt, E.; Xu, Z.; Sfeir, C. Biomimetic Coating of Magnesium Alloy for Enhanced Corrosion Resistance and Calcium Phosphate Deposition. *Acta. Biomater.* **2013**, *9*, 8650–8659. [CrossRef]
35. Bigi, A.; Boanini, E.; Bracci, B.; Facchini, A.; Panzavolta, S.; Segatti, F.; Sturba, L. Nanocrystalline Hydroxyapatite Coatings on Titanium: A New Fast Biomimetic Method. *Biomater.* **2005**, *26*, 4085–4089. [CrossRef] [PubMed]
36. Braun, K.; Pochert, A.; Beck, M.; Fiedler, R.; Gruber, J.; Lindén, M. Dissolution Kinetics of Mesoporous Silica Nanoparticles in Different Simulated Body Fluids. *J. Solgel. Sci. Technol.* **2016**, *79*, 319–327. [CrossRef]
37. Chai, Y.; Maruko, Y.; Liu, Z.; Tagaya, M. Design of Oriented Mesoporous Silica Films for Guiding Protein Adsorption States. *J. Mater. Chem. B* **2021**, *9*, 2054–2065. [CrossRef]
38. Kim, S.R.; Lee, J.H.; Kim, Y.T.; Riu, D.H.; Jung, S.J.; Lee, Y.J.; Chung, S.C.; Kim, Y.H. Synthesis of Si, Mg Substituted Hydroxyapatites and Their Sintering Behaviors. *Biomaterials* **2003**, *24*, 1389–1398. [CrossRef]
39. Eliad, L.; Salitra, G.; Soffer, A.; Aurbach, D. Ion Sieving Effects in the Electrical Double Layer of Porous Carbon Electrodes: Estimating Effective Ion Size in Electrolytic Solutions. *J. Phy. Chem. B* **2001**, *105*, 6880–6887. [CrossRef]
40. Banda, H.; Daffos, B.; Périé, S.; Chenavier, Y.; Dubois, L.; Aradilla, D.; Pouget, S.; Simon, P.; Crosnier, O.; Taberna, P.L.; et al. Ion Sieving Effects in Chemically Tuned Pillared Graphene Materials for Electrochemical Capacitors. *Chem. Mater.* **2018**, *30*, 3040–3047. [CrossRef]
41. Porter, D.W.; Wu, N.; Hubbs, A.F.; Mercer, R.R.; Funk, K.; Meng, F.; Li, J.; Wolfarth, M.G.; Battelli, L.; Friend, S.; et al. Differential Mouse Pulmonary Dose and Time Course Responses to Titanium Dioxide Nanospheres and Nanobelts. *Toxicol. Sci.* **2013**, *131*, 179–193. [CrossRef]
42. Oberdorster, G.; Ferin, J.; Gelein, R.; Soderholm, S.C.; Finkelstein, J. Role of the Alveolar Macrophage in Lung Injury: Studies with Ultrafine Particles. *Environ. Health Perspect.* **1992**, *97*, 193–199. [PubMed]
43. Yu, J.C.; Zhang, L.; Zheng, Z.; Zhao, J. Synthesis and Characterization of Phosphated Mesoporous Titanium Dioxide with High Photocatalytic Activity. *Chem. Mater.* **2003**, *15*, 2280–2286. [CrossRef]
44. Huang, X.; Wang, P.; Yin, G.; Zhang, S.; Zhao, W.; Wang, D.; Bi, Q.; Huang, F. Removal of Volatile Organic Compounds Driven by Platinum Supported on Amorphous Phosphated Titanium Oxide. *Wuji Cailiao Xuebao/J. Inorg. Mater.* **2020**, *35*, 482–490.
45. Aizawa, M.; Nosaka, Y.; Fujii, N. FT-IR Liquid Attenuated Total Reflection Study of TiO_2-SiO_2 Sol-Gel Reaction. *J. Non-Cryst. Solids* **1991**, *128*, 77–85. [CrossRef]

46. Schraml, M.; Walther, K.; Wokaun, A.; Handy, B.; Baiker, A. Porous Silica Gels and TiO_2/SiO_2 Mixed Oxides Prepared via the Sol-Gel Process: Characterization by Spectroscopic Techniques. *J. Non-Cryst. Solids* **1992**, *143*, 93–111. [CrossRef]
47. Thommes, M.; Kaneko, K.; Neimark, A.; Olivier, J.P.; Rodriguez-Reinoso, F.; Rouquerol, J.; Sing, K.S.W. Physisorption of Gases, with Special Reference to the Evaluation of Surface Area and Pore Size Distribution (IUPAC Technical Report). *Pure. Appl. Chem.* **2015**, *87*, 1051–1069. [CrossRef]
48. Kokubo, T.; Kushitani, H.; Sakka, S.; Kitsugi, T.; Yamamuro, T. Solutions Able to Reproduce in Vivo Surface-Structure Changes in Bioactive Glass-Ceramic A-W3. *J. Biomed. Mater. Res.* **1990**, *24*, 721–734. [CrossRef] [PubMed]
49. Atkinson, D.; McLeod, A.I.; Sing, K.S.W. Adsorptive Properties of Microporous Carbons: Primary and Secondary Micropore Filling. *J. Chim. Phys.* **1984**, *81*, 791–794. [CrossRef]
50. Mikhail, R.S.; Brunauer, S.; Bodor, E.E. Investigations of a Complete Pore Structure Analysis I. *Anal. Micropores; J. coll. Interf Sci.* **1968**, *26*, 45–53. [CrossRef]
51. Setoyama, N.; Suzuki, T.; Kaneko, K. A Molecular Simulation Study on Empirical Determination Method of Pore Structures of Activated Carbons. *Tanso 1997* **1997**, 159–166. [CrossRef]

Disclaimer/Publisher's Note: The statements, opinions and data contained in all publications are solely those of the individual author(s) and contributor(s) and not of MDPI and/or the editor(s). MDPI and/or the editor(s) disclaim responsibility for any injury to people or property resulting from any ideas, methods, instructions or products referred to in the content.

Review

The Story, Properties and Applications of Bioactive Glass "1d": From Concept to Early Clinical Trials

Dilshat U. Tulyaganov [1], Simeon Agathopoulos [2], Konstantinos Dimitriadis [3], Hugo R. Fernandes [4], Roberta Gabrieli [5] and Francesco Baino [5,*]

1 Department of Natural-Mathematical Sciences, Turin Polytechnic University in Tashkent, Tashkent 100095, Uzbekistan; tulyaganovdilshat@gmail.com
2 Department of Materials Science and Engineering, University of Ioannina, 451 10 Ioannina, Greece; sagat@uoi.gr
3 Division of Dental Technology, Department of Biomedical Sciences, University of West Attica, 122 43 Athens, Greece; dimitriadiskonstantinos@hotmail.com
4 Department of Materials and Ceramic Engineering, CICECO, University of Aveiro, 3810-193 Aveiro, Portugal; h.r.fernandes@ua.pt
5 Department of Applied Science and Technology, Institute of Materials Physics and Engineering, Politecnico di Torino, 10129 Torino, Italy; roberta.gabrieli@polito.it
* Correspondence: francesco.baino@polito.it

Abstract: Bioactive glasses in the CaO–MgO–Na$_2$O–P$_2$O$_5$–SiO$_2$–CaF$_2$ system are highly promising materials for bone and dental restorative applications. Furthermore, if thermally treated, they can crystallize into diopside–fluorapatite–wollastonite glass-ceramics (GCs), which exhibit appealing properties in terms of mechanical behaviour and overall bone-regenerative potential. In this review, we describe and critically discuss the genesis, development, properties and applications of bioactive glass "1d" and its relevant GC derivative products, which can be considered a good example of success cases in this class of SiO$_2$/CaO-based biocompatible materials. Bioactive glass 1d can be produced by melt-quenching in the form of powder or monolithic pieces, and was also used to prepare injectable pastes and three-dimensional porous scaffolds. Over the past 15 years, it was investigated by the authors of this article in a number of in vitro, in vivo (with animals) and clinical studies, proving to be a great option for hard tissue engineering applications.

Keywords: bioactive glass; glass-ceramic; oxide system; diopside; fluorapatite; wollastonite; bone regeneration

Citation: Tulyaganov, D.U.; Agathopoulos, S.; Dimitriadis, K.; Fernandes, H.R.; Gabrieli, R.; Baino, F. The Story, Properties and Applications of Bioactive Glass "1d": From Concept to Early Clinical Trials. *Inorganics* **2024**, *12*, 224. https://doi.org/10.3390/inorganics12080224

Academic Editors: Roberto Nisticò and Silvia Mostoni

Received: 29 June 2024
Revised: 7 August 2024
Accepted: 8 August 2024
Published: 17 August 2024

Copyright: © 2024 by the authors. Licensee MDPI, Basel, Switzerland. This article is an open access article distributed under the terms and conditions of the Creative Commons Attribution (CC BY) license (https:// creativecommons.org/licenses/by/ 4.0/).

1. Introduction

The first bioactive glass composition, trade named as 45S5 Bioglass®, was designed by Larry Hench [1] in the early 1970s and addressed to bone replacement applications. After the invention of 45S5 glass, many other bioactive glasses have been reported for various medical applications like drug delivery [2], cancer treatment [3] or even soft tissue applications for organ repair [4]. The original 45S5 composition (45SiO$_2$-24.5CaO-24.5Na$_2$O-6P$_2$O$_5$ in wt.%) is based on silica (SiO$_2$) as the main glass-forming oxide and could create bonds with the living bone after in vivo implantations. A sequence of 11 reaction steps is involved in the bonding process of silicate bioactive glasses to living tissue [5], where the steps 1 to 5 are key for the formation of a hydroxyapatite-like layer on the surface of glasses.

Biomedical glasses are conventionally classified as "bioactive" based on these two mechanisms: (i) the formation of a calcium phosphate (hydroxyapatite-like) layer on the surface of the glass when it dissolves in a physiological environment (also in vitro in simulated body fluid (SBF)), and (ii) the release of biologically active ions during in vitro and in vivo testing. Hence, bioactive silicate glasses are both osteoconductive (mechanism (i)) and osteoinductive (mechanism (ii)). The interaction of bioactive glass surfaces with body fluids begins with an exchange of ions that leads to an increment of the pH in the medium, resulting in the development of a SiO_2-rich layer and then the growth of a CaO/P_2O_5-rich layer on the glass surface. This layer is further enriched with carbonates and then crystallizes to form hydroxycarbonate apatite, which mimics the mineral phase of natural bone and ultimately helps to bond with the osseous tissue [6].

The hydroxyapatite-like layer also promotes the next biological reaction stages, including cell migration, proliferation and differentiation to form new bone with good mechanical bonds to the implant surface. The hydroxyapatite layer thickness plays a major role in the bone-bonding ability of the glass as well as on the interfacial shear strength. It was reported that an interface thickness of 20 μm yields adequate shear strength and interfacial bonding [7].

Borate and borosilicate glasses are more reactive than silicate materials when in contact with body fluids or, in general, aqueous media; hence, they are less durable and can convert faster to hydroxyapatite compared to SiO_2-based glasses. In principle, the bioactivity mechanism is very similar except for the formation of a borate-rich gel layer, analogous to the silica gel layer in silicate systems. The apatite-formation rate can be controlled by changing the glass composition, thus varying the reaction time from hours to months [8].

Phosphate glasses have also been proposed for bone tissue engineering applications; their tendency to rapidly dissolve is aqueous media depending on the composition—and, especially, on the metal oxide content—has pushed their use towards advanced therapies mediated by controlled ions release [9].

When high-strength and/or load-bearing applications are major goals, the controlled crystallization of silicate glasses yielding bioactive glass-ceramic (GC) materials is an attractive option. For example, the commercial apatite/wollastonite (A–W)-containing Cerabone (SiO_2-CaO-MgO-P_2O_5-F parent system) has been produced by controlled heat treatment, obtaining 38 wt.% apatite, 34 wt.% of wollastonite and 38 wt.% of residual glass phase, and used as a coating for titanium alloys, artificial vertebrae and bone fillers [10,11]. The bending strength of these A–W glass-ceramics (GCs) is typically higher than that of human cortical bone (160 MPa), but the fracture toughness is three times lower (6 MPa m$^{1/2}$) [12]. Interestingly, it was observed that hydroxyapatite can form on the surface of these GCs even if the silica gel on the surface is absent, as the apatite and wollastonite crystals act as sites for direct nucleation of calcium phosphate phases [13].

Other common examples of bioactive GCs include A–W Ceravital and diopside-containing products, such as apatite–diopside (AD), wollastonite–diopside (WD) and diopside–combeite (DC) GCs. Ceravital (SiO_2-CaO-MgO-Na_2O-K_2O-P_2O_5 parent system) [14] has an analogous bioactivity mechanism to Hench's 45S5 Bioglass®, along with good mechanical properties and better stability in the long term. After the implantation of Ceravital material, an initial degradation of the surface caused by ionic exchange was observed, followed by the formation of reaction layers that protect the implant from further chemical attacks.

The phase in common for the other GC types mentioned above is diopside, which is very appealing for biomedical applications due to its attractive mechanical performance. For example, diopside was combined with hydroxyapatite in order to increase the fracture resistance of the latter [15]. In this case, the fracture strength of the AD bioceramic was found to be 2–3 times higher compared to hydroxyapatite alone; moreover, the material was non-toxic to the body cells and promoted bone regeneration.

In wollastonite–diopside GCs, wollastonite and diopside phases form when the parent glass is thermally treated above 900 °C. Both wollastonite and diopside have good mechanical properties and the latter has a slower dissolution rate upon contact with body fluids. These bioactive GCs have been used in bulk, granular and porous forms for bone graft applications [16].

DC GCs have also been proposed for bone-contact applications; it is worth mentioning that combeite is a highly biocompatible phase, which is typically found in sintered 45S5 Bioglass® treated above 550 °C, too [17].

Therefore, all these studies witnessed that diopside was an attractive crystalline phase to have inside GC biomaterials. These results pushed scientists to design new glass-derived formulations that could originate GC products with superior biological and mechanical properties, of which the "1d" composition is a valuable example.

2. The Genesis of the 1d Composition

The glass composition 1d (Table 1) was designed so that, upon thermal devitrification, a final GC material containing diopside, fluorapatite and wollastonite could be obtained with high mechanical strength and excellent bioactive properties [18].

Table 1. Chemical composition of 1d glass and some of its derivatives (in wt.%) [18–20].

Composition	SiO_2	CaO	MgO	P_2O_5	Na_2O	K_2O	CaF_2
Parent glass composition							
1d	46.1	28.7	8.8	6.2	4.5	0	5.7
1d-a	41.8	32.85	8.85	6.24	4.54	0	5.72
1d-b	37.51	37.07	8.88	6.26	4.55	0	5.73
1e	43.5	30.4	8.8	7.2	4.5	0	5.6
1e-a	43.09	30.13	8.67	9.17	4.44	0	4.47
1e-b	42.71	29.85	8.59	11.1	4.41	0	3.33
1e-c	42.33	29.59	8.52	13.00	4.37	0	2.19
K_2O for Na_2O substitution							
1d-k	45.0	28.0	8.60	6.1	0	6.7	5.6
1e-k	42.5	29.7	8.5	7.1	0	6.7	5.5
MgO for CaO partial substitution							
1d-m	46.6	24.8	11.9	6.3	4.6	0	5.8
1e-m	44.00	26.7	11.8	7.3	4.5	0	5.7

The 1d composition relies on the primary crystallization field of pseudowollastonite in the CaO–MgO–SiO_2 ternary system, in which P_2O_5, Na_2O and CaF_2 were added. Following a melt-quenching route, 1d products can be obtained in an amorphous form (glass) but the development of the three crystalline phases mentioned above can be induced by applying a proper thermal treatment.

The 1d glass was the most promising member of a family of compositions, which were originally designed and studied by a multidisciplinary international research team and resulted in numerous publications since 2006; other sister formulations include 1e glass—having the same components as 1d in different amounts [18–20]—and 1b glass [21,22], also containing B_2O_3 as an additional oxide.

Similarly, other glasses based on the 1d and 1e compositions have been created by replacing the components and changing the ratio of percentages by weight, such as the following:

- 1d-a and 1d-b: in which the CaO/SiO_2 ratio has been progressively increased [18] (Table 1);
- 1e-a, 1e-b, 1e-c: in which the P_2O_5/CaF_2 ratio has been progressively increased [18] (Table 1);
- 1d-k, 1e-k: where Na_2O was replaced with K_2O [19,20] (Table 1);
- 1d-m, 1e-m: where CaO was partially replaced with MgO [19,20] (Table 1).

3. Material Preparation and Basic Properties

The 1d glass is typically synthesized by following a melt-quenching route; several publications can be found in the literature where the basic preparative process has been applied with some modifications, as summarized in Table 2, from the reagent powders to frit production. Examples of 1d glass frit and glass powders after milling are given in Figure 1.

Table 2. Processes to obtain 1d glass powder from precursors powders.

Article	Powders	Batch and Milling	Heat Treatment	Frits and Glass Powders
I. Kansal et al. [18]	SiO_2 (purity > 99.5%), $CaCO_3$ (>99.5%), $MgCO_3$ (>99%), Na_2CO_3(>99%), CaF_2 (>99.9%), $NH_4H_2PO_4$ (>99%)	Homogenous mixture of precursors of about 100 g obtained by ball milling.	Pre-heating at 900 °C for 1 h for calcination. Melting in Pt crucible at 1450–1550 °C for 1 h.	Frits are obtained by quenching of melted glass in water. Frits are dried and then milled in agate mill. Powders are sieved to obtain a mean particle size of about 10 μm.
D. U. Tulyaganov et al. [22]	SiO_2 (purity>99.5%), $CaCO_3$ (>99.5%), $4MgCO_3·Mg(OH)_2·5H_2O$ (>99%), Na_2CO_3 (>99%), CaF_2 (>99%), $NH_4H_2PO_4$ (>99%)	Homogenous mixture of precursors of about 100 g obtained by ball milling.	Pre-heating at 1000 °C for 1 h for decarbonization. Melting in Pt-crucible at 1400 °C for 1 h.	Frits are obtained by quenching of melted glass in water. Frits are dried and then milled in a porcelain mill. Powders are sieved to obtain a mean particle size of 11–14 μm.
S. I. Schmitz et al. [23]	SiO_2 (purity > 99.5%), $CaCO_3$ (>99.5%), $4MgCO_3·Mg(OH)_2·5H_2O$ (>99%), Na_2CO_3 (>99%), CaF_2 (>99%), $NH_4H_2PO_4$ (>99%)	Homogenous mixture of precursors.	Pre-heating at 1000 °C for 1 h for decarbonization. Melting in Pt-crucible at 1400 °C for 1 h.	Frits are obtained by quenching of melted glass in deionized water. Frits are dried and then milled in a planetary mill. Powders are sieved to obtain a final particle size <32 μm.
D. U. Tulyaganov et al. [24]	SiO_2 (purity > 99.5%), $CaCO_3$ (>99.5%), $MgCO_3$ (>99%), Na_2CO_3 (>99%), CaF_2 (>99%), $NH_4H_2PO_4$ (>99%)	Homogenous mixture of precursors.	Pre-heating at 1000 °C for 1 h for decarbonization. Melting in Pt crucible at 1400 °C for 1 h.	Frits are obtained by quenching of melted glass in water. Frites are dried and then milled. Powders are sieved to obtain a mean particle size of 10–15 μm.
K. Dimitriadis et al. [20]	SiO_2 (purity > 99.5%), $CaCO_3$ (>99.5%), $Mg(NO_3)_2·6H_2O$ (>99%), Na_2CO_3 (>99%), CaF_2 (>99%), $(NH_4)_2HPO_4$ (>99%)	Homogenous mixture of precursors of about 100 g.	Pre-heating at 900 °C for 1 h for decarbonization. Melting in Pt crucible at 1400 °C for 1 h.	Frits are obtained by quenching of melted glass in water. Frits are dried and then milled in a planetary mill. Powders are sieved to obtain a final particle size <32 μm.
F. Baino et al. [25]	SiO_2 (purity > 99.5%), $CaCO_3$ (>99.5%), $MgCO_3$ (>99%), Na_2CO_3 (>99%), CaF_2 (>99.9%), $NH_4H_2PO_4$ (>99%)	Homogenous mixture of precursors of about 100 g by ball milling.	Pre-heating at 850 °C for 1 h in an Al_2O_3 at heating rate of 2.5 °C/min. Melting in Pt crucible at 1420 °C for 1 h.	Frits are obtained by quenching of melted glass in water. Frits are dried and then milled in a planetary mill. Powders are sieved to obtain a final particle size <56 μm.
K. Dimitriadis et al. [19]	SiO_2 (purity > 99.8%), $CaCO_3$ (>99%), $Mg(NO_3)_2·6H_2O$ (>99%), Na_2CO_3 (>99.6%), CaF_2 (>99%), $(NH_4)_2HPO_4$ (>99%)	Homogenous mixture of precursors of about 100 g by ball milling.	Pre-heating at 900 °C for 1 h in an Al_2O_3 crucible at heating rate of 1.5 °C/min. Melting in Pt crucible at 1400 °C for 1 h.	Frits are obtained by rapid pouring of melted glass in water. Frits are dried and then milled in a planetary ball-mill at 400 rpm for 45 min in a YSZ milling jar. Powders are sieved to obtain a final particle size <32 μm.

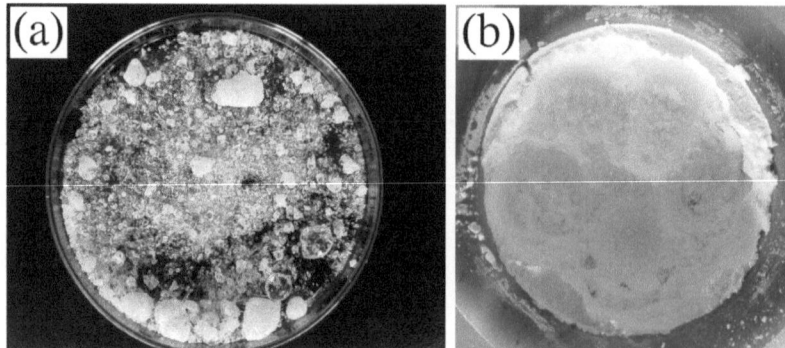

Figure 1. Example of (**a**) 1d glass frit, (**b**) glass powders after ball milling.

The density of 1d glass was reported to be 2.57 ± 0.13 g/cm^3, the characteristic temperatures, i.e., glass transition (Tg), onset of crystallization (Tc) and peak of crystallization (Tp), were assessed in various publications (also by using different experimental methods) and are collected in Table 3.

Table 3. Characteristic temperatures of 1d glass: glass transition (Tg), onset of crystallization (Tc) and peak of crystallization (Tp).

Article	T_g (°C)	T_c (°C)	T_p (°C)
K. Dimitriadis et al. [19]	649 ± 9	783 ± 2	815 ± 13
F. Baino et al. [25]	640	785	830
K. Dimitriadis et al. [20]	655 ± 5	783 ± 2	845 ± 13
D. U. Tulyaganov et al. [24]	607 ± 7	-	815 ± 13
D. U. Tulyaganov et al. [22]	590 ± 10	-	-

4. Crystalline Phases and Mechanical Properties

Powders of 1d glass have been used as-is, even in clinical trials (as discussed later [26]), or as starting materials to fabricate other products, such as 3D porous scaffolds for bone tissue engineering. To obtain these 1d-derived implants, a thermal treatment is necessary to consolidate and join the glass particles together, during which sinter-crystallization may take place. In other words, the formation of crystalline phases is promoted upon heating, leading to the material's transformation from purely amorphous to a GC state with the development of wollastonite, fluorapatite and diopside. Figure 2 shows the microstructure of 1d-derived GCs, in which the part formed by prisms corresponds to diopside, wollastonite crystals have a needle-like shape, and flakes refer to fluorapatite [19].

Table 4 illustrates the influence of heat-treatment temperatures on the mechanical properties of various 1d-derived GCs. It is known that crystallization yields an improvement in the mechanical properties of GCs compared to parent glass; accordingly, the highest increments of flexural strength, hardness, elastic modulus, hardness and fracture toughness, were achieved in the GC produced by heat treatment at 850 °C, i.e., a temperature close to Tp (see Table 3). It was also determined that the brittleness indexes of the produced 1d- and 1e-derived GCs ranged between 3.6 and 3.7 and 3.3 and 3.5 μm$^{-0.5}$, respectively. Qualitatively, machinability reflects the easiness of a material to be cut, and it can be quantified by the magnitude of brittleness [27]. The 1d- and 1e-derived GCs exhibited a brittleness index higher than 3 μm$^{-0.5}$ [28], thus being in the preferred range as the brittleness index for glasses and ceramics typically ranges from 3 to 9 μm$^{-0.5}$ [27]. This discovery holds significance for dental materials production, as these materials are often shaped using specialized cutting tools. Therefore, the aforementioned brittleness index values suggest a reduced risk of fractures or cracks occurring during these processes especially for the

1d-derived GC, since it shows a higher value on the brittleness index compared to the 1e-derived GC [28].

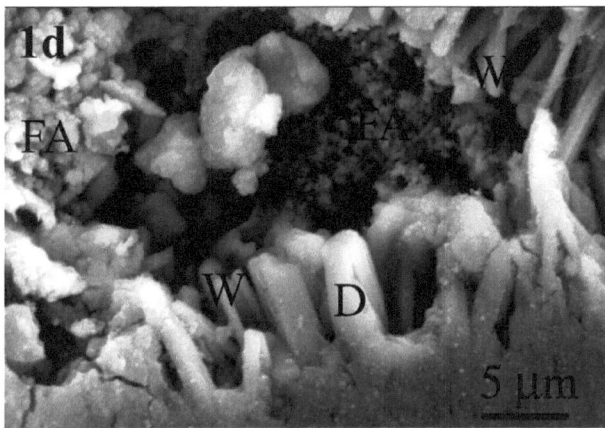

Figure 2. Typical microstructure of 1d-derived GCs produced by heat treatment at 850 °C for 1 h. Observation was performed after etching the polished surface with 2% HF solution (D: diopside; FA: fluorapatite; W: wollastonite).

Table 4. Mechanical properties of different 1d-derived GCs compared to other GCs containing diopside, wollastonite and fluorapatite as crystalline phases.

Materials	Heat Treatment (°C)	Flexural Strength (σ, MPa)	Elastic Modulus (E, GPa)	Vickers Microhardness (HV, GPa)	Fracture Toughness (K_{IC}, MPa·m$^{0.5}$)
1d-derived GCs [28]	800	119 ± 10	24 ± 6	6.0 ± 0.4	1.6 ± 0.1
	850	171 ± 11	27 ± 5	6.1 ± 0.5	1.7 ± 0.1
	900	141 ± 6	22 ± 4	5.2 ± 0.7	1.4 ± 0.1
GCs containing diopside and fluorapatite [29]	850	120–195	-	-	-
GCs containing wollastonite and quartz [30]	900–1000	98 ± 6	-	5.9–6.7	-
GCs containing wollastonite, hydroxyapatite and fluorite [31,32]	700–1000	-	89–100	-	4.6–5.6
GCs containing wollastonite [33]	3100 (flame-spraying)	-	37–56	2.6–5.4	-
GCs containing wollastonite and diopside [33]	3100 (flame-spraying)	-	62–77	2.2–6.5	-

The main characteristics of these three crystalline phases contributing to mechanical properties of GCs (Table 4) are described in the following sections.

4.1. Fluorapatite

Fluorapatite is a mineral that is part of the apatite family with the chemical formula $Ca_{10}(PO_4)_6F_2$. It is a double salt resulting from the bond between calcium phosphate and calcium fluoride. It is a highly biocompatible material and fits well with bone repair applications. In fact, it is present in nature and is part, for example, of the mineralized phase of bones and teeth in mammals [34].

Moreover, having the fluoride ion instead of the hydroxyl ion, fluorapatite exhibits peculiar characteristics that differentiate it from hydroxyapatite; in fact, the former is much less soluble in an acidic environment, such as that of the human mouth, compared to hydroxyapatite. However, despite few differences in some physicochemical properties, the structure of both calcium phosphates is substantially the same. If in hydroxyapatite there are calcium phosphate tetrahedra arranged around hydroxyl ion columns, in fluorapatite there are tetrahedra which develop around fluoride columns (Figure 3).

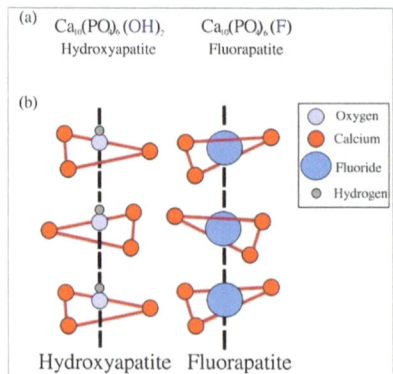

Figure 3. Similarities between fluorapatite and hydroxyapatite: (**a**) chemical formulas and (**b**) chemical structures. Reproduced from [35] with permission.

The crystalline structure of fluorapatite, characterised by flake-like crystals, can be visualized in the SEM image displayed in Figure 4. These morphological observations are also in agreement with theoretical expectations (see Figure 3).

Figure 4. SEM image of fluorapatite. Reproduced from [36] under CC license.

4.2. Wollastonite

Wollastonite is a mineral with interesting characteristics for biomedical applications, including biocompatibility, biodegradability, thermal stability/low thermal expansion, low thermal conductivity and high mechanical properties [37,38]. Given these appealing properties, in the early 1980s, Kokubo et al. [39] first produced a bioactive GC containing both apatite and wollastonite. This new biomaterial belonged to a more complicated system than that invented by Hench one decade earlier (45S5 Bioglass®) [40]. This GC contained apatite (38%), wollastonite (24%) and residual amorphous phase (38%) [39] and has been since marketed under the commercial name of Cerabone®.

Wollastonite is a simple calcium silicate with the chemical formula $CaSiO_3$; alternatively, it can be seen as a mixture of silica (SiO_2) and lime (CaO) having a theoretical percentage of 51.7% and 48.3%, respectively [37]. Due to its crystal structure (Figure 5), wollastonite belongs to the class of minerals known as pyroxenoids. It was reported that pyroxenoid chains are more kinked and have a great repeat distance [37].

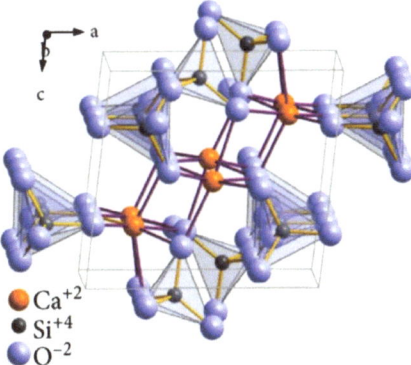

Figure 5. Crystalline structure of wollastonite. Reproduced from [37] under CC license.

Two polymorphs of wollastonite, α-phase and β-phase, potentially exist in nature: the β-phase is the stable state at low temperatures, while the α-phase is found above 1125 °C [41]. Therefore, wollastonite used in biomedical implants is typically the β type and has been proposed as filler in composite fabrication in orthopaedics as well as for dental restoration [37]. The needle-like shape of β-$CaSiO_3$ crystals is shown in Figure 6.

Figure 6. SEM image of wollastonite. Reproduced from [42] with permission.

4.3. Diopside

Diopside ($CaMgSi_2O_6$) is a prismatic monoclinic mineral, but it is not always possible to have this structure; on the contrary, it is easier to find granular and globular structures [43]. The crystalline structure of diopside is very similar to pyroxene, as displayed in Figure 7.

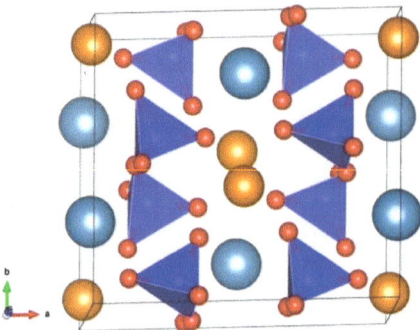

Figure 7. Crystalline structure of diopside found by Warren and Bragg: light blue spheres are calcium atoms, orange spheres are magnesium atoms, each tetrahedron consists of 1 silicon atom and 4 oxygen atoms (red spheres) [44].

Diopside was found to exhibit superior mechanical properties compared to other commonly used bioceramics. For example, it presents a bending strength of 300 MPa and a fracture toughness of 3.5 MPa m$^{1/2}$, which are higher with respect to wollastonite-containing ceramics with similar densities and exceed those of hydroxyapatite by 2–3 times [45].

Diopside was also found to be able to bond with living bone tissue. In this regard, Nonami and Tsutsumi [15] conducted a high-resolution electron microscopy study revealing the continuity between diopside implants and bone tissue in monkeys and rabbits, which was possible due to growth of new tissue at the interface. The same authors also reported that diopside had a much longer degradation time than hydroxyapatite, which was not optimal for bone regeneration; however, this can be useful in those cases where more chemical stability is required or for dental roots.

5. Comparison between 1d Formulation and 45S5 Bioglass®

As previously stated, bioactive glasses in the CaO-MgO-SiO$_2$ ternary system as well as their relevant GC derivatives can indeed be designed for potential use in bone repair, but it is necessary to compare the major characteristics of these new materials with a "gold standard" reference, such as 45S5 Bioglass®. A series of studies were carried out for this specific purpose, as discussed in the next sections.

5.1. Chemical Composition

Table 5 highlights the quantitative differences in the composition of 1d and 45S5 glasses, in terms of amount of ingredients.

Table 5. Nominal compositions of 1d glass and 45S5 Bioglass®.

Glasses	SiO$_2$	CaO	MgO	P$_2$O$_5$	CaF$_2$	Na$_2$O
1d glass (wt.%)	46.1	28.7	8.8	6.2	5.7	4.5
45S5 Bioglass® (wt.%)	45	24.5	-	6	-	24.5

The two bioactive glasses have a similar number of former oxides (SiO$_2$ and P$_2$O$_5$); the content of CaO is higher for 1d glass.

The main differences between the two bioactive glasses can be summarized in two points:
1. The content of Na$_2$O in 45S5 is more than five times greater when compared to 1d;
2. The 1d glass contains additional MgO and CaF$_2$, which are not present in the case of the 45S5 Bioglass®.

The presence of magnesium within the glass composition is useful because Mg^{2+} ions are naturally contained in bone tissue and play an active role in human bone metabolism.

Magnesium ions promote cell adhesion, proliferation and differentiation of osteoblasts [23]. Moreover, Dietrich et al. [46] noted that having an amount of MgO in the glass composition between 0.4 and 1.2 wt.% accelerates glass dissolution, consistent with its role as network modifier.

Instead, the partial substitution of Na_2O with CaF_2 has the effect of decreasing the pH in the surrounding solution during in vitro tests, as well as the melting and glass transition temperatures of the glass; furthermore, a change in cell response was also reported [23]. In this regard, in vitro results with mesenchymal stem cells suggested an advantage of 1d glass concerning cell viability and proliferation. On the other hand, the ions released from 45S5 material appeared to have a stronger osteoinductive effect, whereas no clear superiority of either of the bioactive glasses was observed upon direct cell–glass contact [23]. Further tests are necessary to elucidate these issues, also considering that different experimental conditions could have played a role (e.g., the particle sizes used in that study: <32 μm for 1d and <56 μm for 45S5 glass).

5.2. Advanced Microstructural Analysis (Q^n Units)

Figure 8 shows the Fourier-transformed infra-red (FTIR) spectra of 1d and 45S5 glasses acquired in the wavenumber range of 300 to 1300 cm^{-1}. In order to study the distribution of Q^n (i.e., the degree of polymerization of the structure inside the glass, where n indicates the number of bridging oxygens), Kansal et al. [18] focused on the range within 900–1100 cm^{-1}, which corresponds to SiO_4 with a different number of bridging oxygens.

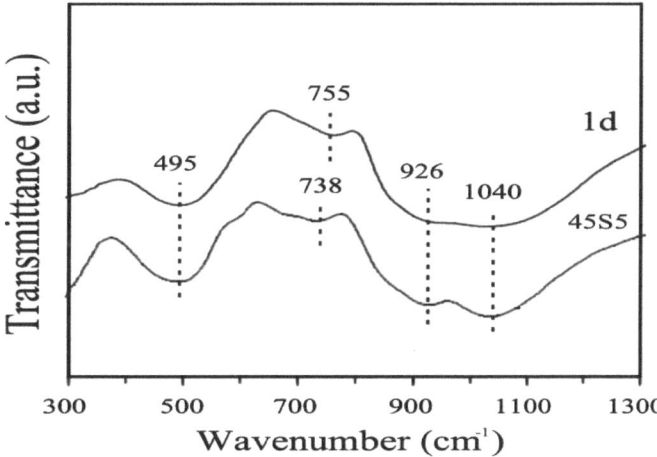

Figure 8. FTIR spectra of bioactive glasses 1d and 45S5. Reproduced from [18] with permission.

There were two interesting bands in this range, around 1040 cm^{-1}, which indicate the presence of silicate Q^3 units, and around 930 cm^{-1}, which indicate the polymerization of silicate Q^2 units along with some Q^1 units. As reported by Tilocca [47], the highest bioactivity of phosphosilicate glasses can be expected if Q^n units are dominated by chains of Q^2 metasilicates, which are sporadically cross-linked through Q^3 units, whereas the chains are terminated by Q^1 units. The structures of 1d and 45S5 glasses are similar since the predominant units are Q^2 and Q^3 for both materials; hence, a high apatite-forming ability is expected. The number of non-bridging oxygens per each tetrahedron was calculated for both 45S5 and 1d glasses, resulting to be 1.99 and 1.88, respectively. These values suggest that 1d glass has a more cross-linked structure than 45S5 Bioglass® because, although predominantly containing Q^2 units, it has a larger fraction of Q^3 units compared to the other glass [21].

5.3. pH In Vitro

Tulyaganov et al. [26] conducted immersion studies in SBF to evaluate the effect of 1d and 45S5 glasses on the pH of the solution. Over the first 300 h, there was a noticeable rise in pH in the case of 45S5 Bioglass® from 7.10 to about 7.75. This increase was due to the ion exchange mechanism already proposed by Hench [5,6,10,13] and explained above. On the contrary, a more moderate increment in pH was observed in the experiment with 1d glass, which was associated with the presence of fluoride ions that were exchanged with OH^- ions from the SBF (from the dissociation of water into H^+ and OH^-), eventually leading to a pH decrease. This was also consistent with other findings about fluoride-containing bioactive glasses [48].

Having a moderately alkaline pH around the implant (up to 7.8–8.0) carries some advantages, such as the accelerated formation of the apatite layer on the glass surface and the stimulation of the viability of osteoblasts; however, if the pH value is too high, damage to tissues and bone cells may occur as well as the inhibition of endothelial cell proliferation [26].

5.4. Mass Loss

Because of the different chemical compositions, there is also a different dissolution rate of 1d and 45S5 glasses in testing solutions. Tulyaganov et al. [26] reported that the mass loss was much higher in the case of Hench's glass (3.7 wt.%) when compared to the 1d composition (2 wt.%) after soaking for 120 h in Tris-HCl. For completeness, it was also reported that the pH value inside this solution increased more significantly after immersion of 45S5 when compared to 1d glass (9.7 versus 8.1, from a starting value of 7.2).

5.5. In Vitro Bioactivity

XRD patterns of the two different bioactive glass compositions, before and after immersion in SBF for different time frames, are reported in Figure 9.

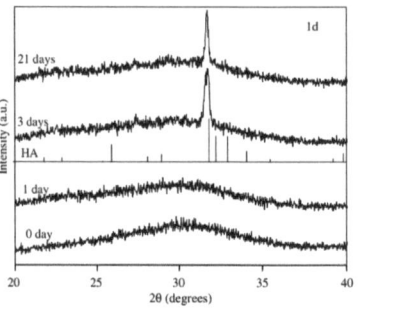
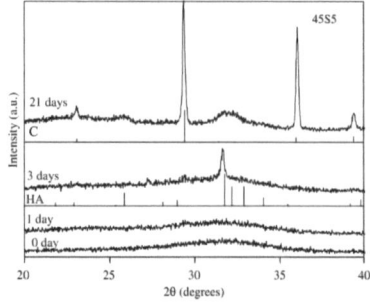

Figure 9. XRD patterns of 1d glass (**left**) and 45S5 Bioglass® (**right**). HA = hydroxyapatite; C = calcite. Reproduced from [26] with permission.

The common feature in both as-produced materials is the presence of the amorphous halo in the range of 28 to 35°, which is typical of silicate glasses. No detectable modifications can be seen after 1 day of immersion in SBF but, after 3 days, the characteristic peaks of hydroxyapatite—especially the main reflection (211) at ~32°—begin to develop. At this time point, the (211) diffraction peak of hydroxyapatite was sharper in the case of the 1d glass compared to 45S5 [26].

The results after 21 days show a different behaviour for the two bioactive glasses; in fact, while only hydroxyapatite was found to grow on the 1d surface, in the case of 45S5 glass, there were some peaks related to calcium carbonate (calcite), too. The coexistence of apatite and calcite was reported to occur in some in vitro studies, which may depend on multiple factors (e.g., glass particle size, volume of solution used, glass composition) not fully elucidated yet, as comprehensively discussed elsewhere [49].

5.6. Direct and Indirect Cell Culture

Schmitz et al. [23] performed an experiment to investigate the biological responses elicited by 45S5 and 1d glasses in two different cell-culture settings with a focus on cell proliferation, viability and osteogenic differentiation:

(a) Indirect culture: the bioactive glasses were immersed in a solution for 24 h at body temperature and shaken to promote the release of ionic dissolution products. At the end of this first phase, the glasses were removed, and the solution was filtered and used as a culture medium for mesenchymal stem cells (MSCs). This setting is used to verify the effects of the ions released.

(b) Direct culture: the bioactive glasses removed from the solution in the point (a) were placed in direct contact with MSCs.

The use of both settings was very useful to describe in detail both the action of the ions that were released by the bioactive glasses through the indirect culture setting and the reactions that take place once the physical contact of the materials with the cells occurs through the direct culture setting.

The major results from the two different settings can be summarized as follows:

- Indirect culture setting: 45S5 glass elicited a better osteogenic action compared to 1d glass. This behaviour was because, in the former case, there was a higher concentration of P and Si ions in the cell culture medium having osteostimulatory effects. In fact, phosphorus stimulates osteogenic differentiation and bone mineralization [50], and Si ions also activate gene families in bone cells, ultimately promoting osteogenic differentiation [51].

- Direct culture setting: there was a reversal in the trend compared to the indirect culture setting. In fact, the osteogenic action was no longer so different between the two bioactive glasses, although the expression of the specific protein OCN (marker of osteogenic differentiation) was greater in the 1d cultures. The concentration of magnesium, which was released by the 1d glass, in the first 24 h was one tenth of that detected after a week, and the concentration of Ca increased over time in the medium. This progressive, increasing release of beneficial ions is important as magnesium ions have the ability to increase cell viability [52], and calcium ions improve cell viability and proliferation [53]. These promising in vitro results on 1d glass were corroborated by in vivo tests.

The obtained results demonstrated an advantage of 1d glass in regard to cell viability and proliferation. Owing to its good osteogenic potential compared to the benchmark 45S5 glass and its higher biocompatibility, 1d glass was proposed to be an interesting alternative to 45S5 glass for bone tissue engineering applications [23].

5.7. Antibacterial Properties

The antibacterial properties of some experimental compositions based on 1d glass were tested against *Escherichia coli* (*E. coli*) [54]. Silver from 0.035 M, 0.077 M, 0.150 M and 0.220 M $AgNO_3$ aqueous solutions have been incorporated into the surface of 1d bioactive glass through the ion exchange approach assisted with ultrasonic treatment. Antibacterial tests showed that the silver-containing glasses inhibited the growth of *E. coli*, which exhibited a rapid decrease in its viability, reaching the limit of detection after a maximum of 2 h. Parent 1d glass induced a slight decrease in bacterial number after one hour compared with the bacterial inoculum. However, after two hours of incubation at 37 °C, the number of bacteria increased again, being comparable to the inoculum. It was observed that, according to the mid IR spectra, the structure of silver-modified glasses was similar to that of the parent 1d glass, which indicates that the treatment performed did not significantly alter the structure of the glass network and, thus, was not expected to interfere with its bioactivity mechanisms.

6. In Vivo Experiments and Clinical Trials

Tulyaganov et al. [26] studied the effects of bioactive glass 1d on both animals and human patients. In the first set of experiments, 1d glass particles were inserted directly into osseous defects produced in rabbit femora after being properly sterilized. Reactions in rabbits were monitored at different time points, i.e., 1 week, 2 weeks, 1 month, 2 months, 3 months, 4 months and 6 months (Figure 10).

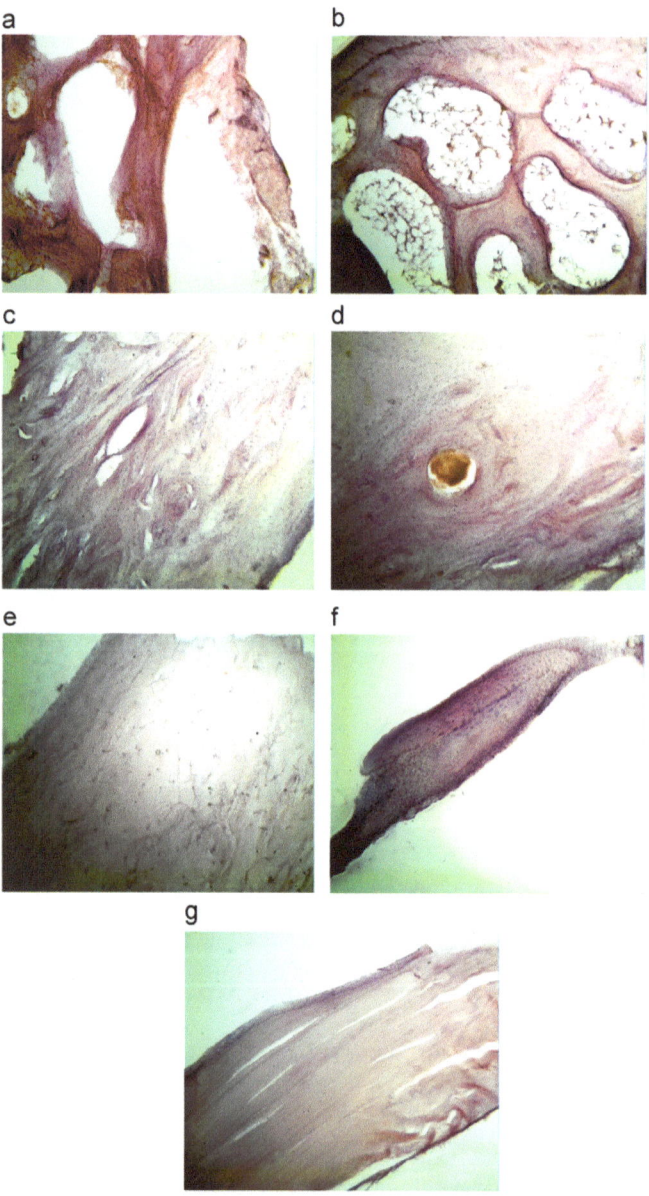

Figure 10. Histopathological sections of the bone in cortical area of bone (magnification 400×) after implantation for different periods: (**a**) 1 week; (**b**) 2 weeks; (**c**) 1 month; (**d**) 2 months; (**e**) 3 months; (**f**) 4 months; and (**g**) 6 months. Reproduced from [26] with permission.

Since the first week, the presence of low inflammatory infiltrate and an increase in the thickness of blood vessels was observed. From the first month onwards, the formation of new bone began, starting from the outside and spreading inwards. As the months passed, the bone trabeculae became thicker and thicker, and the implanted material was progressively embedded inside the newly formed bone. After 6 months, the surgically created cavities were completely filled with regenerated bone, which was mature and homogeneous. Overall, 1d glass powders were fully compatible with the surrounding tissues without eliciting any significant adverse reaction throughout the experiment duration.

The same 1d glass particulate was clinically tested over 8 months to treat jawbone defects in human patients mainly after a cystectomy operation. The test was conducted in 45 volunteers (21 males and 24 females) aged between 19 and 60. Patients were examined before and after surgery at 2 weeks, 2 months and 6 months. In this study, glass particles were inserted where there was a defect in the alveolar bone in order to avoid the progressive loss of bone over time (resorption) and assure the stability of the patient's teeth. These early clinical trials showed that the glass formed a cohesive mass with the patient's blood, thus demonstrating a homeostatic effect. Figure 11 shows that new, regenerated bone was formed after 2 months where there was the empty space of a jawbone defect before surgery. Similar results were found in all patients and, thus, these early clinical trials support the suitability of 1d glass for the conventional treatment of bone defects (highly biocompatible and bioactive filler). Figure 12 shows two other radiographs of the patient showing that after a cystectomy operation, the lesions were filled up with newly formed cancellous bone. However, in order to gain regulatory approval and consider this bioactive glass for routine surgery, it will be necessary to increase the number of patients during a further phase of clinical trials.

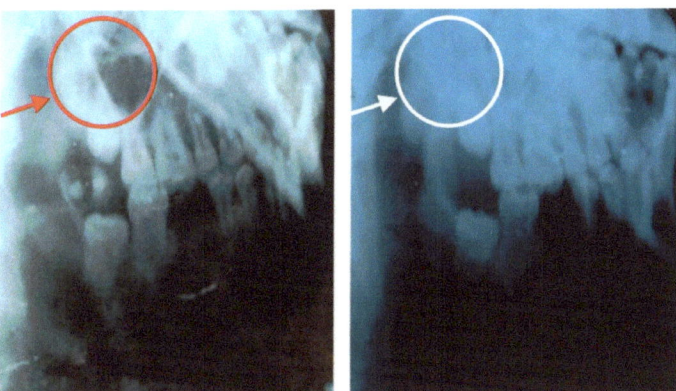

Figure 11. Radiographic images of jawbone defects before ((**left**) side) and after surgical intervention (2 months of follow-up) with implantation of 1d glass particulates ((**right**) side). Reproduced from [26] with permission.

In an attempt to avoid undesired losses of glass particulates during the grafting procedure and to further improve the handling properties, composites based on bioactive glass 1d and organic carriers, i.e., glycerol and polyethylene glycol (PEG), were synthesized [55]. Homogeneous mixtures were obtained that could be handled as mouldable pastes, demonstrating cohesive injectability. All pastes exhibited high apatite-forming rates after immersion in SBF, consistent with the in vitro bioactivity of 1d particles. The potential suitability of these materials for osteostimulatory bone healing was recently confirmed in vivo through implantation experiments in rabbit femoral defects [24].

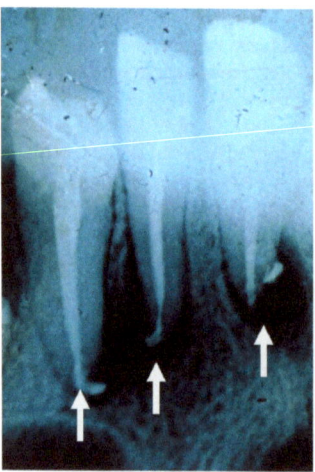 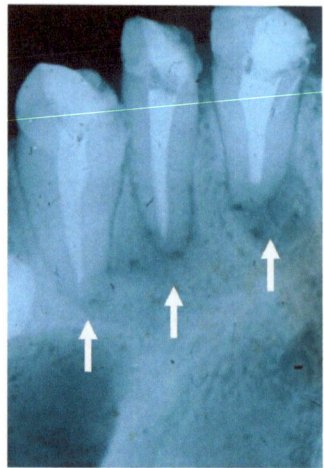

Figure 12. High-magnification radiographs of the patient before ((**left**) side) and after 2 months from operation ((**right**) side) showing that the lesions were filled up with trabeculae of new bone.

7. Towards the Future: 1d-Derived GC Dental Implants and Porous Scaffolds

7.1. GC Dental Implants

Dental implants rely on osteointegration through osteoinduction (whereby osteogenesis is induced) and osteoconduction, which involves the formation of hydroxyapatite [56–58]. The mechanical properties of natural tissues like the jawbone and dental hard tissue play a crucial role in the longevity of dental implants in the oral cavity [58,59]. After the placement of a prosthetic restoration, the dental implant receives the loads from the occlusal forces. When the elastic modulus and fracture toughness of the implant material align closely with those of the jawbone, the implant can effectively distribute the load to the adjacent bone, maintaining its density. However, if these properties differ significantly from those of the jawbone, then the dental implant is the only one that is loaded mechanically by the occlusal forces; as a result, the implant does not transfer the occlusal forces to the jawbone. This phenomenon is called stress shielding [60], whereby the osteocytes lose their main role (i.e., the preservation of the extracellular matrix), resulting in a reduction in the bone density of the jawbone, and eventually in the failure of the dental implant. The high elastic modulus of commonly-used dental implant materials like titanium alloys and zirconia (110 and 220 GPa, respectively) compared to the jawbone (7–30 GPa) and dentine (15–30 GPa) [31,61–64] is a common cause of implant failure and decreased bone density post-implantation. Bone grafts are utilized in various clinical scenarios when a patient's jawbone fails to meet the necessary criteria for optimal dental implant placement (specifically, due to insufficient bone quantity resulting from tooth loss) [31,62–64]. Initially, autogenous and allograft jawbone grafts were employed for their osteogenic and osteoinductive/osteoconductive properties, respectively. Subsequently, researchers redirected their focus towards synthetic materials (hydroxyapatite, tricalcium phosphates, bioactive glasses and GCs) to overcome the drawbacks associated with traditional grafts (such as patient discomfort, infection, complex surgical procedures, non-simultaneous absorption of the graft/new bone formation) and reduce costs [61–67].

According to Dimitriadis et al. [20,60], 1d-derived GCs display an excellent bioactive behaviour, yielding to spontaneous formation of hydroxyapatite on their surface after immersion in SBF at 37 °C. Besides having an adequate bioactivity, these GCs exhibited a well-sintered, dense microstructure embedding biocompatible crystalline phases that affected their mechanical properties (Table 4). These features, along with the attractive aesthetics (a white colour) [19,20], encourage further studies on such highly promising dental implant materials. More specifically, heat-treatment at 850 °C yielded GCs with mechanical properties comparable to those of dentine (which is the biological tissue to be replaced by a dental implant) and the jawbone (which is the biological tissue put in direct contact with the material of a dental implant). Therefore, these appealing characteristics warrant further investigation regarding the suitability of 1d-based GCs in the production of dental implants.

The range of mechanical properties could be further modulated and finely tuned by combining the bioactive glass with biocompatible polymers, thus obtaining glass/polymer composites. These multiphasic biomaterials allow combining the peculiar features of bioactive glasses, such as bioactivity and osteostimulatory properties, with the added values of polymers, including flexibility [68,69]. These composites show great promise in overcoming some traditional limits of bioactive glasses such as low fracture toughness, which is a particularly critical issue in glass-derived porous scaffolds [70].

7.2. Glass-Derived Porous Scaffolds

Biomaterials addressed to bone repair, including bioactive glasses or GCs, are often produced in the form of porous templates with architectural characteristics mimicking the trabecular structure of cancellous bone [71]. Following this "biomimicry-guided" criterion is thought to improve the regenerative properties of the implants, which are thus dictated not only by the inherent characteristics of the material (e.g., apatite-forming ability, biocompatibility with bone cells) but also by the porous geometry, allowing biofluids to flow in and out, cells to colonize the scaffold walls and blood vessel to grow in. Macroporous 1d-derived scaffolds have been recently fabricated for the first time using the sponge replica method [25]. Upon high-temperature thermal treatment at 800 °C for 3 h, the 1d glass particles underwent sinter-crystallization, leading to the consolidation of a scaffold structure and the concurrent development of diopside, fluorapatite and wollastonite, as expected from the material design. The sintered 1d-derived GC scaffolds exhibited a 3D pore-strut architecture and total porosity (68 vol.%) comparable to those of cancellous bone, while the compressive strength (29.7 MPa) and elastic modulus (1.4 GPa) were even superior to those of trabecular bone tissue (50–500 MPa), suggesting suitability for application in load-bearing sites. The scaffolds were also highly bioactive in vitro as demonstrated by the formation of a calcium phosphate layer after immersion in SBF for just 48 h. In an attempt to improve the scaffold reproducibility and the scalability of the whole fabrication process, early trials using additive manufacturing technologies to process 1d glass powders are currently ongoing in the context of a research collaboration among the authors of this review article.

8. Conclusions

Since its invention, 45S5 Bioglass® has been implanted in millions of human patients and is currently being marketed for various dental and orthopaedic applications. This has led to a considerable effort towards understanding the fundamentals that govern the physical, chemical and biological properties of bioactive glasses based on—or inspired by—45S5 glass. On the other hand, some limitations of the 45S5 composition—e.g., the high pH environment created by a high sodium content and poor sinterability—pushed scientists to develop new bioactive glass formulations. In this regard, bioactive glasses belonging to the $CaO-MgO-SiO_2$ ternary system as well as their relevant GC derivatives that feature low sodium oxide contents can indeed be used in bone repair as an alternative to the "gold standard" reference 45S5 Bioglass®. This was demonstrated through promising in vitro results on 1d glass that were corroborated by in vivo tests. The experimental data collected over the past 15 years supports the suitability of 1d glass in a variety of clinical applications for the repair of periodontal defects, ridge preservation and sinus augmentation. The full potential of bioactive glasses and GCs based on the $CaO-MgO-SiO_2$ ternary system, with special reference to the 1d composition, is still to be fully exploited and indeed deserves further investigation in the near future.

Author Contributions: Conceptualization, D.U.T. and F.B.; methodology, D.U.T., S.A., K.D., H.R.F., R.G. and F.B.; validation, D.U.T.; investigation, D.U.T., S.A., K.D., H.R.F., R.G. and F.B.; writing—original draft preparation, D.U.T. and F.B.; writing—review and editing, S.A., K.D., H.R.F. and R.G.; supervision, D.U.T. and F.B.; project administration, F.B.; funding acquisition, F.B. All authors have read and agreed to the published version of the manuscript.

Funding: This study was partially carried out within the project "Artificial Intelligence-based design of 3D PRINTed scaffolds for the repair of critical-sized BONE defects—I-PRINT-MY-BONE" funded by the European Union—Next Generation EU within the PRIN 2022 program (D.D. 104—02/02/2022 Ministero dell'Università e della Ricerca). This manuscript reflects only the authors' views and opinions, and the Ministry cannot be considered responsible for them.

Data Availability Statement: No new data were created in this study, which is a review article. Original data can be found in the cited references.

Acknowledgments: D.T. thanks J.M.F. Ferreira from the University of Aveiro (Portugal) whose support, guidance and great expertise were instrumental in shaping the research direction.

Conflicts of Interest: The authors declare no conflicts of interest.

References

1. Hench, L.L. The story of Bioglass. *J. Mater. Sci. Mater. Med.* **2006**, *17*, 967978. [CrossRef]
2. Wu, C.; Chang, J. Mesoporous bioactive glasses: Structure characteristics, drug/growth factor delivery and bone regeneration application. *Interface Focus* **2012**, *2*, 292–306. [CrossRef]
3. Moeini, A.; Hassanzadeh Chinijani, T.; Malek Khachatourian, A.; Fook, M.V.L.; Baino, F.; Montazerian, M. A critical review of bioactive glasses and glass-ceramics in cancer therapy. *Int. J. Appl. Glass. Sci.* **2023**, *14*, 69–87. [CrossRef]
4. Miguez-Pacheco, V.; Greenspan, D.; Hench, L.; Boccaccini, A. Bioactive glasses in soft tissue repair. *Am. Ceram. Soc. Bull.* **2015**, *94*, 27–31.
5. Hench, L.L.; Greenspan, D. Interactions between bioactive glass and collagen: A review and new perspectives. *J. Aust. Ceram. Soc.* **2013**, *49*, 1–40.
6. Hench, L.L. Bioactive ceramics. *Ann. N. Y. Acad. Sci.* **1988**, *523*, 54–71. [CrossRef]
7. Roy, M.; Bandyopadhyay, A.; Bose, S. Ceramics in Bone Grafts and Coated Implants. In *Materials for Bone Disorders*, 1st ed.; Bandyopadhyay, A., Bose, S., Eds.; Elsevier: Philadelphia, PA, USA, 2017; pp. 265–314.
8. Bi, L.; Rahaman, M.N.; Day, D.E.; Brown, Z.; Samujh, C.; Liu, X.; Mohammadkhah, A.; Dusevich, V.; Eick, J.D.; Bonewald, L.F. Effect of bioactive borate glass microstructure on bone regeneration, angiogenesis, and hydroxyapatite conversion in a rat calvarial defect model. *Acta Biomater.* **2013**, *9*, 8015–8026. [CrossRef]
9. Abou Neel, E.A.; Pickup, D.M.; Valappil, S.P.; Newport, R.J.; Knowles, J.C. Bioactive functional materials: A perspective on phosphate-based glasses. *J. Mater. Chem.* **2009**, *19*, 690–701. [CrossRef]
10. McEntire, B.; Bal, B.S.; Rahaman, M.; Chevalier, J.; Pezzotti, G. Ceramics and ceramic coatings in orthopaedics. *J. Eur. Ceram. Soc.* **2015**, *35*, 4327–4369. [CrossRef]
11. Balasubramanian, S.; Gurumurthy, B.; Balasubramanian, A. Biomedical applications of ceramic nanomaterials: A review. *Int. J. Pharma Sci. Res.* **2017**, *8*, 4950–4959.
12. Shanmugam, K.; Sahadevan, R. Bioceramics—An Introductory Overview. In *Fundamental Biomaterials: Ceramics*, 1st ed.; Thomas, S., Balakrishnan, P., Sreekala, M.S., Eds.; Elsevier Ltd.: Philadelphia, PA, USA, 2018; pp. 1–46.
13. Montazerian, M.; Zanotto, E.D. History and trends of bioactive glass-ceramics. *J. Biomed. Mater. Res. A* **2016**, *104*, 1231–1249. [CrossRef] [PubMed]
14. Blayney, A.W.; Bebear, J.P.; Williams, K.R.; Portmann, M. Ceravital in ossiculoplasty: Experimental studies and early clinical results. *Am. J. Otol.* **1986**, *100*, 1359–1366. [CrossRef] [PubMed]
15. Nonami, T.; Tsutsumi, S. Study of diopside ceramics for biomaterials. *J. Mater. Sci. Mater. Med.* **1999**, *10*, 475–479. [CrossRef] [PubMed]
16. Salinas, A.J.; Vallet-Regí, M. Bioactive ceramics: From bone grafts to tissue engineering. *RSC Adv.* **2013**, *3*, 11116–11131. [CrossRef]
17. Peitl, O.; LaTorre, G.P.; Hench, L.L. Effect of crystallization on apatite layer formation of bioactive glass 45S5. *J. Biomed. Mater. Res.* **1996**, *30*, 509–514.
18. Kansal, I.; Tulyaganov, D.U.; Goel, A.; Pascual, M.J.; Ferreira, J.M.F. Structural analysis and thermal behavior of diopside fluorapatite wollastonite based glasses and glass ceramics. *Acta Biomater.* **2010**, *6*, 4380–4388. [CrossRef] [PubMed]
19. Dimitriadis, K.; Vasilopoulos, K.C.; Vaimakis, T.C.; Karakassides, M.A.; Tulyaganov, D.U.; Agathopoulos, S. Synthesis of glass ceramics in the Na_2O/K_2O-CaO-MgO-SiO_2-P_2O_5-CaF_2 system as candidate materials for dental applications. *Int. J. Appl. Ceram. Technol.* **2020**, *17*, 2025–2035. [CrossRef]
20. Dimitriadis, K.; Moschovas, D.; Tulyaganov, D.U.; Agathopoulos, S. Development of novel bioactive glass ceramics in the Na_2O/K_2O-CaO-MgO-SiO_2-P_2O_5-CaF_2 system. *J. Non-Cryst. Solids* **2020**, *533*, 119936. [CrossRef]

21. Tulyaganov, D.U.; Agathopoulos, S.; Ventura, J.M.G.; Karakassides, M.A.; Fabrichnaya, O.; Ferreira, J.M.F. Synthesis of glass–ceramics in the CaO–MgO–SiO$_2$ system with B$_2$O$_3$, P$_2$O$_5$, Na$_2$O and CaF$_2$ additives. *J. Eur. Ceram. Soc.* **2006**, *26*, 1463–1471. [CrossRef]
22. Tulyaganov, D.U.; Agathopoulos, S.; Valerio, P.; Balamurugan, A.; Saranti, A.; Karakassides, M.A.; Ferreira, J.M.F. Synthesis, bioactivity and preliminary biocompatibility studies of glasses in the system CaO-MgO-SiO$_2$-Na$_2$O-P$_2$O$_5$-CaF$_2$. *J. Mater. Sci. Mater. Med.* **2011**, *22*, 217–227. [CrossRef]
23. Schmitz, S.I.; Widholz, B.; Essers, C.; Becker, M.; Tulyaganov, D.U.; Moghaddam, A.; Gonzalo de Juan, I.; Westhauser, F. Superior biocompatibility and comparable osteoinductive properties: Sodium-reduced fluoride-containing bioactive glass belonging to the CaO–MgO–SiO$_2$ system as a promising alternative to 45S5 bioactive glass. *Bioact. Mater.* **2020**, *5*, 55–65. [CrossRef] [PubMed]
24. Tulyaganov, D.U.; Akbarov, A.; Zyyadullaeva, N.; Khabilov, B.; Baino, F. Injectable bioactive glass-based pastes for potential use in bone tissue repair. *Biomed. Glas.* **2020**, *6*, 23–33. [CrossRef]
25. Baino, F.; Tulyaganov, D.U.; Kahharov, Z.; Rahdar, A.; Vernè, E. Foam-Replicated Diopside/Fluorapatite/Wollastonite-Based Glass–Ceramic Scaffolds. *Ceramics* **2022**, *5*, 120–130. [CrossRef]
26. Tulyaganov, D.U.; Makhkamov, M.E.; Urazbaev, A.; Goel, A.; Ferreira, J.M.F. Synthesis, processing and characterization of a bioactive glass composition for bone regeneration. *Ceram. Int.* **2013**, *39*, 2519–2526. [CrossRef]
27. Boccaccini, A.R. Machinability and brittleness of glass-ceramics. *J. Mater. Process Technol.* **1997**, *65*, 302–304. [CrossRef]
28. Dimitriadis, K.; Tulyaganov, D.U.; Agathopoulos, S. Production of Bioactive Glass-Ceramics for Dental Application through Devitrification of Glasses in the Na$_2$O/K$_2$O–CaO–MgO–SiO$_2$–P$_2$O$_5$–CaF$_2$ System. In *Bioactive Glasses and Glass-Ceramics: Fundamentals, Applications, and Advances*; Baino, F., Kargozar, S., Eds.; John Wiley and Sons Ltd.: Hoboken, NJ, USA, 2022; pp. 431–457.
29. Kapoor, S.; Goel, A.; Filipa, A.; Pascual, M.J.; Lee, H.; Kim, H.; Ferreira, J.M.F. Influence of ZnO/MgO substitution on sintering, crystallisation, and bio-activity of alkali-free glass-ceramics. *Mater. Sci. Eng. C* **2015**, *53*, 252–261. [CrossRef]
30. Soares, V.O.; Daguano, J.K.M.B.; Lombello, C.B.; Bianchin, O.S.; Gonçalves, L.M.G.; Zanotto, E.D. New sintered wollastonite glass-ceramic for biomedical applications. *Ceram. Int.* **2018**, *44*, 20019–20027. [CrossRef]
31. Saadaldin, S.A.; Rizkalla, A.S. Synthesis and characterization of wollastonite glassceramics for dental implant applications. *Dent. Mater.* **2014**, *30*, 364–371. [CrossRef]
32. Saadaldin, S.A.; Dixon, S.J.; Rizkalla, A.S. Bioactivity and biocompatibility of a novel wollastonite glass-ceramic biomaterial. *J. Biomater. Tissue Eng.* **2014**, *4*, 939–946. [CrossRef]
33. Garcia, E.; Miranzo, P.; Sainz, M.A. Thermally sprayed wollastonite and wollastonite-diopside compositions as new modulated bioactive coatings for metal implants. *Ceram. Int.* **2018**, *44*, 12896–12904. [CrossRef]
34. Dorozhkin, S.V. Calcium orthophosphates as bioceramics: State of the art. *J. Funct. Mater.* **2010**, *1*, 22–107. [CrossRef]
35. Ramadoss, R.; Padmanaban, R.; Subramanian, B. Role of bioglass in enamel remineralization: Existing strategies and future prospects-A narrative review. *J. Biomed. Mater. Res. B Appl. Biomater.* **2022**, *110*, 45–66. [CrossRef] [PubMed]
36. Wiglusz, K.; Dobrzynski, M.; Gutbier, M.; Wiglusz, R.J. Nanofluorapatite Hydrogels in the treatment of Dentin Hypersensitivity: A study of physiochemical Proprieties and Fluoride Realease. *Gels* **2023**, *9*, 271–287. [CrossRef]
37. Zenebe, C.G. A Review on the Role of Wollastonite Biomaterial in Bone Tissue Engineering. *BioMed Res. Int.* **2022**, *2022*, 996530. [CrossRef]
38. Tulyaganov, D.U.; Dimitriadis, K.; Agathopoulos, S.; Baino, F.; Fernandes, H.R. Wollastonite-containing glass-ceramics from the CaO–Al$_2$O$_3$–SiO$_2$ and CaO–MgO–SiO$_2$ ternary systems. *Open Ceram.* **2024**, *17*, 100507. [CrossRef]
39. Kokubo, T.; Shigematsu, M.; Nagashima, Y.; Tashiro, M.; Nakamura, T.; Yamamuro, T.; Higash, S. Apatite-and wollastonite-containing glass-ceramics for prosthetic application. *Bull. Inst. Chem. Res. Kyoto Univ.* **1982**, *60*, 260–268.
40. Hench, L.L.; Splinter, R.J.; Allen, W.C.; Greenlee, T.K. Bonding mechanisms at the interface of ceramic prosthetic materials. *J. Biomed. Mater. Res.* **1971**, *5*, 117–141. [CrossRef]
41. Vichaphund, S.; Kitiwan, M.; Atong, D.; Thavorniti, P. Microwave synthesis of wollastonite powder from eggshells. *J. Eur. Ceram. Soc.* **2011**, *31*, 2435–2440. [CrossRef]
42. He, Z.; Shen, A.; Lyu, Z.; Li, Y.; Wu, H.; Wang, W. Effect of wollastonite microfibers as cement replacement on the properties of cementitious composites: A review. *Constr. Build. Mater.* **2020**, *261*, 119920. [CrossRef]
43. "Chrome Diopside", Chemistry Views. 22 May 2014. Available online: https://www.chemistryviews.org/details/ezine/6047511/Chrome_Diopside/#:~:text=Diopside%20has%20the%20composition%20CaMgSi,the%20crystals%20are%20often%20twinned (accessed on 1 November 2023).
44. "A Surprisingly Important Structure-Diopside", Chrystallography365. 23 August 2014. Available online: https://crystallography365.wordpress.com/2014/08/23/a-surprisingly-important-structure-diopside/ (accessed on 1 November 2023).
45. Wang, G.C.; Lu, Z.F.; Zreiqat, H. Bioceramics for Skeletal Bone Regeneration. In *Bone Substitute Biomaterials*, 1st ed.; Mallick, K., Ed.; Woodhead Publishing: Cambridge, UK, 2014; pp. 180–186.
46. Dietrich, E.; Oudadesse, H.; Lucas Girot, A.; Le Gal, Y.; Jeanne, S.; Cathelineau, G. Effects of Mg and Zn on the surface of doped melt derived glass for biomaterials applications. *Appl. Surf. Sci.* **2008**, *255*, 391–395. [CrossRef]
47. Tilocca, A. Structural models of bioactive glasses from molecular dynamics simulations. *Proc. R Soc. A* **2009**, *465*, 1003–1027. [CrossRef]

48. Brauer, D.S.; Karpulthina, N.; O'Donnell, M.D.; Law, R.V.; Hill, R.G. Fluoride-containing bioactive glasses: Effect of glass design and structure on degradation, pH and apatite formation in simulated body fluid. *Acta Biomater.* **2010**, *6*, 3275–3282. [CrossRef] [PubMed]
49. Mozafari, M.; Banijamali, S.; Baino, F.; Kargozar, S.; Hill, R.G. Calcium carbonate: Adored and ignored in bioactivity assessment. *Acta Biomater.* **2019**, *91*, 35–47. [CrossRef] [PubMed]
50. Khoshniat, S.; Bourgine, A.; Julien, M.; Petit, M.; Pilet, P.; Rouillon, T.; Masson, M.; Gatius, M.; Weiss, P.; Guicheux, J.; et al. Phosphate-dependent stimulation of MGP and OPN expression in osteoblasts via the ERK1/2 pathway is modulated by calcium. *Bone* **2011**, *48*, 894–902. [CrossRef] [PubMed]
51. Reffitt, D.M.; Ogston, N.; Jugdaohsingh, R.; Cheung, H.F.; Evans, B.A.; Thompson, R.P.; Powell, J.J.; Hampson, G.N. Orthosilicic acid stimulates collagen type 1 synthesis and osteoblastic differentiation in human osteoblast-like cells in vitro. *Bone* **2003**, *32*, 127–135. [CrossRef]
52. Zheng, J.; Mao, X.; Ling, J.; Chen, C.; Zhang, W. Role of magnesium transporter subtype 1 (MagT1) in the osteogenic differentiation of rat bone marrow stem cells. *Biol. Trace Elem. Res.* **2016**, *171*, 131–137. [CrossRef] [PubMed]
53. Hench, L.L. Genetic design of bioactive glasses. *J. Eur. Ceram. Soc.* **2009**, *29*, 1257–1265. [CrossRef]
54. Gonzalo-Juan, I.; Xie, F.; Becker, M.; Tulyaganov, D.U.; Fischer, A.; Riedel, R. Synthesis of silver modified bioactive glassy materials with antibacterial properties via facile and low-temperature route. *Materials* **2020**, *13*, 5115. [CrossRef] [PubMed]
55. Gonzalo-Juan, I.; Tulyaganov, D.U.; Balan, C.; Linser, R.; Ferreira, J.M.F.; Riedel, R.; Ionescu, E. Tailoring the viscoelastic properties of injectable biocomposites: A spectroscopic assessment of the interactions between organic carriers and bioglass particles. *Mater. Des.* **2016**, *97*, 45–50. [CrossRef]
56. Jones, J.R.; Brauer, D.S.; Hupa, L.; Greenspan, D.C. Bioglass and bioactive glasses and their impact on healthcare. *Int. J. Appl. Glass Sci.* **2016**, *7*, 423–434. [CrossRef]
57. Fernandes, H.R.; Gaddam, A.; Rebelo, A.; Brazete, D.; Stan, G.E.; Ferreira, J.M.F. Bioactive glasses and glass-ceramics for healthcare applications in bone regeneration and tissue engineering. *Materials* **2018**, *11*, 2530. [CrossRef]
58. Albrektsson, T.; Johansson, C. Osteoinduction, osteoconduction and osseointegration. *Eur. Spine J.* **2001**, *10*, S96–S101.
59. Lakatos, É.; Magyar, L.; Bojtár, I. Material properties of the mandibular trabecular bone. *J. Med. Eng.* **2014**, *2014*, 470539. [CrossRef]
60. Dimitriadis, K.; Tulyaganov, D.U.; Agathopoulos, S. Development of novel alumina-containing bioactive glass-ceramics in the CaO-MgO-SiO$_2$ system as possible candidates for dental implant applications. *J. Eur. Ceram. Soc.* **2021**, *41*, 929–940. [CrossRef]
61. Montazerian, M.; Zanotto, E.D. Bioactive and inert dental glass-ceramics. *Biomed. Mater. Res. A* **2017**, *105*, 619–639. [CrossRef] [PubMed]
62. Saadaldin, S.A.; Dixon, S.J.; Costa, D.O.; Rizkalla, A.S. Synthesis of bioactive and machinable miserite glass-ceramics for dental implant applications. *Dent. Mater.* **2013**, *29*, 645–655. [CrossRef] [PubMed]
63. Fillingham, Y.; Jacobs, J. Bone grafts and their substitutes. *Bone Jt. J.* **2016**, *98-B*, 6–9. [CrossRef]
64. Laurencin, C.; Khan, Y.; El-Amin, S.F. Bone graft substitutes. *Expert. Rev. Med. Devices* **2006**, *3*, 49–57. [CrossRef] [PubMed]
65. Kumar, P.; Vinitha, B.; Fathima, G. Bone grafts in dentistry. *J. Pharm. Bioallied. Sci.* **2013**, *5*, S125–S127. [CrossRef]
66. Chen, Q.Z.; Thompson, I.D.; Boccaccini, A.R. 45S5 Bioglass-derived glass-ceramic scaffolds for bone tissue engineering. *Biomaterials* **2006**, *27*, 2414–2425. [CrossRef] [PubMed]
67. Fu, L.; Engqvist, H.; Xia, W. Glass-ceramics in dentistry: A review. *Materials* **2020**, *13*, 1049. [CrossRef]
68. Rezwan, K.; Chen, Q.Z.; Blaker, J.J.; Boccaccini, A.R. Biodegradable and bioactive porous polymer/inorganic composite scaffolds for bone tissue engineering. *Biomaterials* **2006**, *27*, 3413–3431. [CrossRef] [PubMed]
69. Mohamad Yunos, D.; Bretcanu, O.; Boccaccini, A.R. Polymer-bioceramic composites for tissue engineering scaffolds. *J. Mater. Sci.* **2008**, *43*, 4433–4442. [CrossRef]
70. Rehorek, L.; Chlup, Z.; Meng, D.; Yunos, D.M.; Boccaccini, A.R.; Dlouhy, I. Response of 45S5 bioglass® foams to tensile loading. *Ceram. Int.* **2013**, *39*, 8015–8020. [CrossRef]
71. Baino, F.; Fiume, E.; Barberi, J.; Kargozar, S.; Marchi, J.; Massera, J.; Verné, E. Processing methods for making porous bioactive glass-based scaffolds—A state-of-the-art review. *Int. J. Appl. Ceram. Technol.* **2019**, *16*, 1762–1796. [CrossRef]

Disclaimer/Publisher's Note: The statements, opinions and data contained in all publications are solely those of the individual author(s) and contributor(s) and not of MDPI and/or the editor(s). MDPI and/or the editor(s) disclaim responsibility for any injury to people or property resulting from any ideas, methods, instructions or products referred to in the content.

MDPI AG
Grosspeteranlage 5
4052 Basel
Switzerland
Tel.: +41 61 683 77 34

Inorganics Editorial Office
E-mail: inorganics@mdpi.com
www.mdpi.com/journal/inorganics

Disclaimer/Publisher's Note: The title and front matter of this reprint are at the discretion of the Guest Editors. The publisher is not responsible for their content or any associated concerns. The statements, opinions and data contained in all individual articles are solely those of the individual Editors and contributors and not of MDPI. MDPI disclaims responsibility for any injury to people or property resulting from any ideas, methods, instructions or products referred to in the content.

www.ingramcontent.com/pod-product-compliance
Lightning Source LLC
LaVergne TN
LVHW072344090526
838202LV00019B/2472